Effective Field Theories for Nuclei and Compact-Star Matter

Chiral Nuclear Dynamics (CND-III)

Other World Scientific Titles by the Author

Chiral Nuclear Dynamics II: From Quarks to Nuclei to Compact Stars
2nd Edition
ISBN: 978-981-270-588-4

Quarks, Nuclei and Stars:
Memorial Volume Dedicated to Gerald E Brown
ISBN: 978-981-3220-64-5

The Multifaceted Skyrmion
2nd Edition
ISBN: 978-981-4704-40-3

Effective Field Theories for Nuclei and Compact-Star Matter

Chiral Nuclear Dynamics (CND-III)

Yong-Liang Ma
Jilin University, China

Mannque Rho
CEA Saclay, France

World Scientific

NEW JERSEY • LONDON • SINGAPORE • BEIJING • SHANGHAI • HONG KONG • TAIPEI • CHENNAI • TOKYO

Published by

World Scientific Publishing Co. Pte. Ltd.
5 Toh Tuck Link, Singapore 596224
USA office: 27 Warren Street, Suite 401-402, Hackensack, NJ 07601
UK office: 57 Shelton Street, Covent Garden, London WC2H 9HE

British Library Cataloguing-in-Publication Data
A catalogue record for this book is available from the British Library.

EFFECTIVE FIELD THEORIES FOR NUCLEI AND COMPACT-STAR MATTER
Chiral Nuclear Dynamics (CND-III)

ISBN 978-981-3273-31-3

For any available supplementary material, please visit
https://www.worldscientific.com/worldscibooks/10.1142/11072#t=suppl

Desk Editor: Cheryl Heng

Printed in Singapore

To **Gerry Brown**

who shared with some of us Sherlock Holmes' conviction that "When you have eliminated the impossible, whatever remains, however improbable, must be the truth"

– Sir Arthur Conan Doyle

Preface

This monograph is a sequel to the previous two volumes, the first entitled *Chiral Nuclear Dynamics* (CNDI for short) that was co-authored by Maciek Nowak, Ismail Zahed and one of the present coauthors (MR), the second entitled *Chiral Nuclear Dynamics II: From Quarks to Nuclei to Compact Stars* (CNDII for short) authored by MR , and develops further on the main theme of the topics treated in those volumes. The present volume is different in spirit from the previous volumes. The previous volumes exploited what's established in, or inferred from, various developments in strong-interaction dynamics to describe highly correlated dense nuclear medium. In this volume our objective is to arrive at a single unified framework, with a minimum number of unknown parameters, anchored on what's considered consistent in the framework of effective theories for QCD, and make predictions on highly compressed baryonic matter to be observed in terrestrial laboratories and extracted from observables from massive compact stars such as those accurately measured recently and to be revealed in forth-coming gravitational waves from merging neutron stars.

The current activities in this field of research are focused on starting from a theoretical framework that accurately fits the known areas of nuclear matter near the equilibrium density n_0 and then extrapolating to high density relevant to compact stars, presently the only source of such extreme matter. There are presently no terrestrial laboratories available for such a density regime. In doing this, one constructs a model, mostly phenomenologically or chiral-symmetry-implemented, that is "accurately" fit, with certain number of parameters, to available experimental data in the vicinity of normal nuclear matter and then extrapolated to high density. For most of the workers in this field, much effort is currently paid to meet a number of constraints of symmetric matter and asymmetric nuclear matter empirically determined at near n_0. These "constraints" are imposed as the essential quantities to be satisfied by *any* approach for accessing compact stars, the densities involved there being in the order of (5-7) times n_0. In the absence of both experimental data and reliable theoretical tools, how to extrapolate to higher density is very much unknown. Therefore what is usually done with such a model is to adjust whatever additional freedoms there may be in the model to reproduce the star properties, such as maximum mass, radius and other observables consistent with certain global constraints such as causality etc. In this class of approaches, the objective is to come up with a model to explain or postdict the given star properties, with the assumption that the given theory or model is continuous from low density, say, n_0, to high density $\sim$ (5-7) n_0. Accurately satisfying the "constraints" set a n_0 which is *imposed* as the *necessary* criterion for a model to be considered valid. A model that may adequately explain star properties but does not meet accurately the n_0 constraints is then "ruled out."

In this volume, we are adopting a drastically different attitude. Our view is that a theory or a model that can explain correctly dense matter of compact stars need not necessarily reproduce the nuclear-density constraints. And vice-versa. As in the previous volumes, we start with an effective field theory Lagrangian matched at a certain matching scale Λ_M with QCD. The effective Lagrangian is implemented with certain symmetries which are "hidden," hence not visible, in QCD in the vacuum but assumed to manifest in medium at increasing density. The emergent symmetries could encode phase changes of the matter which need not belong, i.e., topology change, to the standard Ginzburg-Landau-Wilsoninan paradigm of phase transitions. Consequently one Lagrangian can encompass several phases with no changes in degrees of freedom in going from the density regime n_0 to the compact-star-density regime, with the parameters of the Lagrangian controlled by — *i.e.*, sliding with the symmetries emergent in the medium. This is a novel approach, only recently developed, to be confronted with Nature although the basic ingredients of the reasoning have been around. As such, it involves a variety of assumptions, though deemed robust but yet to be confirmed.

This monograph draws largely from the works done, and discussions, with a large number of collaborators. In particular, most of the ideas developed in the five-year (2008-2013) World-Class University Program at Hanyang University directed by Hyun Kyu Lee form the crucial part of the volume. MR participated in the Program as a Guest Scholar and YLM was a frequent visitor there, collaborating with the members of the program. It is highly proper to emphasize that the principal objective of the WCU/Hanyang Program was to presage the potential novel development in the forthcoming nuclear physics project at the RIB (rare-ion-beam) accelerator called "RAON" in the Institute of Basic Science in Korea. This was the only WCU program among many with the future in view of exploring a totally uncharted domain of nuclear astrophysics. The topic of the WCU/Hanyang Program was "From Dense Matter to Compact Stars" with the aim to link what will be studied at RAON to the CBM (compressed baryonic matter) program at FAIR/Darmstadt and ultimately to the gravity-wave observables of coalescing neutron stars to be measured at various LIGOs. It is appropriate to note that while completing this volume, the first gravity-wave signal from coalescing neutron stars GW170817 was detected, offering a direct window for the first time into dense baryonic matter discussed in this volume.

Above all, this volume goes a long way towards supporting what Gerry Brown wrote in his last, unpublished, paper: "Landau-Migdal fixed point interaction G_0' (a.k.a., 'Ericson-Ericson-Lorentz-Lorenz term'), nuclear tensor force and 'Brown-Rho scaling' run the show and make all nuclear forces equal."

We would like to acknowledge invaluable discussions and collaborations with the core members and participants of the Program, Tom Kuo, Hyun Kyu Lee, Won-Gi Paeng and Sang-Jin Sin. Our views are strongly influenced by the works of, and by numerous correspondences with, Rod Crewther, Masayasu Harada, Lewis Tunstall and Koichi Yamawaki.

Yong-Liang Ma and Mannque Rho

Notation and Conventions

bsHLS	scale-invariant baryonic hidden local symmetry
sHLS	scale-invariant hidden local symmetry
BPS	Bogomoln'yi-Prasad-Sommerfeld
CC	Cheshire Cat
CCP	Cheshire Cat Principle
χ	conformal compensator field
χPT	chiral perturbation theory
$\chi\mathrm{PT}_n$	$SU(n)_L \times SU(n)_R$ chiral perturbation theory
$\chi\mathrm{PT}_\sigma$	scale-chiral perturbation theory
CFL	color flavor lock
DFT	density functional theory
DLFP	dilaton-limit fixed point
EDF	energy density functional
EFT	effective field theory
EoS	equation of state
$\epsilon^{\mu_1\mu_2\cdots\mu_D}$	totally anti-symmetric Levi-Civita tensor with $\epsilon^{12\cdots d} = 1$.
FT	Folk Theorem
f_π	pion decay constant
f_σ	dilaton decay constant
G^a_μ	gluon field
g_s	color gauge coupling
g (g_V)	hidden gauge coupling
$\hbar = c = 1$	units adopted
HBχPT	heavy baryon chiral perturbation theory
HLS	hidden local symmetry
IDD	Intrinsic density dependence
IRFP	Infrared fixed point

k_F	Fermi momentum
λ_a	Gell-Mann matrices
LEC	low energy constant
MFT	mean field theory
n	nuclear matter density
n_0	normal nuclear matter density
$n_{1/2}$	onset density for half-skyrmion phase
n_c	critical density of chiral restoration
n_{VM}	vector manifestation density
NGB	Nambu-Goldstone boson
Ψ	nucleon doublet
q	light quark fields
Q	heavy quark fields
QCD	quantum chromodynamics
RG	renormalization group
RMF	relativistic mean field
SkEF	Skyrme energy functional
σ	dilaton field
SNPA	standard nuclear physics approach
τ_a	Pauli matrices
$\theta_{\mu\nu}$	energy-momentum tensor
VeV	vacuum expectation value
VM	vector manifestation
VMD	vector meson dominance

Contents

Chapter 1

Introduction

Where the proton (or generally nucleon) mass comes from remains, even after the discovery of Higgs boson, as one of the great mysteries of Nature. In this volume, we explore the possibility that the answer could perhaps be found in nuclear physics by way of unraveling the structure of highly dense matter present in the interior of massive compact stars.

The problem can be stated as follows. Given that the mass of the proton is very accurately measured, 938.272081$\pm$0.000006 MeV [Patrignani (2016)], more than 99.9% of the "visible" mass around us — all the way down to atoms — can be accounted for in great accuracy by adding the number of nucleons involved in the system. Even that of the nucleus which is in the core of atoms is accounted for up to 98%. The strong interactions taking place inside the nucleus are now quantitatively described by quantum chromodynamics (QCD). So the mass of a nucleus is nearly completely given by the sum of the mass of the nucleons in the nucleus with only a small correction of binding energy, less than 2% of the proton mass, and even that small value can be fairly well explained, though in a highly intricate way, by QCD. This additive accountability, however, ends abruptly — and singularly — at this point. The proton mass, the source for the nuclear mass, is no longer accounted for as a sum of "something." In QCD, the relevant constituents are quarks, and for the nucleon the (current) quarks involved, "up" and "down," are very light, less than 5 MeV. Thus how the proton mass arises from its constituents must be drastically different from how the mass of a nucleus comes about. Effectively, QCD for the proton is a theory for "mass without masses" [Wilczek (2006)].

For most of particle physicists, however, the proton mass is satisfactorily "explained." Hence the end of the story. In Wilczek's words [Wilczek (2006)], "Here there are no uncontrolled approximations, no perturbative theory, no cutoff, and no room for fudge factors. A handful of parameters, inserted into a highly constrained theory of extraordinary symmetry, either will or won't account for the incredible wealth of measured phenomena in the strong interactions. And it certainly appears that they do." Indeed lattice simulation with QCD with no mass terms and with the heavy quarks c, b and t ignored, termed "QCDLite" by Wilczek, predicts $\sim 95\%$ of the proton mass. If pressed to explain in more detail, they will then say it comes from the massless gluons interacting with the nearly massless, high-momentum quarks, winding up confined in a "bag." In a nutshell, one may say the mass is generated due to the breaking of chiral symmetry associated with the nearly massless quarks by the confinement [Casher (1979)]. In the standard paradigm, the chiral symmetry spontaneously broken leading to the mass generation is characterized by the non-zero quark condensate, $\langle\bar{q}q\rangle \neq 0$, a potential chiral order parameter.

To nuclear physicists, ironically, this explanation raises more questions than answers. This may be due to the fact that confinement is very poorly — if at all — understood. For mathematicians, it is "One-Million Dollar Clay Millenium Problem", still to be solved. It may even be, as some argue, that "deconfinement", corollary to confinement, does not take place within the framework of QCD. For nuclear physicists, it presents numerous puzzles, some of which bear directly on observables as we will describe below.

How to decipher the mystery of the proton mass is the most fundamental issue in nuclear physics.

In this volume, we address this issue by means of "un-breaking" the spontaneously broken — or, more precisely, "unhiding" the hidden — chiral symmetry. The idea is to "dial" the quark condensate to zero. This could be done by heating hadronic matter to high temperature and/or crushing baryonic matter to high density. Hadrons have been heated in heavy-ion collisions to high temperature, say, several hundreds of MeV, allowing us to have a glimpse of the Universe at the age of $\sim 10^{-6}$ second after the Big Bang. Baryonic matter is being, and will be soon more strongly, compressed to a density several times the normal nuclear matter density $n_0 \simeq 0.16\ \mathrm{fm}^{-3}$, mimicking the conditions that are supposed to be present in the interior of compact stars. These processes bring hadronic matter to a temperature or density at which the spontaneously broken chiral symmetry is restored and/or the quarks and gluons get deconfined. Along the way the matter experiences a variety of conditions in which the mass and coupling constants get modified significantly, influencing crucially the properties of hadrons living in these environments. Lattice QCD, the only trustful tool known for highly correlated strongly-interacting matter, has, backed by heavy-ion accelerators, beautifully shed light on hadronic matter at high temperature. The near perfect quark-gluon liquid has been discovered, exposing a picture drastically different from, and far more exciting than, what was naively expected, weakly interacting quark-gluon plasma. But the situation at high density is totally different. Since the lattice technique cannot yet access — and there are no other known nonperturbative tools to describe — highly dense matter, the strongly compressed baryonic matter relevant in the interior of massive compact stars remains largely unknown.

The assertion in this volume is that effective field theories (EFTs for short) offer the most promising approach to access dense baryonic matter.

The approach adopted in this volume is anchored on the "Folk Theorem (FT)" on EFTs championed by Steven Weinberg [Weinberg (1997)]. This theorem, briefly stated in Chapter 2, will be the underlying strategy throughout the volume. In fact nuclear physics offers a surprisingly convincing "folk proof" of this theorem, perhaps even better than in particle physics. The early "proof" of this theorem was recounted in the volume CNDII [Rho (2008)] and there have been more recent impressive developments, in various nuclear structure phenomena, that reconfirm the early vindication of the FT in nuclear physics. In this volume the aim is to venture further, boldly, into that direction for accessing the uncharted domain of high density. It will inevitably involve certain uncertainties and complications that need to be lifted by future experiments and of course theoretical progress. The import of this development is that EFT, not only making a connection to QCD, can supplement and improve on — not just replace — the standard nuclear physics approach (SNPA for short). What makes EFT more powerful than the SNPA in its framework is that it can make predictions that go beyond what was feasible in standard nuclear physics approaches. In doing this, we revisit certain concepts treated in CNDI and CNDII, sharpening them — and correcting errors committed therein — with more recent theoretical insight and contact with nature.

We define the appropriate languages to address this issue in Chapter 2. The notations that will figure in the volume are defined in this chapter. We revisit in Chapter 3 the thesis treated in both CNDI and CNDII on the Cheshire Cat phenomenon that addresses the hadron-quark continuity in nuclear processes, one of the principal themes of this volume. The reason for this is that that notion, so crucial for making contacts with the topological aspect of nucleon structure, reappears in holographic QCD and in a much broader context in the equation of state (EoS) that enters in the structure of compact stars. How EFT works in nuclear processes at a density in the vicinity of normal nuclear matter is outlined in Chapter 4 with the aim to extrapolate to high-density regime where the EFT has not been tested. Its connection to phenomenological approaches, such as "energy density functional (EDF)" popular in nuclear community, is pointed out there. Various symmetries hidden in QCD proper, such as scale symmetry, hidden local symmetry (HLS), "mended symmetries" are discussed in Chapter 5 with the suggestion that their tell-tale signals can emerge in strongly-correlated baryonic systems at high density. To apply the EFTs to dense matter to explore the pattern of the emerging symmetries, effective Lagrangians are constructed in Chapters 6-8. The notable case is the scale symmetry. Scale symmetry, manifest at the classical level in gauge theories, is broken explicitly at the quantum level by a quantum anomaly and can also be broken spontaneously. Whether and how it can manifest in QCD is highly controversial, and figures importantly at large number of flavors (N_f) for dilatonic Higgs for going beyond the Standard Model in particle physics. In this volume, it is argued to be a crucial element *also* in the EoS for dense matter and the origin of the nucleon mass.

Baryons as topological solitons, skyrmions, treated in both CNDI and CNDII, are revisited in Chapter 7. Skyrmions in lower dimensions are found to describe a variety of fascinating phenomena in condensed matter physics. And they are also figuring, in an intriguing way, in nuclear structure as one can see in [Rho and Zahed (2016)]. It turns out as one finds in this chapter that it plays an extremely important role in the EoS for dense matter with a hint at the source of the proton mass. There is a phase structure of baryonic matter that is not visible in models that do not resort to topology that appears at a density 2 or 3 times the normal nuclear matter density. There emerges, at high density, parity-doubling that indicates that the proton mass has an important component that is not *directly* associated with the spontaneous breaking of chiral symmetry. At this density, nothing much is known experimentally and hence whether such a phase change is real or not is not known. What is remarkable is that it has a significant qualitative impact on baryonic matter at a density relevant to massive compact stars. How the topology change influences the structure of the EFT Lagrangian in the wide range of densities involved is discussed in Chapter 8. In Chapter 9, the EFT so determined is applied, nearly parameter-free, to massive compact stars. The predictions made are the maximum star mass, the radius, the sound velocity and the tidal deformability in gravitational waves with an initial contact with the recently observed gravitational waves coming from merging neutron stars GW170817 [Abbott *et al.* (2017)]. The connection between the source of the proton mass and the emergence of hidden symmetries is inferred from these results. The notion that the hadron-quark continuity can be applied to the EoS of compact stars is put forward in this chapter. All the way to massive compact stars with the central density $\sim 6n_0$, strangeness did not figure in the description. This matter will be treated in Chapter 10 where various arguments are given to the effect that *explicit* strangeness degrees of freedom need not figure in the density regime involved in the massive compact stars so far observed.

Chapter 2

The "Folk Theorem" in Nuclear Physics

2.1 Foreword

As stated in Introduction, our principal theoretical tool will be effective field theory (EFT). It will cover the whole range of nuclear physics, from finite nuclei and infinite nuclear matter to highly compressed baryonic matter relevant to compact stars. In the first part of this Chapter, we will give a precise definition of what EFT that concerns us is all about, then give a short account of how well EFT works in nuclear physics. In the second part of the chapter, we give the minimalistic list of the basic elements in the fundamental theory, put in non-rigorous terms, that are to figure in one or more of EFTs we want to construct. Those ingredients will then be implemented in the following chapters to the best that is feasible to capture a variety of different concepts that we deem indispensable in building the theoretical framework to address the range of phenomena that we are interested in. The ultimate aim is to have one unified approach with as few a set of parameters as feasible.

Quantum Chromodynamics (QCD) is undoubtedly the correct theory for strong interactions, as it is now widely recognized, so it should apply not only to elementary hadrons but also to nuclear physics as well as compact-star physics. But it cannot at present directly access the latter two with any degree of accuracy, particularly the complex structure of compact-star matter. It will be some time before it will be able to do so. EFTs in a variety of different guises are currently the only tools available to explore the wide variety of strong interactions that take place in going from quarks to hadrons to superdense matter involved. And the only promising one.

So what is effective quantum field theory?

This question in a broad context was raised in 1970's and addressed by Steven Weinberg in what he called "Folk Theorem" on effective quantum field theories [Weinberg (1979, 1997, 2016, 2001)].[1] The folk theorem (FT for short) states: "If one writes down the most general possible Lagrangian, including all terms consistent with assumed symmetry principles, and then calculates matrix elements with this Lagrangian to any given order of perturbation theory, the result will simply be the most general possible S-matrix consistent with perturbative unitarity, analyticity, cluster decomposition, and the assumed symmetry properties."

This theorem would apply to *all* interactions in Nature going beyond the Standard Model even if there existed a confirmed "Theory of Everything" with which everything could possibly be calculated. The most striking cases are found in condensed matter physics. In strongly correlated condensed matter systems a large variety of EFTs are uncovering a wealth of fascinating phenomena, exposing emergent symmetries without visible connection to QED, the arguably "fundamental theory" for condensed matter systems. In particle physics the Standard Model (SM) is a preeminent case of EFT in search of more fundamental theory or "ultimate theory" attempting to go beyond. Unlike in QED, QCD, the "fundamental theory" for strong interactions, cannot yet be solved in the nonperturbative regime and so is far from being able to directly address nuclear phenomena. Thus EFT is the only approach available in nuclear physics. And it has proven to be highly successful.

In this chapter it will be argued that this impressive success of EFT in nuclear physics could be taken as a "proof" of the FT and in our view one of the most solid case of how the theorem works in strong interaction physics. In fact the initial development dates back to 1960's with Skyrme's pioneering work on (a) unified theory of mesons and baryons [Skyrme (1962)] and (b) energy density functional (EDF) in nuclear physics [Skyrme (1959)], and it has evolved with the recognition of chiral symmetry in nuclear dynamics via the current algebras in 1970's before the advent of QCD, and then chiral perturbation theory (χPT) in 1990's for nuclear potentials and electroweak (EW) currents.

We first illustrate with a few selected, albeit non-exhaustive, cases of how the "proof" works out in nuclear processes. We take the stand that the FT will be more quantitatively vindicated in nuclear physics under normal conditions, that is, up to the normal nuclear matter density n_0 or

[1] We put the quotation mark on the folk theorem to distinguish it from the conventional theorem in mathematics and theoretical physics. One possible "folk" proof offered is "how could it go wrong?"

slightly above, as higher-order corrections are calculated with more powerful numerical techniques within the well-defined EFT scheme and more and higher-accuracy experimental date become available.

This illustrative discussion will be followed by what's in QCD that the effective theory should embody. Those ingredients that will figure will be introduced, with, in some cases, detailed formulas that will be later referred to but not written down due to their lengthy structure.

We then outline the scheme to go beyond what's established by the FT to the domain for which very little is established, both theoretically and experimentally. We will give the premises on which our arguments will be built, outlining the core ideas that will be developed in as much detail as feasible in the chapters that will follow. Our ultimate aim is to confront the extreme condition of density that is considered to be relevant in massive compact stars, and make predictions to be checked in the future observations to be made in both terrestrial and spatial laboratories. We admit that some of the assumptions we are forced to make to make progress are highly controversial or conjectural and will require to be given a rigorous scrutiny.

2.2 How nuclear effective field theory fares

2.2.1 *Nuclear proof: "Chiral filter" mechanism*

The current breakthrough in EFT in nuclear physics initially germinated from Weinberg's 1979 article on phenomenological Lagrangians for pion dynamics [Weinberg (1979)] and got materialized in 1990's in the formulation from effective chiral Lagrangians of nuclear forces [Weinberg (1990, 1991); Ordónez and van Kolck (1992); Ordónez, Ray and van Kolck (1994)] and of electroweak responses [Rho (1991)]. A remarkable progress has been made since then on formulating systematic approaches to wide-ranging nuclear dynamics, ranging from nuclear forces to nuclear structure.[2] The EFT approach as it stands now is established as the standard tool to probe strongly-interacting correlated matter under normal conditions, say, up to $\sim n_0 \approx 0.16$ fm^{-3} or slightly above. It is also the *only* tool available to go beyond the normal towards the extreme conditions met in compact stars and merging of neutron stars giving rise to gravity waves.

[2]There are numerous reviews on the current status of nuclear EFT, among which are [Meissner (2016); Machleidt and Sammarruca (2016)]. For a recent status on how nuclear EFT works for the spectra of light nuclei, see [Piarulli *et al.* (2018)].

To see how this development constitutes a proof of the FT in nuclear physics, we go back to the pre-QCD period. For this we focus on nuclear response functions to external fields instead of nuclear forces. This is because while the nuclear currents and spectra are intimately connected to each other, the nuclear response functions can be more accurately calculated and compared with experiments than the nuclear spectra.

That chiral symmetry could play an important role in nuclear physics was recognized already in early 1970's. It was implicit in considerations of pion-mediated nuclear potentials, but the first realization of its potential impact in nuclear physics was in nuclear response functions to the electroweak (EW) external potential. There was no EFT then, but just phenomenological Lagrangians. Both the nuclear potentials and response functions were then calculated only at tree order with the Lagrangian since higher order had no meaning. It was in calculating the EW response functions that the soft-pion theorems were applied to pionic exchange currents. Given that the pion-exchange represented dominant contribution, with massive meson degrees of freedom suppressed by their heavy mass, the soft-pion contribution with constraints from the current algebras were calculated reliably parameter-free. This represented the first indication that chiral symmetry could figure significantly in nuclear processes [Chemtob and Rho (1971)]. The further development in this line of reasoning led to the prediction that the $M1$ matrix element and the weak axial-charge matrix element should receive an important contribution from the soft-pion theorems. It was recognized on the ground of the soft-pion theorems that the axial-charge transitions can have an extremely clean meson-exchange effects [Kubodera, Delorme and Rho (1978)]. This pre-eminent role of soft pions was dubbed then as "chiral filter hypothesis" and later given a justification in χPT [Rho (1991)]. Indeed the nuclear EFT so formulated led to high precision calculations of electroweak processes in light nuclei, a brief summary of which is given in [Kubodera and Rho (2011)]. Among them highly notable are the solar pp fusion $p + p \to d + e^+ + \nu_e$, the νd reactions $\nu_e + d \to e^- + p + p$, $\nu + d \to \nu + n + p$, the solar Hep process $^3\mathrm{He} + p \to^4 \mathrm{He} + \nu_e + e^+$ etc. The first two precesses have been computed parameter-free with a theoretical precision of less than 1% error and could be further improved. Together with the thermal np capture $n + p \to d + \gamma$, these are to date the most precise theoretical calculation that has ever been achieved in nuclear physics. As for the Hep process, the first calculation performed by Salpeter in 1950's and the present-day EFT calculation differ by roughly 2 orders of magnitude. Calculating the Hep process with some accuracy was heralded by John Bahcall [Bahcall (2000); Park *et al.* (2003)] as the outstanding challenge in nuclear structure physics. At present it can be calculated within an error bar of $\sim 15\%$ [Park *et al.* (2003)], a giant progress since 1950's. An even more accurate calculation, unquestionably doable given the currently available techniques, remains to be done.

An extremely interesting process, both in the context of nuclear EFT, double soft pions and going beyond the standard chiral perturbation theory with pions only, dubbed SχPT, is the first forbidden process involving the axial-charge operator $J_5^{0\pm}$,

$$A(0^\pm) \to B(0^\mp) + e + \nu, \quad \Delta T = 1. \tag{2.1}$$

In nuclear EFT, the leading — in the chiral power counting — one-body axial-charge operator has the form

$$J_5^{0\pm}(\vec{x})_{1-\text{body}} = -g_A^{\text{t}} \sum_i \tau_i^\pm \vec{\sigma}_i \cdot (\vec{p}_i - \vec{k}/2)/m_N \delta(\vec{x} - \vec{x}_i) \tag{2.2}$$

where $\vec{p}$ is the initial momentum of the nucleon making the transition and $\vec{k}$ is the momentum carried by the axial current. The nucleon in nuclei is taken to be non-relativistic in writing down the single-particle operator. This is justified as it will be seen that the nucleon mass remains at all densities involved near its vacuum value. Since Lorentz invariance is spontaneously broken in nuclear systems, the axial coupling constant g_A has the time component g_A^t which is different from the space component g_A^s. How to determine g_A in medium will be discussed in Chapter 8. We borrow the result given there

$$g_A^t = g_A \frac{f_\pi}{f_\pi^*} \tag{2.3}$$

where f_π^* is the pion decay constant in nuclei that depends on the average density of the system. The pion decay constant is determined experimentally in deeply bound pionic systems and it has been determined at normal nuclear matter density n_0 [Kienle and Yamazaki (2004)][3]

$$f_\pi^*(n_0)/f_\pi \equiv \Phi(n = n_0) \approx 0.8. \tag{2.4}$$

To the next order in the chiral counting, two-body currents contribute. The most important one is the pion-exchange term depicted in Fig. 2.1. Consider the upper vertex. If one considers the time component of the axial field as soft as it would be in the first-forbidden beta decay, then the

[3]This value of f_π^* at $n = n_0$ will appear frequently in what follows and will be considered as established.

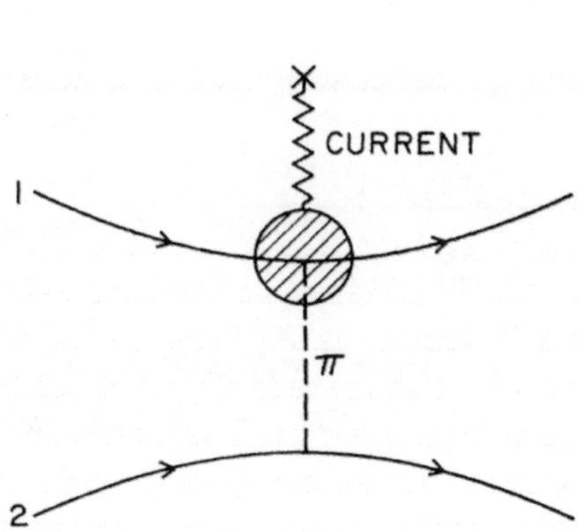

Fig. 2.1 Two-body exchange current. The upper vertex involves two soft pions for the axial charge transition.

vertex effectively involves two soft pions controlled by the current algebra, and the axial-charge contribution gives an $O(1)$ term. This was recognized already in 1978. In terms of chiral Lagrangian this is automatically encoded at the leading tree order with higher order terms manifestly suppressed. It is easy to work out the two-body charge operator [Kubodera, Delorme and Rho (1978)]

$$J_5^{0\pm}(\vec{x})_{2-\text{body}} = g_A^t \Phi^{-1} \frac{m_\pi^2}{8\pi f_\pi^2} \times \sum_{1<j} (\tau_i \times \tau_j)^{\pm} [\vec{\sigma}_i \cdot \hat{r} \delta(\vec{x} - \vec{x}_j) + (i \leftrightarrow j)] \mathcal{Y}(r) \quad (2.5)$$

with

$$\mathcal{Y} = \frac{e^{-m_\pi r}}{m_\pi r} \Big(1 + \frac{1}{m_\pi r}\Big).$$

Given accurate wavefunctions, the transition matrix elements of both the one-body and two-operators can be calculated accurately. If one defines R as

$$\frac{R}{\Phi(n)} = \frac{\langle B | J_5^{0\pm}(\vec{x})_{2-\text{body}} | A \rangle}{\langle B | J_5^{0\pm}(\vec{x})_{1-\text{body}} | A \rangle} \quad (2.6)$$

it is found to be highly insensitive to the wave functions and hence to the density of the system. It is also surprisingly big as a "correction"

$$R = 0.5 \pm 0.1 \quad (2.7)$$

ranging in mass number from $A = 12$ to $A = 280$. This extreme insensitivity to nuclear structure (and density) can be understood by the fact that the accurately calculable two-body operator is long-ranged given by a soft-pion exchange.

The effective axial-charge operator is then given by replacing g_A^{t} by the effective one

$$g_A^{t\ \mathrm{eff}} = g_A^t\left(1 + \frac{R}{\Phi}\right). \tag{2.8}$$

Precise data are available in the Pb region $A = 205 - 212$ [Warburton (1991)]. The analysis in [Warburton (1991)] is given in terms of the medium-renormalization of the free-space axial-coupling constant g_A

$$\epsilon = g_A^{t\ \mathrm{eff}}/g_A. \tag{2.9}$$

The prediction for $n = n_0$ [Kubodera and Rho (1991)] and the experiment in the Pb region [Warburton (1991)] are

$$\epsilon(n_0)_{theory} = 2.0 \pm 0.2, \quad \epsilon_{exp} = 2.01 \pm 0.05. \tag{2.10}$$

This quantitative agreement represents one of most successful check of EFT in nuclear physics. Similar enhancements are also found in the lighter nuclei $A = 12 - 15$. There is a great advantage with these night nuclei, offering more powerful tests. In this mass range, an accurate calculation in full consistency with χPT [Rho (1991)] should be feasible since higher-order corrections to the two-body axial-charge operator, e.g., multi-body currents, are strongly suppressed and *ab initio* nuclear structure calculations should be doable.

There is the other side of the coin to this phenomenon termed "chiral filter hypothesis" [Kubodera, Delorme and Rho (1978); Rho (1991)] that when soft-pions are suppressed, there are no good reasons to think that corrections appearing at higher chiral orders are not suppressed at increasing order. This meant that the exchange current that enters in the Gamow–Teller transitions, controlled by the space component of the axial current, for which soft-pion terms are suppressed, cannot be reliably computed at a few next orders in chiral perturbation expansion. It is necessary to go to very high orders. This has been confirmed very recently in *ab initio* quantum Monte Carlo calculations of axial-vector transitions in light nuclei [Pastore *et al.* (2018)]. It was shown that to get the meson exchange correction to the leading-order one-body operator, one has to go at least to N^4LO and the correction amounts to less than $\sim 3\%$.

Together with the enhanced axial-charge transitions, this observation offers a highly compelling proof of the FT.

2.2.2 *Soft theorems and the "quenched g_A problem" in nuclei*

An important spin-off of the FT in action described above is the resolution of the four-decades-old problem of quenched g_A in nuclear Gamow–Teller transitions. It has been observed since 1970's [Wilkinson (1978); Engel and Menèndez (2017); Suhonen (2017)] that Gamow–Teller beta decay transitions could be explained in simple shell models only if one took the effective $g_A^* \approx 1$, that is, $\sim 20\%$ quenched from the current experimental value in the vacuum, $g_A = 1.27$. This mystery can be very simply resolved in terms of the soft-pion theorems and hidden scale and local symmetries implemented in the spirit of nuclear FT as developed in Chapters 6-8. This stunning development is recounted in Chapter 4. There the soft-pion theorems via chiral filter enter into the nuclear FT and hint at the possibility that the deep issue of the way a variety of soft theorems figure in the infrared structure of gravity and gauge theories [Strominger (2017)] may also manifest in nuclear dynamics.

2.3 Essentials of QCD

In going to the uncharted domain of high density, our ultimate goal being massive compact-star matter, we will be going beyond the regime where effective quantum theories are more or less well-formulated as discussed in the preceding section. This process of going beyond entails, unavoidably, various assumptions which may not be always justifiable. In this section we summarize as concisely as possible the aspects of QCD that are accepted as "facts," hence should underlie the theoretical arguments, and will be invoked whenever needed for formulating the EFT. Details will be avoided.

For completeness and for clarity in notations that will be adopted in the volume, we write down the QCD Lagrangian (found in *all* physics textbooks)

$$\mathcal{L}_{\rm QCD} = \sum_l \left\{ \bar{q}_{l,\alpha} i \left(\partial_\mu \gamma^\mu - i g_s G_\mu^a T^a \gamma^\mu\right)_{\alpha\beta} q_{l,\beta} - m_l \bar{q}_l q_l \right\} - \frac{1}{4} G_{\mu\nu}^a G^{a\mu\nu} \tag{2.11}$$

where T^a is the generator of $\mathrm{SU}(3)_c$ group satisfying $\mathrm{tr}(T^a T^b) = (1/2)\delta^{ab}$, the subscripts α, β stand for the color indices, the summation is over the flavor index l, $G_{\mu\nu}^a = \partial_\mu G_\nu^a - \partial_\nu G_\mu^a - i g_s f_{abc} G_\mu^b G_\nu^c$ is the field strength tensor of gluon fields, g_s is the QCD gauge coupling constant and f_{abc} is the structure constant of $\mathrm{SU}(3)_c$ group. In QCD, the fundamental parameters are the coupling constant g_s (or $\alpha_s = g_s^2/4\pi$) and the current quark masses m_l.

What we will deal with is, formally, the generating functional. For QCD, it is of the form

$$Z[J]_{\text{QCD}} = \int \mathcal{D}q\mathcal{D}\bar{q}\mathcal{D}G_\mu \exp\left\{\int d^4x \left[\mathcal{L}_{\text{QCD}}^{m_q=0} + \bar{q}Jq\right]\right\}, \quad (2.12)$$

in the presence of the external fields $J(x)$ with $J(x) = -s(x) + ip(x)\gamma_5 + \not{v}(x) + \not{a}\gamma_5$ for the sources for the scalar (s), pseudoscalar (p), vector (v_μ) and axial-vector (a_μ) currents.

Along the line of the FT, we will focus on EFTs that embody QCD in the sense described above. For standard chiral perturbation theory (SχPT),

$$\mathcal{L}_{S\chi\text{PT}} = \mathcal{L}_{S\chi\text{PT}}(U, \partial_\mu U, \cdots), \quad (2.13)$$

the generating functional becomes

$$Z[J]_{S\chi\text{PT}} = \int \mathcal{D}U \exp\left\{\int d^4x \mathcal{L}_{S\chi\text{PT}}(U, J; \partial_\mu U, \cdots)\right\} \quad (2.14)$$

where U is the chiral field to be defined precisely below. The objective of this volume is to go beyond the $S\chi$PT, incorporating hidden gauge fields and hidden dilaton field in the effective field action so constructed, and access dense baryonic matter systems. For this objective, we take into account as faithfully as possible the (generally) accepted properties of QCD. They are:

(1) **Confinement**:
We will take it for granted that at zero temperature and zero density, quarks and gluons are confined and that at some temperature $T_{c_1} > 0$ and some density $n_{c_1} > n_0$ (where $n_0 \simeq 0.16$ fm^{-3} is the normal nuclear matter density), quarks and gluons are deconfined. How this happens, which is still unknown, will not be addressed. In Chapter 3, it will be caricatured in what will be referred to as "Cheshire Cat phenomenon."

(2) **Chiral symmetry**:
We will be mainly dealing with the up (u) and down (d) quarks, but the strange (s) quark will figure as well. We will be mostly considering the "chiral limit" where $m_u \approx m_d \approx m_s \approx 0$ although in nature, the u- and d-quark masses are a few MeV and the s-quark mass ~ 100 MeV. Unless otherwise stated, these will be referred to as "light quarks."

The QCD in this limit is referred to as "QCDLite" [Wilczek (2006)]. The QCD Lagrangian is invariant under chiral transformations. This is the $SU(3)_L \times SU(3)_R$ chiral symmetry for $N_f = 3$.[4] Thus both the left (L) and right (R) currents or alternatively the vector and axial vector currents are conserved. The conservation of the vector current is not affected when the quark masses are turned on because the vector ($V = L + R$) symmetry is unbroken, but the axial current will have a term proportional to the quark mass if the quark mass is not zero.
The chiral symmetry can, however, be broken or rather "hidden" by the vacuum. This is referred to as "spontaneous breaking." The vector symmetry is proven to remain unbroken [Vafa and Witten (1984)], but the axial symmetry is broken by the vacuum. According to the Nambu–Goldstone (NG) theorem, associated with each broken generator, there is a zero-mass boson (in QCDLite), i.e., NG boson[5] except for the axial $U(1)$ (see below). It is this NG field that U in (2.14) represents. It is now understood — and supported by nature — that the symmetry associated with the light quarks in the zero-mass limit is spontaneously broken below some temperature T_{c_2} and some density n_{c_2}.

(3) **Anomalies**:
Certain symmetries of QCD present at the classical level are broken *explicitly* by quantum effects. Such a phenomenon breaks the current conservation unlike the spontaneous breaking which leaves the relevant currents conserved. This is what is called "quantum anomaly" or "anomaly" in short. Among a variety of anomalies associated with QCD proper and effective theories thereof, what will concern us is the scale invariance in the chiral limit. The axial anomaly is also relevant to certain processes in hadron physics but will not figure directly in our discussions.

- *Trace anomaly:*
  The QCD Lagrangian with no quark masses is scale invariant. So the energy-momentum tensor $\theta_{\mu\nu}$ is traceless at the classical level. However at the quantum level, due to an anomaly, scale invariance is broken and the trace of the energy-momentum tensor is not

[4] There are in addition an unbroken $U(1)_V$ symmetry corresponding to baryon charge and anomalous $U(1)_A$ which will be mentioned below.

[5] In systems with Lorentz invariance spontaneously broken, i.e., in medium, this one-to-one correspondence between broken generators and NG bosons may not hold. We will come to this matter in later chapters.

zero[6]

$$\theta^\mu_\mu = \frac{\beta(\alpha_s)}{4\alpha_s} G^a_{\mu\nu} G^{a\mu\nu} \tag{2.15}$$

with $\beta(\alpha_s)$ being the beta function for the QCD gauge coupling constant. The non-vanishing trace of $\theta_{\mu\nu}$ means that the scale invariance present in the Lagrangian is explicitly broken quantum mechanically. The divergence of the dilation current D_μ associated with scale symmetry is non-zero, proportional to the trace of $\theta_{\mu\nu}$, θ^μ_μ. This matter will be discussed in detail in Chapter 6.

- *$U(1)_A$ anomaly*:
  While the $SU(N_f)$ axial currents are conserved, spontaneously broken at low temperature and low density, and considered to be restored at some temperature, T_{c_2}, and some density, n_{c_2}, the $U(1)$ axial current is not conserved due to a triangle[7]

$$\partial_\mu j^{\mu 5} = \frac{N_f \alpha_s}{4\pi} \epsilon^{\mu\nu\alpha\beta} \mathrm{tr} G_{\mu\nu} G_{\alpha\beta} \,. \tag{2.16}$$

  In the limit $N_c \to \infty$, the anomaly vanishes — since $\alpha_s \sim O(N_c^{-1})$ — with the $U_A(1)$ symmetry spontaneously broken. There emerges a NG boson. In nature, the would-be NG boson, η', is massive given that $N_c \ll \infty$. The question raised in nuclear physics is whether the anomaly can be made to disappear at high temperature and/or density. In this volume, we will not deal directly with this anomaly.

(4) **Large N_c**:
QCD in the large N_c limit provides useful, thus far only known analytic, information on nonperturbative aspects of the strong interactions at high density. In nature, N_c is 3, but in some sense which cannot be given a precise meaning, $N_c = 3$ can be considered much bigger than the number of flavors $N_f = 2, 3$ that we are dealing with. In terms of the power of N_c factors, the counting goes as follows:

- *Gluons*
  The counting in the gluon sector is done in terms of the 't Hooft constant $\lambda = g_s^2 N_c$. This is taken to be $\mathcal{O}(1)$.[8] Thus

[6] With non-zero quark mass there is an additional term on the RHS of the form $+\sum_f m_f(1+\gamma_m^f)\bar{q}_f q_f$ with γ_m^f the anomalous dimension for the bilnear quark field $\bar{q}_f q_f$.

[7] When the quark is massive, there is an additional term of the form $+2m_f \bar{q}_f \gamma^5 q_f$ on the RHS.

[8] In holographic QCD discussed in several chapters that follow, λ will be taken to be large in the sense that an expansion is made in $1/\lambda$.

$g_s^2 \sim \mathcal{O}(1/N_c)$. Planar diagrams dominate, with non-planar graphs suppressed by $1/N_c$ or higher.

- *Mesons*

  The leading contribution to the correlation functions of quark-antiquark currents is the sum of planar graphs, which is $\mathcal{O}(N_c)$. Thus, the decay constant of a meson made of a quark and an anti-quark scales as $\sqrt{N_c}$ and the meson-meson couplings are down by powers of $1/N_c$. For instance, the pion decay constant $f_\pi \sim \sqrt{N_c}$, the 3-meson interactions scale as $1/\sqrt{N_c}$, the 4-meson interactions $1/N_c$, *etc.*. The mass of the $\bar{q}q$-type mesons is $\mathcal{O}(N_c^0)$. They appear as sharp resonances at this order. Exotics are suppressed by $\mathcal{O}(1/N_c)$ or higher.

- *Baryons*

  Baryon mass scales as $\mathcal{O}(N_c)$. This follows from the fact that N_c quarks are needed to make "color" singlet. This plays an important role in what follows in this volume. Roughly

  $$m_B \sim N_c \sim 1/g_s^2. \tag{2.17}$$

  The mass being inversely proportional to g_s^2 indicates that the baryon can be considered as a soliton in a weak coupling theory. Indeed the skyrmion description of baryons discussed below is intimately tied to this scaling as we will see later. Simple counting shows that $g_A \sim N_c$ and $g_{\pi NN} \sim N_c^{3/2}$. There are a variety of relations between the skyrmion model and non-relativistic quark models in the large N_c, which we will invoke frequently.

(5) **Anomaly matching**:

In order for an effective theory to be consistent with a fundamental theory, the anomaly in EFT should be identical to the anomaly in the fundamental theory. This is the 't Hooft anomaly matching condition ['t Hooft (1980)]. In QCD with chiral symmetry realized in NG mode, the condition requires that in the hadronic world where hadrons are composites of quarks and gluons, there be a massless, colorless object which couples to the axial current and photon and reproduces exactly the triangle anomaly. The NG bosons and zero mass quarks do the matching. The same applies to the trace anomaly in terms of the gluons and the dilaton.

2.4 "QCDLite"

In this section, we explain the basic aspects of QCD that will be taken for granted in what follows. They will not be recalled unless absolutely necessary. For simplicity we will stick to the chiral limit, i.e., called "QCDLite", where the quark masses are taken to be zero [Wilczek (2006)]. Whenever chiral symmetry breaking effects, i.e., NG boson masses, are called for, they will be specifically mentioned. Otherwise our discussions will be in QCDLite.

2.4.1 *Chiral symmetry*

When a fermion is massless, its helicity cannot flip. The helicity states can be projected out by using the R(ight)- and L(eft)-handed projection operators

$$P_R \equiv \frac{1}{2}(1+\gamma_5); \qquad P_L \equiv \frac{1}{2}(1-\gamma_5). \tag{2.18}$$

Using them, one can decompose a generic fermion field as

$$\psi = \psi_R + \psi_L, \qquad \bar\psi = \bar\psi_R + \bar\psi_L, \tag{2.19}$$

with

$$\begin{aligned} \psi_R(x) &= P_R\psi(x), & \psi_L(x) &= P_L\psi(x), \\ \bar\psi_R(x) &= \bar\psi(x)P_L, & \bar\psi_L(x) &= \bar\psi(x)P_R. \end{aligned}$$

In terms of the right- and left-handed fermion fields, specifically quark fields $q_{L,R}$, omitting the flavor and color indices, the quark sector of QCDLite is of the form

$$\mathcal{L}_{\rm q} = \bar q_R i\not{D} q_R + \bar q_L i\not{D} q_L \tag{2.20}$$

where $q^T = (q_1, q_2, \cdots, q_N)$ with N-flavor light quarks and the pure gluon part has been omitted. The covariant derivative D_μ is in the color space, so the QCD Lagrangian (2.20) is invariant under the $\mathrm{U}(N)_L \times \mathrm{U}(N)_R$ flavor transformation.

Corresponding to the $\mathrm{U}(N)_L \times \mathrm{U}(N)_R$ transformation there are the Noether currents

$$\begin{aligned} J_L^{\mu,a} &= \bar q_L \gamma^\mu T^a q_L, \quad & J_R^{\mu,a} &= \bar q_R \gamma^\mu T^a q_R, \quad a = 1, 2, \cdots, N, \\ J_L^\mu &= \bar q_L \gamma^\mu q_L, & J_R^\mu &= \bar q_R \gamma^\mu q_R \end{aligned} \tag{2.21}$$

which can be written as vector and axial-vector currents,

$$J^{\mu,a} = J_R^{\mu,a} + J_L^{\mu,a} = \bar q \gamma^\mu T^a q, \quad J_5^{\mu,a} = J_R^{\mu,a} - J_L^{\mu,a} = \bar q \gamma^\mu \gamma_5 T^a q, \tag{2.22}$$

$$J^\mu = J_R^\mu + J_L^\mu = \bar q \gamma^\mu q, \qquad J_5^\mu = J_R^\mu - J_L^\mu = \bar q \gamma^\mu \gamma_5 q. \tag{2.23}$$

All these currents are conserved at the classical level, but the conservation of the color-singlet $U_A(1)$ current $J_{\mu 5}$ is explicitly broken by the triangle anomaly due to the quantum corrections [Adler (1969); Bell and Jackiw (1969)]. So, in the chiral limit, QCD Lagrangian (2.20) has global $\mathrm{SU(N)}_L \times \mathrm{SU(N)}_R \times \mathrm{U(1)}_V$ symmetry.

It is generally considered that chiral symmetry, as a global symmetry, can be realized in either of two ways: Wigner–Weyl (WW) or Nambu–Goldstone (NG). In the former, the symmetry is manifest with the ground-sate of the theory invariant under the symmetry and particles appearing in degenerate (parity) multiplets, and in the latter, the symmetry is "hidden" with the ground state non-invariant under the symmetry and particles are not in multiplet, but there appear massless spinless bosons, called NG bosons. This latter manifestation is referred to as "spontaneously broken chiral symmetry" but the term "broken" is actually a misnomer since the symmetry is not really broken but hidden. In Nature, there are pions, massless in the chiral limit, so chiral symmetry is realized in NG mode in QCD.

The systems we will consider can, in principle, be realized in a variety of ways, either singly or mixed. A state in one symmetry, say, NG, can be transformed into the other symmetry, i.e., WW. In what follows, it will be the baryonic matter at high density that is thought to induce the symmetry transformation. This phenomenon is referred to as phase transition. Such a transition, typically involving local order-parameter fields with symmetry change, is addressed by what's known as Ginzburg–Landau–Wilson (GLW) paradigm. It will turn out, however, in what seems to appear in dense baryonic medium that a changeover of state from one form to other takes place without involving well-defined order parameters to which the GLW paradigm does not seem to apply. In fact, what we will encounter involves a topology change entailing no symmetry change.

In terms of the bilinear quark fields only, i.e., the scalar and pseudoscalar quark-antiquark densities, the non-vanishing quark condensation is the sufficient condition for the spontaneous chiral symmetry breaking but is not the necessary condition because of the possibility that multi-quark condensates can also intervene. It is this situation that will occur in dense skyrmion matter as discussed in Chapter 7.

With the chiral symmetry spontaneously broken, we have

$$\langle 0|J^a_{5\mu}(0)|\phi^b(p)\rangle = ip_\mu f_\pi \delta^{ab} \tag{2.24}$$

where f_π is the pion decay constant. For relativistic systems with the Lorentz invariant ground state, it is well-known through the NG theorem that there must exist massless NGBs for each broken generators, so the number of NG bosons (NBs) is the same as the number of broken generators. However, as alluded above, for non-relativistic systems, the counting is more complicated and less clear-cut. The case we will be principally concerned with, i.e., compressed baryonic matter, at least at low densities, is non-relativistic with many-body excitations involving spinless bosonic modes. Thus there may be certain complications, as in condensed matter physics, in dealing with NG modes in matter. In the past these have not been taken into account. Our point of view adopted in what follows is that our system, specially at high density, is nearly relativistic, with Lorentz symmetry spontaneously broken and we can disregard whatever complications there may arise in NG modes. This point of view will apply equally to the putative dilaton scalar degree of freedom associated with the spontaneous breaking of scale symmetry discussed below.

2.4.2 *Scale symmetry*

Scale symmetry is a part of conformal symmetry with 15 generators. For our purpose we do not need other than scale symmetry.[9] The scale transformation is defined through the dilatation of coordinate

$$x_\mu \to x'_\mu = \lambda^{-1} x_\mu, \tag{2.25}$$

where λ is the transformation parameter. For infinitesimal transformation, $\lambda = 1 + \epsilon$ with ϵ being an infinitesimal parameter. Along the standard procedure, one can derive the generator of the scale transformation

$$D = -ix^\mu \partial_\mu. \tag{2.26}$$

For a system invariant under scale transformation, one can easily show that a field $\Phi(x)$ transforms as

$$\Phi(x) \to \lambda^\Delta \Phi(\lambda^{-1} x), \tag{2.27}$$

with $\Delta = 1$ and $3/2$ for a boson field and a fermion field, respectively. Explicitly, in QCDLite, the quark field q and gluon field G^a_μ scale-transform as

$$\begin{aligned} q(x) &\to \lambda^{3/2} q(\lambda^{-1} x), \\ G^a_\mu(x) &\to \lambda G^a_\mu(\lambda^{-1} x). \end{aligned} \tag{2.28}$$

[9]In D-dimensional conformal field theory, (D+1) generators including the dilatation can be spontaneously broken, but contrary to global symmetries here there is only one NG boson, i.e., the dilaton. This may be found in textbooks but it is currently a topical matter in the literature [Di Vecchia, Marotta and Mojaza (2017)].

QCDLite is scale-invariant at the classical level, so the Noether current, the dilation current, associated with the scale transformation

$$J^{\mu}_{D} = x_{\nu}\Theta^{\mu\nu}, \tag{2.29}$$

where $\Theta^{\mu\nu}$ is the improved energy-momentum tensor, is divergenceless and traceless

$$\partial_{\mu}J^{\mu}_{D} = \Theta^{\mu}_{\mu} = 0. \tag{2.30}$$

Though classically conserved, the quantization necessitates regularization, bringing in a scale and leads to an anomaly that breaks *explicitly* the scale symmetry [Adler, Collins and Duncan (1977); Collins, Duncan and Joglekar (1977)]. This is the trace anomaly which for QCDLite takes the form

$$\partial_{\mu}J^{\mu}_{D} = \Theta^{\mu}_{\mu} = \frac{\beta(\alpha_s)}{4\alpha_s}G^{a}_{\mu\nu}G^{a\mu\nu}, \tag{2.31}$$

where $\beta = \mu\partial\alpha_s/\partial\mu$ is the beta function for the gauge coupling constant g_s. In the presence of chiral symmetry explicit breaking, it is

$$\partial_{\mu}J^{\mu}_{D} = \Theta^{\mu}_{\mu} = \frac{\beta(\alpha_s)}{4\alpha_s}G^{a}_{\mu\nu}G^{a\mu\nu} + \sum_{q}\left(1+\gamma^{q}_{m}(\alpha_s)\right)m_q\bar{q}q, \tag{2.32}$$

where $\gamma^{q}_{m} = \mu\partial\ln m_q/\partial\mu$ is the anomalous dimension of the quark mass operator.

2.4.3 *Spontaneous breaking of symmetries*

In the case of chiral symmetry, it makes sense to talk about the "chiral limit" and hence turning off the quark mass term, i.e., QCDLite, has a well-defined meaning. Thus we can talk about "turning off" the symmetry breaking. The explicit symmetry breaking due to the quark mass, originated at a scale higher than the Standard Model, if small as for the light quarks relevant to nuclear physics, can be taken into account as perturbation. This is exemplified by χPT. *This sets the stage of 'proving' the FT.*

In contrast, the scale symmetry seems to be profoundly different. Unlike chiral symmetry, there seems to be no sense of "turning off" the explicit symmetry breaking, that is, the trace anomaly. The trace anomaly is a renormalization-group invariant quantity, so if non-zero at a given scale, it cannot be made to go to zero at any other scale. Furthermore there are arguments that say that in the absence of an explicit symmetry breaking, spontaneous symmetry breaking of scale symmetry is not possible [Freund and Nambu (1968); Fubini (1976); Serra (2013)]. This makes the NG mode of scale symmetry extremely subtle, accounting for the fact that either it is unjustifiably ignored as in nuclear physics or it generates tremendous confusion as it is in all other areas of physics ranging from condensed matter to cosmology. In Chapter 6 we will finesse this subtle issue in terms of a possible nonperturbative infrared fixed point [Crewther and Tunstall (2015); Golterman and Shamir (2016)] and the notion of emergent symmetries in medium like the compact-star matter.

2.5 Nonlinear realization of chiral symmetry and its topology

In Nature, at low energy and low density, chiral symmetry is captured in nonlinear sigma model. But at high density it seems to be more reliably realized in linear sigma model. Thus it seems that the two ways of realizing chiral symmetry in hadron dynamics can be related by "tweaking" the condition. In this section, limiting to two flavors u and d for convenience, we describe how to go from the linear sigma model to the nonlinear sigma model by tuning a parameter. This is also relevant to how scale symmetry is exposed from an EFT Lagrangian which does not exhibit scale symmetry. In later chapters where we confront Nature, we will discuss how this "tweaking" can be done in nuclear physics.

In this section we will be brief, returning to detailed discussions in Chapter 6. We will start with nonlinear sigma model because this model leads naturally to the lowest vector mesons $V = (\rho, \omega)$ emerging as local gauge fields and to the baryons emerging as topological solitons, "skyrmions," from the pion field impacted by the cloud of vector mesons. This phenomenon constitutes the principal mechanism that leads to the equation of state (EoS) of baryonic matter stiff enough to produce ~ 2 solar-mass star. These are covered in detail in Chapters 7, 8 and 9.

2.5.1 *From linear sigma model to nonlinear sigma-model and vice-versa*

We schematically write the meson fields as quark-antiquark bound states as

$$M_{ba} = (q_{b\alpha})^\dagger \gamma_0 \frac{1-\gamma_5}{2} q_{a\alpha} = \bar{q}_{R,b\alpha} q_{L,a\alpha}, \tag{2.33}$$

where a and α are, respectively, the flavor and color indices. Under the chiral transformation the matrix M transforms as

$$M \to g_L M g_R^\dagger, \tag{2.34}$$

where $g_{L,R} \in \mathrm{SU}(2)_{L,R}$ when two flavors are considered. We can decompose the matrix M in terms of the isosinglet field ϕ_0 and the isotriplet pseudoscalar $\boldsymbol{\phi}$ as

$$M = \phi_0 \mathbb{I} + i\boldsymbol{\tau} \cdot \boldsymbol{\phi}, \tag{2.35}$$

with τ_i as the Pauli matrix. In terms of the meson matrix M one can write the linear sigma model Lagrangian in the form

$$\begin{aligned}\mathcal{L}_{\mathrm{L\sigma M}} &= \frac{1}{4}\mathrm{Tr}(\partial_\mu M \partial^\mu M^\dagger) - V_0(M, M^\dagger) - V_{SB} \\ &= \frac{1}{2}\left(\partial_\mu \phi_0 \partial^\mu \phi_0 + \partial_\mu \boldsymbol{\phi} \cdot \partial^\mu \boldsymbol{\phi}\right) - V_0(\phi_0^2 + \boldsymbol{\phi}^2) - V_{SB},\end{aligned} \tag{2.36}$$

where $V_0(M)$ is the potential term which is invariant under $\mathrm{SU}(2)_L \times \mathrm{SU}(2)_R \times \mathrm{U}(1)_V$ transformation, and V_{SB} stands for the explicit chiral symmetry breaking due to the current quark mass. Since $\mathrm{SU}(2)_L \times \mathrm{SU}(2)_R \sim \mathrm{SO}(4)$, the Lagrangian (2.36), except for the V_{SB} term, has SO(4) symmetry with the basis $(\phi_0, \boldsymbol{\phi})$. We refer to this as "chiral four-vector."

Now, let us introduce α_L and α_R to parameterize the chiral transformation matrices

$$g_L = e^{\boldsymbol{\alpha}_L \cdot \boldsymbol{\tau}/2}, \qquad g_R = e^{\boldsymbol{\alpha}_R \cdot \boldsymbol{\tau}/2}. \tag{2.37}$$

Under the infinitesimal vector transformation, $\alpha_L = \alpha_R = \alpha$, the matrix M transforms as

$$M \to M' = g_L M g_R^\dagger\Big|_{\alpha_L = \alpha_R = \alpha} = M - \left[i\frac{1}{2}\boldsymbol{\alpha} \cdot \boldsymbol{\tau}, i\boldsymbol{\tau} \cdot \boldsymbol{\phi}\right] + \cdots, \tag{2.38}$$

which leads to

$$\delta_V \phi_0 = 0, \qquad \delta_V \boldsymbol{\phi} = (\boldsymbol{\alpha} \times \boldsymbol{\phi}). \tag{2.39}$$

This means that ϕ_0 is a scalar under the vector transformation while ϕ^a is a vector. Similarly, under the axial-vector transformation, $\alpha_R = -\alpha_L = \beta$, the meson matrix M transforms as

$$M \to M' = g_L M g_R^\dagger\Big|_{\alpha_R = -\alpha_L = \beta} = M + i\boldsymbol{\beta} \cdot \boldsymbol{\tau}\phi_0 + \left\{i\frac{1}{2}\boldsymbol{\beta} \cdot \boldsymbol{\tau}, i\boldsymbol{\tau} \cdot \boldsymbol{\phi}\right\} + \cdots, \tag{2.40}$$

which yields

$$\delta_A \phi_0 = -\boldsymbol{\beta} \cdot \boldsymbol{\phi}, \qquad \delta_A \boldsymbol{\phi} = \boldsymbol{\beta}\phi_0. \tag{2.41}$$

The Noether currents associated with the left- and right-handed transformations from Lagrangian (2.36) are

$$J^i_{L,\mu} = \frac{\partial \mathcal{L}_{\mathrm{L}\sigma\mathrm{M}}}{\partial \partial_\mu \alpha_{L,i}} = -\frac{i}{8}\mathrm{Tr}\left[\tau_i M \partial_\mu M^\dagger - \tau_i \partial_\mu M M^\dagger\right],$$
$$J^i_{R,\mu} = \frac{\partial \mathcal{L}_{\mathrm{L}\sigma\mathrm{M}}}{\partial \partial_\mu \alpha_{R,i}} = -\frac{i}{8}\mathrm{Tr}\left[\tau_i M^\dagger \partial_\mu M - \tau_i \partial_\mu M^\dagger M\right] \tag{2.42}$$

from which follow

$$J^i_\mu = J^i_{R,\mu} + J^i_{L,\mu} = \epsilon^{ijk}\phi^j\partial_\mu\phi^k,$$
$$J^i_{\mu 5} = J^i_{R,\mu} - J^i_{L,\mu} = \phi_0\partial_\mu\phi^i - \phi^i\partial_\mu\phi_0\,. \tag{2.43}$$

For a special choice of the potential V_0 in (2.36), for example $V_0 = -\mu^2(\phi_0^2+\boldsymbol{\phi}^2)+(\phi_0^2+\boldsymbol{\phi}^2)^2$ with μ^2 a positive parameter with the dimension of mass squared, in the NG mode, the vacuum expectation value (VeV) of the ϕ_0 field is non-zero. Thus we have

$$\langle 0|J^i_{\mu 5}(x)|\pi^k(p)\rangle = \langle 0|\phi_0\partial_\mu\phi^i|\pi^k(p)\rangle = \langle 0|\phi_0|0\rangle i p_\mu e^{ip\cdot x}\delta^{ik} \tag{2.44}$$

where we have assumed in the vacuum, $\partial_\mu\phi_0 = 0$. From Eq. (2.44) with Eq. (2.24) one finds ϕ_i to be an interpolating field of pion

$$\langle 0|\phi^i(x)|\pi^k(p)\rangle = e^{ip\cdot x}\delta^{ik}, \tag{2.45}$$

and the pion decay constant f_π can be defined as

$$f_\pi = \langle 0|\phi_0|0\rangle. \tag{2.46}$$

This shows that in the present model, f_π is proportional to the two-quark condensate and thus the two-quark condensate could be regarded as the order parameter of chiral symmetry breaking.[10] In sum, the model (2.36) encodes the intrinsic symmetry property of QCD for the light-quark systems, chiral symmetry and its breaking.

At low energy — and at low density in nuclear medium we will be encountering in this volume, the linear sigma model, renormalizable in principle, does not, at low enough an order for practical calculations, work at the elementary level as well as in nuclear medium. At the tree order, it fails to describe low-energy $\pi\pi$ and πN interactions and does not give a stable nuclear matter [Kerman and Miller (1974)]. The Lee–Wick state [Lee and Wick (1974)] predicted in linear sigma-model-type model is not realized in experiments. Going beyond the low orders is highly problematic given the

[10]It is important for later to note that in general, not only the two-quark condensate but also multi-quark condensates can be the order parameter of chiral symmetry breaking.

strong coupling involved. It is now well-established that it is in the nonlinear realization of chiral symmetry that the current algebras are naturally encoded in and amenable to systematic χPT. This means that one has to go over to the nonlinear representation. How to do this is as follows [Weinberg (2001)].

For simplicity, let us ignore the explicit chiral symmetry breaking term V_{SB}, that is, go to "QCDLite" and write (2.36) in terms of $\phi_n (n = 0, 1, 2, 3)$,

$$\mathcal{L}_{\mathrm{L}\sigma\mathrm{M}} = \frac{1}{2}\left(\sum_{n=0}^{3} \partial_\mu \phi_n \partial^\mu \phi_n\right) - V_0\left(\sum_{n=0}^{3} \phi_n^2\right). \tag{2.47}$$

Given the $SO(4)$ invariance, one can get ϕ_n from a special vector $(\sigma, 0, 0, 0)$ using a local orthogonal matrix R

$$\phi_n = R_{n0}\sigma \tag{2.48}$$

with $R^T R = 1$. Then, the Lagrangian $\mathcal{L}_{\mathrm{L}\sigma\mathrm{M}}$ becomes

$$\mathcal{L}_{\mathrm{L}\sigma\mathrm{M}} = \frac{1}{2}\left(\partial_\mu\sigma\partial^\mu\sigma + \sigma^2 \partial_\mu R_{n0} \partial^\mu R_{n0}\right) - V_0\left(\sigma^2\right). \tag{2.49}$$

Now parameterize the orthogonal matrix R as

$$R_{00} = \cos\left(\frac{\pi^a \tau^a}{f_\pi}\right), \quad R_{i0} = \frac{\pi_i}{|\pi|}\sin\left(\frac{\pi^a \tau^a}{f_\pi}\right) \tag{2.50}$$

which can be recast into a compact form

$$R_{00} + i\tau_i R_{i0} = \exp\left(\frac{i\pi^a \tau^a}{f_\pi}\right) \equiv U(x). \tag{2.51}$$

Then we have

$$\mathcal{L}_{\mathrm{L}\sigma\mathrm{M}} = \frac{1}{2}\partial_\mu\sigma\partial^\mu\sigma + \frac{1}{2}\sigma^2 \mathrm{Tr}\big(\partial_\mu U(x) \partial^\mu U^\dagger(x)\big) - V_0\left(\sigma^2\right). \tag{2.52}$$

It is important to note here that in this form the σ in the Lagrangian is a chiral scalar, not to be confused with the fourth-component of the chiral four-vector in the usual notation of linear sigma model. When the chiral symmetry is realized in the NG mode that can be dialed in V_0, banishing the fourth-component of the chiral four-vector to infinity to agree with the soft-pion theorems, i.e., the current algebras, one finds that $\sigma \to f_\pi$ and arrives at the non-linear sigma model

$$\mathcal{L}_{\mathrm{NL}\sigma} = \frac{f_\pi^2}{4}\mathrm{Tr}\big(\partial_\mu U(x) \partial^\mu U^\dagger(x)\big), \tag{2.53}$$

in which each interaction of the NGBs, here pions, is accompanied with a spacetime derivative.

Note that if the potential is ignored, the Lagrangian (2.52) is invariant under the scale transformation (2.25). This is because the σ field and U field have scale dimension one and zero, respectively. As shown in Chapter 5, the opposite limit in the potential to that of nonlinear realization (2.53) makes the σ become massless. This can be identified as a NG boson of spontaneously broken scale symmetry, namely a dilaton. The VeV of the dilaton, $\langle\sigma\rangle$, can then be identified with the dilaton decay constant. This shows, as pointed out by Yamawaki [Yamawaki (2015, 2017)], that *the linear sigma model (2.52) has not only the non-linearly realized chiral symmetry but also hidden scale symmetry.* This point, one of the key observations that plays an important role in this volume, is taken up in detail in Chapters 5 and 6.

In what follows, we will not be dealing with nuclear electroweak (EW) responses, but for completeness, we write the basic structure of the EW interaction in the chiral Lagrangian. External-gauging the EW symmetry in the Lagrangian (2.53) one has

$$\mathcal{L}_{\mathrm{NL}\sigma}^{\mathrm{gauged}} = \frac{f_\pi^2}{4}\mathrm{Tr}\left[\mathscr{D}_\mu U \mathscr{D}_\mu U^\dagger\right], \tag{2.54}$$

where the covariant derivative $\mathscr{D}_\mu$ is defined as

$$\mathscr{D}_\mu U = \partial_\mu U - i\mathscr{L}_\mu U + iU\mathscr{R}_\mu, \tag{2.55}$$

with

$$\begin{aligned}
\mathscr{R}_\mu &= -eQA_\mu - g\frac{\sin^2\theta_W}{\cos\theta_W}QZ_\mu, \\
\mathscr{L}_\mu &= -eQA_\mu + gQ_Z Z_\mu + \frac{g}{\sqrt{2}}(W_\mu^+ Q_W + W_\mu^- Q_W^\dagger),
\end{aligned} \tag{2.56}$$

where Q is the charge matrix of quarks which in the two-flavor case is $Q = \mathrm{diag}(2/3, -1/3)$ and e is given by $e = g\sin\theta_W$. The matrices Q_W and Q_Z are defined as

$$Q_W = \begin{pmatrix} 0 & V_{ud} \\ 0 & 0 \end{pmatrix}; \qquad Q_Z = \frac{1}{\cos\theta}\begin{pmatrix} 1/2 & 0 \\ 0 & -1/2 \end{pmatrix} - \frac{\sin^2\theta_W}{\cos\theta_W}Q, \tag{2.57}$$

where V_{ud} is the appropriate Cabibbo–Kobayashi–Maskawa matrix elements.

2.5.2 *Topology in the nonlinear realization of chiral symmetry*

In accessing dense nuclear matter relevant to compact stars, we will find that the crucial ingredient in formulating the appropriate EFT comes from topology change in the soliton description of dense matter. This matter will be covered in detail in Chapters 7, 8 and 10. Here we give the essential aspect of topology involved in the pion field, the NG mode of chiral symmetry in the non-linear sigma model obtained from a linear sigma model. To be as simple as possible while comprehensible, our discussion will follow the review [Holzwarth and Schwesinger (1986)].

Consider the unitary chiral field $U(x)$ representing the NG pion field. Since it is unitary, at any fixed time, the matrix $U(\boldsymbol{x})$ defines a map from R^3 to the manifold S^3. In the low energy limit, QCD goes to the vacuum that is accounted for by the condition

$$U(|\boldsymbol{x}| \to \infty) = \mathbf{1}. \tag{2.58}$$

This limit tells us that all the points at $|\boldsymbol{x}| \to \infty$ are mapped onto the north pole of S^3 and the energy of the system is finite. We therefore have the nontrivial map

$$U(\boldsymbol{x}) : R^3 \to S^3, \tag{2.59}$$

for the static configuration $U(\boldsymbol{x})$. We learn from mathematics on topology that one can classify all the maps into homotopically distinct classes according to the times that the sphere S^3 is covered while $\boldsymbol{x}$ takes all values of coordinate space. In the language of topology, these maps constitute the third homotopy group $\pi_3(S^3) \sim Z$ with Z being the additive group of integers which accounts for the times that S^3 is covered by the mapping $U(\boldsymbol{x})$, i.e., winding numbers Z

$$B^0 : \pi_3(S^3) = Z. \tag{2.60}$$

Given that a change of the time coordinate can be regarded as homotopy transformation which forbids transit between the field configurations in homotopically distinct classes, the winding number is a conserved quantity in the homotopy transformation by the unitary condition of the field $U(x)$ and the condition (2.58). In the skyrmion description of baryon, the number of which is a conserved quantity in QCD, the winding number is identified with the baryon number.

For constructing the mapping B^0, we introduce the notation

$$U(x) = \varphi_0 + i\tau \cdot \boldsymbol{\varphi}, \tag{2.61}$$

which is the vector representation of $\mathrm{SU}(2)_L \times \mathrm{SU}(2)_R$. In the nonlinear case, using the constraint $U^\dagger U = 1$, the $\varphi^i(x)$ can be regarded as an angular variable.

In the group manifold, in terms of ϕ^i, a fundamental surface element is characterized by

$$d^3\Sigma = \epsilon^{ijkl}\varphi^i\partial_1\varphi^j\partial_2\varphi^k\partial_3\varphi^l dx^1 dx^2 dx^3, \tag{2.62}$$

where x^i are the corresponding coordinates on R^3 obtained by stereographic projection from S^3. In affine geometry, (2.62) is the Jacobian associated to the transformation: $S^3 \to R^3$. Hence, one can construct the normalized topological density as

$$B^0 = \frac{1}{12\pi^2}\epsilon^{ijkl}\epsilon^{\nu\alpha\beta}\varphi^i\partial_\nu\varphi^j\partial_\alpha\varphi^k\partial_\beta\varphi^l. \tag{2.63}$$

By using (2.61), one can write Eq. (2.63) in terms of $U(x)$ as

$$B^0 = \frac{i}{24\pi^2}\epsilon^{0\nu\alpha\beta}\mathrm{Tr}\left[\partial_\nu U\cdot U^\dagger\partial_\alpha U\cdot U^\dagger\partial_\beta U\cdot U^\dagger\right]. \tag{2.64}$$

Notice that B^0 does not vanish if and only if all the 3 pion degrees of freedom, π^0, $\pi^\pm$, are excited. The expression (2.64) clearly shows that

$$\int_{R^3} B^0 d^3x = \text{winding number}. \tag{2.65}$$

The topological charge (2.64) can be easily written into a Lorentz covariant form

$$B^\mu = \frac{i\epsilon^{\mu\nu\alpha\beta}}{24\pi^2}\mathrm{Tr}\left[\partial_\nu U\cdot U^\dagger\partial_\alpha U\cdot U^\dagger\partial_\beta U\cdot U^\dagger\right], \tag{2.66}$$

which is conserved in R^3. Note that the conservation of the current (2.66) is independent of the EoM of the pion.

In accessing baryons and baryonic matter in terms of topological object, i.e., skyrmion [Skyrme (1962)], which will be discussed in Chapter 7, this conserved topological number will be identified with the conserved baryon number in QCD. This allows us, after an appropriate arrangement of terms of basic blocks in the nonlinear realization of chiral symmetry, to use the mesonic theory to describe elementary baryon properties as well as dense baryonic matter. A brief resume will be given below, with details relegated to later chapters.

2.6 Chiral effective field theory of hadrons

In the nonlinear realization of chiral symmetry, as we have pointed out, the interactions involving pseudo-scalar NG bosons — apart from symmetry breaking terms — are all derivative-coupling and hence at low energy scale, one can have a well-defined power counting that allows a systematic calculation. The NG boson mass, away from the chiral limit, is counted the same as the derivative. Denoting the leading-order derivative (or momentum) term as "LO," next-to the leading order will be denoted as NLO, next-to-next to the leading order as N^{2}LO and so on.

Generically, an EFT in particle physics requires two inputs: The energy/momentum scale, "cut-off," below which the theory is applicable and the consistent power counting, which can be used to order terms of decreasing importance. In χPT, the scale of the theory can be estimated through physical processes, such as π-π scattering to one-loop. In such a way, the scale of the χPT is found to be around $\Lambda_\chi \simeq 4\pi f_\pi \simeq 1$ GeV. This is usually referred to as characteristic chiral scale.

The degrees of freedom in chiral EFT relevant to nuclear dynamics that will figure in what follows are the NG bosons (triplet for 2 flavors), the vector mesons (ρ and ω), the scalar dilaton (σ) and the baryons (proton and neutron), either arising as solitons — skyrmions — from mesonic theory or put in explicitly (by hand). We specify each of them briefly below. Some lengthy Lagrangians involving them, referred to in the chapters that follow but not written down explicitly, will be written explicitly here. They will serve the purpose of definitions.

2.6.1 *Nambu–Goldstone bosons: Pseudoscalar mesons*

Due to Lorentz invariance, the chiral effective theory for mesons including only pseudoscalar mesons has the general form of power expansion

$$\mathcal{L}_{\chi\mathrm{PT}} = \sum_{n=1}^{\infty} \mathcal{L}_{eff}^{2n}. \tag{2.67}$$

The mass of the lowest-lying pseuscalar meson, pion, $m_\pi \simeq 140$ MeV, is very small compared with the chiral scale Λ_χ, so the expansion parameter in χPT[11] is $\mathcal{O}(m_\pi^2) \sim \mathcal{O}(\partial^2) \sim \mathcal{O}(p^2)$. The external sources $\mathscr{L}_\mu$ and $\mathscr{R}_\mu$ always appear in company with derivative, so they can be taken as $\mathcal{O}(p)$. When the scalar $\mathscr{S}$ and pseudoscalar $\mathscr{P}$ sources are introduced in the generating functional in company with the light quark mass, they can be regarded as $\mathcal{O}(p^2)$. With these criteria, the chiral perturbation perturbation series can be systematically set up. We summarize the counting rules of the operators in χPT in Table. 2.1. The leading order (LO) in the chiral

[11]The standard χPT where the boson degrees of freedom are just pions is referred to as SχPT, to be distinguished form "generalized chiral perturbation" GχPT that will include vector and scalar mesons.

Table 2.1 Counting rules of the operators and fields in χPT.

Operator/field	U	∂_μ	$\mathscr{L}_\mu$	$\mathscr{R}_\mu$	$\mathscr{S}$	$\mathscr{P}$	m_q
Counting rule	$\mathcal{O}(1)$	$\mathcal{O}(p)$	$\mathcal{O}(p)$	$\mathcal{O}(p)$	$\mathcal{O}(p^2)$	$\mathcal{O}(p^2)$	$\mathcal{O}(p^2)$

expansion with NG bosons only is then, generically, $\mathcal{O}(p^2)$ and takes the form

$$\mathcal{L}_2 = \frac{f_\pi^2}{4}\mathrm{Tr}\left(\mathscr{D}_\mu U \mathscr{D}^\mu U^\dagger\right) + \frac{f_\pi^2}{4}\mathrm{Tr}\left(\mathcal{M}^\dagger U + U^\dagger \mathcal{M}\right), \tag{2.68}$$

where the source (or spurion) field $\mathcal{M} = 2B(\mathscr{S} + i\mathscr{P})$ can be taken as the quark mass matrix with $\mathcal{M} = \mathrm{diag}(m_\pi^2, m_\pi^2, 2m_K^2 - m_\pi^2)$.

To the next-to-leading order (NLO), the Lagrangian for three flavors takes the form [Weinberg (1979); Gasser and Leutwyler (1984, 1985)]

$$\begin{aligned}
\mathcal{L}_4 = {} & L_1 \left\{\mathrm{Tr}[\mathscr{D}_\mu U (\mathscr{D}^\mu U)^\dagger]\right\}^2 + L_2 \mathrm{Tr}\left[\mathscr{D}_\mu U (\mathscr{D}_\nu U)^\dagger\right] \mathrm{Tr}\left[\mathscr{D}^\mu U (\mathscr{D}^\nu U)^\dagger\right] \\
& + L_3 \mathrm{Tr}\left[\mathscr{D}_\mu U (\mathscr{D}^\mu U)^\dagger \mathscr{D}_\nu U (\mathscr{D}^\nu U)^\dagger\right] \\
& + L_4 \mathrm{Tr}\left[\mathscr{D}_\mu U (\mathscr{D}^\mu U)^\dagger\right] \mathrm{Tr}\left(\mathcal{M} U^\dagger + U \mathcal{M}^\dagger\right) \\
& + L_5 \mathrm{Tr}\left[\mathscr{D}_\mu U (\mathscr{D}^\mu U)^\dagger (\mathcal{M} U^\dagger + U \mathcal{M}^\dagger)\right] + L_6 \left[\mathrm{Tr}\left(\mathcal{M} U^\dagger + U \mathcal{M}^\dagger\right)\right]^2 \\
& + L_7 \left[\mathrm{Tr}\left(\mathcal{M} U^\dagger - U \mathcal{M}^\dagger\right)\right]^2 + L_8 \mathrm{Tr}\left(U \mathcal{M}^\dagger U \mathcal{M}^\dagger + \mathcal{M} U^\dagger \mathcal{M} U^\dagger\right) \\
& - i L_9 \mathrm{Tr}\left[f_{\mu\nu}^R \mathscr{D}^\mu U (\mathscr{D}^\nu U)^\dagger + f_{\mu\nu}^L (\mathscr{D}^\mu U)^\dagger \mathscr{D}^\nu U\right] + L_{10} \mathrm{Tr}\left(U f_{\mu\nu}^L U^\dagger f_R^{\mu\nu}\right) \\
& + H_1 \mathrm{Tr}\left(f_{\mu\nu}^R f_R^{\mu\nu} + f_{\mu\nu}^L f_L^{\mu\nu}\right) + H_2 \mathrm{Tr}\left(\mathcal{M}\mathcal{M}^\dagger\right),
\end{aligned} \tag{2.69}$$

where the covariant derivative $\mathscr{D}^\mu U$ is defined by Eq. (2.55). In principle, these low energy coefficients L_i could be derived from QCD in the large N_c expansion in terms of the correlation functions, e.g., [Wang, Kuang, Xiao and Wang (2000)] or extracted in lattice simulations [Aoki *et al.* (2017)]. However, these correlation functions are very difficult to calculate. In practice, they are estimated from experimental data for various processes. Here we quote the list given in [Harada and Yamawaki (2003)] in Table. 2.2. Although what's given in the table will not explicitly figure in the development given in this volume, it is illustrative of what's involved.

With the counting rules so defined, one can then expand the SχPT amplitude in derivative expansion following [Weinberg (1979); Harada and Yamawaki (2003)]. The series is given by the power $\bar{D}$

$$\bar{D} = 2 + \sum_{d,j} N_{d,j}(2d + 2j - 2) + 2N_L\ , \tag{2.70}$$

with N_L the number of meson loops in the matrix, d the number of derivatives in the vertex and j the number of quark mass in the vertex.

Table 2.2 Renormalized low-energy constants L_i^r in unit of 10^{-3} at scale m_η [Gasser and Leutwyler (1985)] and m_ρ [Ecker, Gasser, Pich and de Rafael (1989)].

	$L_i^r(\mu = m_\eta)$	$L_i^r(\mu = m_\rho)$	source
$L_1^r(\mu)$	0.9 ± 0.3	0.7 ± 0.3	$\pi\pi$ D-waves, Zweig rule
$L_2^r(\mu)$	(1.7 ± 0.7)	1.3 ± 0.7	$\pi\pi$ D-waves
$L_3^r(\mu)$	$-\,4.4 \pm 2.5$	$-\,4.4 \pm 2.5$	$\pi\pi$ D-waves, Zweig rule
$L_4^r(\mu)$	0 ± 0.5	$-\,0.3 \pm 0.5$	Zweig rule
$L_5^r(\mu)$	2.2 ± 0.5	1.4 ± 0.5	$F_K{:}F_\pi$
$L_6^r(\mu)$	(0 ± 0.3)	$-\,0.2 \pm 0.3$	Zweig rule
$L_7^r(\mu)$	$-\,0.4 \pm 0.15$	$-\,0.4 \pm 0.15$	Gell-Man–Okubo, L_5, L_8
$L_8^r(\mu)$	1.1 ± 0.3	0.9 ± 0.3	K^0-K^+, R, L_5
$L_9^r(\mu)$	7.4 ± 0.7	6.9 ± 0.7	$\langle r^2\rangle^\pi_{\rm e.m.}$
$L_{10}^r(\mu)$	$-\,6.0 \pm 0.7$	$-\,5.2 \pm 0.7$	$\pi \to e\nu\gamma$

2.6.2 *Vector mesons*

At the next energy scale to the pions in the meson sector come the vector mesons $V \equiv (\rho, \omega)$ with $m_V \sim 770$ MeV and the scalar σ with $m_\sigma \sim 600$ MeV. In this subsection, we deal with the vectors, relegating the scalar to the next subsection. We will give some lengthy expressions in this part for completeness. In applications given in later chapters, they will be greatly simplified with specified approximations.

Although the chiral symmetry breaking scale is estimated to be $\Lambda_\chi \simeq 1$ GeV [Manohar and Georgi (1984)], χPT cannot describe the low energy processes of pions quite well in the whole energy region below Λ_χ: for example the π-π scattering at P-wave, even with the loop corrections included. Moreover, lessons from nuclear physics tell us that the vector mesons, such as the ρ and ω mesons, play essential roles in the nuclear force, for example the tensor force and repulsive force between nucleons (for a recent review, see, e.g., [Rho (2014)]). These observations render the incorporation of the vector mesons natural and often necessary.

In the literature, there are several alternative frameworks for treating hadron dynamics including vector mesons, such as the matter field method [Ecker, Gasser, Leutwyler, Pich and de Rafael (1989)], massive Yang–Mills method [Schwinger (1967, 1970)], anti-symmetric tensor field method [Gasser and Leutwyler (1984); Ecker, Gasser, Pich and de Rafael (1989)], and hidden gauge symmetric approach [Bando, Kugo, Uehara, Yamawaki and Yanagida (1985); Bando, Kugo and Yamawaki (1988); Harada and Yamawaki (2003)] etc. In the matter-free vacuum, although these methods give different results for off-shell amplitudes, they are equivalent for the tree-level on-shell amplitudes [Bando, Kugo and Yamawaki (1988); Harada and Yamawaki (2003)]. However there is a power in hidden local symmetry (HLS) superior to other approaches if the vector meson mass is considered to be "light" in the sense that it can be put on the same footing as the pion mass. In the vacuum, considering the ρ mass, say, on the same footing as the pion mass may look highly problematic. However if it is taken "light" then the notion of HLS leads to such low-energy theorems as the KSRF relation [Kawarabayashi and Suzuki (1966); Riazuddin and Fayyazuddin (1966)] which work remarkably well [Harada and Yamawaki (2003); Komargodski (2011); Yamawaki (2018)]. More significantly when the vector mass can be "dialed" to go down, say, by density as in baryonic matter (described in Chapter 5), it allows to make predictions that cannot be made by other approaches.

The idea of the HLS is the following: A nonlinear sigma model based on the manifold G/H is gauge equivalent to a model having the symmetry $G_{\text{global}} \times H_{\text{local}}$. After spontaneous breaking of the local symmetry H_{local}, the gauge boson of the local symmetry H_{local} is identified as the massive vector mesons. Here we consider the case in which $G_{\text{global}} = [\text{SU}(2)_L \times \text{SU}(2)_R]_{\text{global}}$ and $H_{\text{local}} = [\text{SU}(2)_V]_{\text{local}}$. The extension to other H_{local}, such as $[U(2)_V]_{\text{local}}$ and $[SU(2)_V \times U(1)_V]_{\text{local}}$, is straightforward. Integrating out the vector mesons from this model and keeping only the minimal number derivative terms leads to the two-flavor nonlinear sigma model. This will be shown explicitly below for self-containness.

The strategy to bring in vector mesons V as gauge flavor gauge bosons in nonlinear sigma model will be described in Chapter 5. Here we simply mention that it amounts to exploiting redundancies present in the chiral field $U(x)$

$$U(x) = \xi_L^\dagger(x)\xi_R(x). \tag{2.71}$$

Being a product of two local fields, it has an infinite redundancies. The infinity can be exploited to build a "dimensionally deconstructed" 5D hidden local symmetric model as we will encounter later. Here we take one that brings in a gauge symmetry H_{local} by requiring that $\xi_{L,R}$ have the transformation

$$\xi_{L,R}(x) \mapsto \xi'_{L,R}(x) = h(x)\xi_{L,R}(x)g^\dagger_{L,R}, \tag{2.72}$$

where $h(x) \in H_{\text{local}}$. The variables $\xi_{L,R}$ can be parameterized as

$$\xi_{L,R}(x) = e^{i\sigma(x)/(2f_\sigma)}e^{\pm i\pi(x)/(2f_\pi)}, \quad \pi(x) = \pi^a X^a, \quad \sigma(x) = \sigma^\alpha S^\alpha, \tag{2.73}$$

where S^a is the generator of the unbroken subgroup H of $G_{\rm global}$, here SU(2), while X^a is the broken generator. π and σ denote the NG bosons[12] associated with the spontaneous breaking of $G_{\rm global}$ and $H_{\rm local}$, respectively, and f_π and f_σ are the corresponding decay constants.

The above redundancy is elevated to a propagating gauge field $V_\mu(x)$ that transforms $V_\mu(x) \to V_\mu(x)' = h(x)(V_\mu(x) - i\partial_\mu h(x))h^\dagger(x)$ with $h(x) \in$ SU$(2)_V$ and with the kinetic energy term added.[Harada and Yamawaki (2003)].[13] To simplify notations, we define here two 1-forms which will be used in later chapters:

$$\hat{\alpha}_{\parallel\mu} = \frac{1}{2i}(D_\mu\xi_R \cdot \xi_R^\dagger + D_\mu\xi_L \cdot \xi_L^\dagger),$$

$$\hat{\alpha}_{\perp\mu} = \frac{1}{2i}(D_\mu\xi_R \cdot \xi_R^\dagger - D_\mu\xi_L \cdot \xi_L^\dagger), \tag{2.74}$$

where the covariant derivative is defined as $D_\mu\xi_{R,L} = (\partial_\mu - iV_\mu)\xi_{R,L}$, and both of these quantities transform as $\hat{a}^\mu_{\parallel,\perp} \to h(x)\hat{a}^\mu_{\parallel,\perp}h(x)^\dagger$. We write down for completeness the gauge field strength tensor

$$V_{\mu\nu}(x) = \partial_\mu V_\nu(x) - \partial_\nu V_\mu(x) - i[V_\mu(x), V_\nu(x)], \tag{2.75}$$

that transforms as $V_{\mu\nu}(x) \to h(x)V_{\mu\nu}(x)h(x)^\dagger$.

With the two 1-forms defined by Eq. (2.74) and the field strength tensor (2.75), one can construct a Lorentz invariant Lagrangian with the minimal number of derivatives as

$$\mathcal{L}_{\rm HLS} = f_\pi^2 {\rm Tr}[\hat{a}_{\perp\mu}\hat{a}_\perp^\mu] + f_\sigma^2 {\rm Tr}[\hat{a}_{\parallel\mu}\hat{a}_\parallel^\mu] - \frac{1}{2g^2}{\rm Tr}[V_{\mu\nu}V^{\mu\nu}], \tag{2.76}$$

where g is the gauge coupling of HLS $[SU(2)_V]_{\rm HLS}$.[14] The kinetic energy term makes the vector fields propagate and hence become dynamical. How the kinetic energy term arises in the HLS framework can be seen in [Yamawaki (2018)].

[12]We should warn that the σ here is not to be confused with the fourth component of the chiral four-vector in linear sigma model and more seriously with the dilaton that arises from spontaneous breaking of scale symmetry to which we will come below and in the chapters that follow. The σ here is an would-be NG boson to be higgsed in HLS theory. This unfortunate multiplicity in the notation is due to the fact that it appeared in the original paper of HLS. We will point out this problem whenever there may arise possible confusion.

[13]How the kinetic energy term can arise in the theory is given a detailed treatment in [Yamawaki (2018)].

[14]Whenever there is no cause for confusion between the QCD gauge coupling and the hidden gauge coupling, g will stand for the hidden gauge coupling. When they appear simultaneously, the QCD coupling will be denoted as g_s and the hidden gauge coupling as g_V.

As written, the Lagrangian (2.76) is local-gauge invariant. But the local gauge symmetry is not "visible" in QCD because of the mass. It will be unhidden when the vector mesons become massless, which is not the case in the matter-free vacuum. The point, crucial in the reasoning developed, is that it can happen in dense matter. The mass measured in the vacuum can be generated by Higgs mechanism, namely, by HLS spontaneously broken. To obtain the vector masses, the simple way is to take the unitary gauge, i.e., $\sigma = 0$,

$$\xi_L^\dagger = \xi_R \equiv \xi = e^{i\pi/(2f_\pi)}, \quad U(x) = \xi^2(x). \tag{2.77}$$

The unitary-gauge Lagrangian is

$$\mathcal{L}_{\rm HLS} = f_\sigma^2 \mathrm{Tr}\left[\left(\partial_\mu\xi\cdot\xi^\dagger + \partial_\mu\xi^\dagger\cdot\xi - 2iV_\mu\right)\left(\partial^\mu\xi\cdot\xi^\dagger + \partial^\mu\xi^\dagger\cdot\xi - 2iV^\mu\right)\right] + \cdots . \tag{2.78}$$

from which follows the mass formula[15]

$$m_V^2 = g^2 f_\sigma^2. \tag{2.79}$$

For $f_\sigma^2 = 2f_\pi^2$, this is the well-known KSRF relation first obtained by the current algebras. It also turns out to be valid to all loop orders in HLS with corrections coming at $\mathcal{O}(m_V^2/\Lambda_\chi^2)$ [Harada, Kugo and Yamawaki (1993)].[16]

We show for completeness how the standard χPT Lagrangian can be gotten from HLS Lagrangian. For this, one integrates out the vector mesons from the HLS using the equations of motion derived from Eq. (2.78),

$$2V_\mu^b \mathrm{Tr}(T^aT^b) = -i\mathrm{Tr}\{T^a(\partial_\mu\xi\cdot\xi^\dagger + \partial_\mu\xi^\dagger\cdot\xi)\} + \mathcal{O}(p^3), \tag{2.80}$$

[15]For those who are unfamiliar with the language, we sketch the Higgs mechanism that is involved here. Expand $\xi_{L,R}$ as

$$\xi_L = 1 + i\sigma(x)/(2f_\sigma) + i\pi(x)/(2f_\pi) + \cdots ,$$
$$\xi_R = 1 + i\sigma(x)/(2f_\sigma) - i\pi(x)/(2f_\pi) + \cdots ,$$

which leads to

$$f_\sigma^2 \mathrm{Tr}[\hat{a}_{\parallel\mu}\hat{a}_\parallel^\mu] = -f_\sigma^2 \mathrm{Tr}\left[V_\mu + \frac{1}{f_\sigma}\partial_\mu\sigma\right]^2 + \cdots .$$

Now the mixing term $V_\mu\partial_\mu\sigma$ of V_μ and σ in the Lagrangian can be removed by the gauge fixing, i.e., the unitary gauge $\sigma = 0$ used above.

[16]It will be argued in Chapters 8 and 9 that high density drives the vector mesons towards — but not on top of — the "vector manifestation" fixed point at which the vectors become massless and the local gauge symmetry emerges. This will be seen to be the key mechanism responsible in our approach for massive compact stars.

where the $\mathcal{O}(p^3)$ term is from the vector meson kinetic term.[17] Limiting to the leading order $\mathcal{O}(p^2)$ term, we have

$$2V_\mu = -i(\partial_\mu \xi \cdot \xi^\dagger + \partial_\mu \xi^\dagger \cdot \xi). \tag{2.81}$$

Substituting Eq. (2.81) into the Lagrangian (2.78) with the vector kinetic term ignored, we get the current algebra term

$$\mathcal{L} = \frac{f_\pi^2}{4} \mathrm{Tr}\big(\partial_\mu U \partial^\mu U^\dagger\big), \tag{2.82}$$

which is the leading order of SχPT. It is in this sense that the linear model based on the manifold $G_{\mathrm{global}} \times H_{\mathrm{local}}$ is gauge-equivalent to the nonlinear model based on the manifold G/H.

2.6.2.1 $G\chi PT$

As mentioned, the notion of HLS makes a powerful sense, mysterious though it may seem, when the vector meson mass is considered as "small" [Komargodski (2011); Yamawaki (2018)]. This seems problematic since $\frac{m_\rho^2}{\Lambda_\chi^2} \sim 0.5$. But the actual calculation shows that a chiral perturbative calculation does come out rather reasonable. Let us simply assume that one can think of the vector-meson mass as of the same scale as that of pseudo-NG bosons and then set up a generalized chiral perturbation theory (GχPT) including the vector mesons. In fact this has been discussed before and is found to make sense in confronting experiments [Harada and Yamawaki (2003)]. Here we discuss the power counting following [Harada and Yamawaki (2003)] and generalize (2.70) to the case where vector mesons are present.

In addition to the derivative

$$\partial_\mu \sim \mathcal{O}(p) \tag{2.83}$$

that figures in SχPT, the vector field should be counted as of the same power. It is incorporated in the covariant derivative with the power counted as[18]

$$V_\mu = g\rho_\mu \sim \mathcal{O}(p). \tag{2.84}$$

---

[17] When we integrate out the vector meson fields from the model, it implicitly implies that the vector mesons are heavy objects so that M_V^2 cannot be regarded as an $\mathcal{O}(p^2)$ quantity but a large constant. Thus the vector-meson kinetic energy term is $\mathcal{O}(p^4)$, not $\mathcal{O}(p^2)$.

[18] This corresponds to putting the vector meson mass on the *same footing*, i.e., small, as the pion mass. Repeated once more to bring out the crucial point that is not properly appreciated by nuclear physics community, this is the feature of HLS that is *distinctively different* from the variety of approaches anchored on SχPT.

The KSRF formula implies that

$$m_\rho^2 = g^2 f_\sigma^2 \sim \mathcal{O}(p^2) , \tag{2.85}$$

and

$$g \sim \mathcal{O}(p). \tag{2.86}$$

This counting immediately shows that the Lagrangian (2.76) is the leading order Lagrangian.

Following the same line of reasoning as that which enters in the derivation of Eq. (2.70), one can arrive, in the case of HLS, at

$$\bar{D} = 2 + N_L + \sum_{d,j,l}(2d - 2 + 2j + l), \tag{2.87}$$

where d is the number of derivatives, j is the power of m_π and l is the number of vector mesons in the vertex of the given diagram. With this power counting, one can write down the analog to the power series of SχPT for GχPT with HLS.

For later reference, we write down here the full HLS Lagrangian valid to next-to-leading order (NLO) for the symmetry group $[SU(3)_L \times SU(3)_R]_{\text{global}} \times U(3)_{\text{HLS}}$. It is lengthy, involving more than 30 terms, with most of the parameters of which (theoretically and experimentally) undetermined. In our applications in later chapters we will refer to it, but we will not write it out explicitly. It appears that it can be reduced to a surprisingly simplified form with only four terms without losing quantitative accuracy. This has been verified in skyrmion matter as discussed in Chapter 7.

Witten out fully, the $\mathcal{O}(p^4)$ HLS Lagrangian is[19]

$$\mathcal{L}_{\text{HLS}} = \mathcal{L}_{(2)} + \mathcal{L}_{(4)\text{y}} + \mathcal{L}_{(4)\text{w}} + \mathcal{L}_{(4)\text{z}} + \mathcal{L}_{\text{anom}}, \tag{2.88}$$

where

$$\mathcal{L}_{(2)} = f_\pi^2 \text{Tr}[\hat{\alpha}_{\perp\mu}\hat{\alpha}_\perp^\mu] + a f_\pi^2 \text{Tr}[\hat{\alpha}_{\parallel\mu}\hat{\alpha}_\parallel^\mu] - \frac{1}{2g^2}\text{Tr}[V_{\mu\nu}V^{\mu\nu}] + \frac{1}{4} f_\chi^2 \text{Tr}[\hat{\mathcal{M}} + \hat{\mathcal{M}}^\dagger], \tag{2.89}$$

---

[19]Here we write in full glory all the terms of $\mathcal{O}(p^4)$ including the mass terms from [Harada and Yamawaki (2003)]. It's a luxury that is beyond our reach in our practical calculations. The parameter f_χ is equal to f_π at the tree order but differs from it at higher orders due to quadratic divergences.

$$
\begin{aligned}
\mathcal{L}_{(4)y} = {} & y_1 \mathrm{Tr}\left[\hat{\alpha}_{\perp\mu}\hat{\alpha}_{\perp}^{\mu}\hat{\alpha}_{\perp\nu}\hat{\alpha}_{\perp}^{\nu}\right] + y_2 \mathrm{Tr}\left[\hat{\alpha}_{\perp\mu}\hat{\alpha}_{\perp\nu}\hat{\alpha}_{\perp}^{\mu}\hat{\alpha}_{\perp}^{\nu}\right] \\
& + y_3 \mathrm{Tr}\left[\hat{\alpha}_{\parallel\mu}\hat{\alpha}_{\parallel}^{\mu}\hat{\alpha}_{\parallel\nu}\hat{\alpha}_{\parallel}^{\nu}\right] + y_4 \mathrm{Tr}\left[\hat{\alpha}_{\parallel\mu}\hat{\alpha}_{\parallel\nu}\hat{\alpha}_{\parallel}^{\mu}\hat{\alpha}_{\parallel}^{\nu}\right] \\
& + y_5 \mathrm{Tr}\left[\hat{\alpha}_{\perp\mu}\hat{\alpha}_{\perp}^{\mu}\hat{\alpha}_{\parallel\nu}\hat{\alpha}_{\parallel}^{\nu}\right] + y_6 \mathrm{Tr}\left[\hat{\alpha}_{\perp\mu}\hat{\alpha}_{\perp\nu}\hat{\alpha}_{\parallel}^{\mu}\hat{\alpha}_{\parallel}^{\nu}\right] \\
& + y_7 \mathrm{Tr}\left[\hat{\alpha}_{\perp\mu}\hat{\alpha}_{\perp\nu}\hat{\alpha}_{\parallel}^{\nu}\hat{\alpha}_{\parallel}^{\mu}\right] \\
& + y_8 \left\{\mathrm{Tr}\left[\hat{\alpha}_{\perp\mu}\hat{\alpha}_{\parallel}^{\mu}\hat{\alpha}_{\perp\nu}\hat{\alpha}_{\parallel}^{\nu}\right] + \mathrm{Tr}\left[\hat{\alpha}_{\perp\mu}\hat{\alpha}_{\parallel\nu}\hat{\alpha}_{\perp}^{\nu}\hat{\alpha}_{\parallel}^{\mu}\right]\right\} \\
& + y_9 \mathrm{Tr}\left[\hat{\alpha}_{\perp\mu}\hat{\alpha}_{\parallel\nu}\hat{\alpha}_{\perp}^{\mu}\hat{\alpha}_{\parallel}^{\nu}\right] \\
& + y_{10}\left(\mathrm{Tr}\left[\hat{\alpha}_{\perp\mu}\hat{\alpha}_{\perp}^{\mu}\right]\right)^2 + y_{11}\,\mathrm{Tr}\left[\hat{\alpha}_{\perp\mu}\hat{\alpha}_{\perp\nu}\right]\mathrm{Tr}\left[\hat{\alpha}_{\perp}^{\mu}\hat{\alpha}_{\perp}^{\nu}\right] \\
& + y_{12}\left(\mathrm{Tr}\left[\hat{\alpha}_{\parallel\mu}\hat{\alpha}_{\parallel}^{\mu}\right]\right)^2 + y_{13}\,\mathrm{Tr}\left[\hat{\alpha}_{\parallel\mu}\hat{\alpha}_{\parallel\nu}\right]\mathrm{Tr}\left[\hat{\alpha}_{\parallel}^{\mu}\hat{\alpha}_{\parallel}^{\nu}\right] \\
& + y_{14}\,\mathrm{Tr}\left[\hat{\alpha}_{\perp\mu}\hat{\alpha}_{\perp}^{\mu}\right]\mathrm{Tr}\left[\hat{\alpha}_{\parallel\nu}\hat{\alpha}_{\parallel}^{\nu}\right] + y_{15}\,\mathrm{Tr}\left[\hat{\alpha}_{\perp\mu}\hat{\alpha}_{\perp\nu}\right]\mathrm{Tr}\left[\hat{\alpha}_{\parallel}^{\mu}\hat{\alpha}_{\parallel}^{\nu}\right] \\
& + y_{16}\left(\mathrm{Tr}\left[\hat{\alpha}_{\perp\mu}\hat{\alpha}_{\parallel}^{\mu}\right]\right)^2 + y_{17}\,\mathrm{Tr}\left[\hat{\alpha}_{\perp\mu}\hat{\alpha}_{\parallel\nu}\right]\mathrm{Tr}\left[\hat{\alpha}_{\perp}^{\mu}\hat{\alpha}_{\parallel}^{\nu}\right] \\
& + y_{18}\,\mathrm{Tr}\left[\hat{\alpha}_{\perp\mu}\hat{\alpha}_{\parallel\nu}\right]\mathrm{Tr}\left[\hat{\alpha}_{\parallel}^{\mu}\hat{\alpha}_{\perp}^{\nu}\right],
\end{aligned}
\tag{2.90}
$$

$$
\begin{aligned}
\mathcal{L}_{(4)w} = {} & w_1 \frac{f_\chi^2}{f_\pi^2}\mathrm{Tr}\left[\hat{\alpha}_{\perp\mu}\hat{\alpha}_{\perp}^{\mu}\left(\hat{\mathcal{M}}+\hat{\mathcal{M}}^\dagger\right)\right] \\
& + w_2 \frac{f_\chi^2}{f_\pi^2}\mathrm{Tr}\left[\hat{\alpha}_{\perp\mu}\hat{\alpha}_{\perp}^{\mu}\right]\mathrm{Tr}\left[\hat{\mathcal{M}}+\hat{\mathcal{M}}^\dagger\right] \\
& + w_3 \frac{f_\chi^2}{f_\pi^2}\mathrm{Tr}\left[\hat{\alpha}_{\parallel\mu}\hat{\alpha}_{\parallel}^{\mu}\left(\hat{\mathcal{M}}+\hat{\mathcal{M}}^\dagger\right)\right] \\
& + w_4 \frac{f_\chi^2}{f_\pi^2}\mathrm{Tr}\left[\hat{\alpha}_{\parallel\mu}\hat{\alpha}_{\parallel}^{\mu}\right]\mathrm{Tr}\left[\hat{\mathcal{M}}+\hat{\mathcal{M}}^\dagger\right] \\
& + w_5 \frac{f_\chi^2}{f_\pi^2}\mathrm{Tr}\left[\left(\hat{\alpha}_{\parallel}^{\mu}\hat{\alpha}_{\perp\mu}-\hat{\alpha}_{\perp\mu}\hat{\alpha}_{\parallel}^{\mu}\right)\left(\hat{\mathcal{M}}-\hat{\mathcal{M}}^\dagger\right)\right] \\
& + w_6 \frac{f_\chi^4}{f_\pi^4}\mathrm{Tr}\left[\left(\hat{\mathcal{M}}+\hat{\mathcal{M}}^\dagger\right)^2\right] + w_7 \frac{f_\chi^4}{f_\pi^4}\left(\mathrm{Tr}\left[\hat{\mathcal{M}}+\hat{\mathcal{M}}^\dagger\right]\right)^2 \\
& + w_8 \frac{f_\chi^4}{f_\pi^4}\mathrm{Tr}\left[\left(\hat{\mathcal{M}}-\hat{\mathcal{M}}^\dagger\right)^2\right] + w_9 \frac{f_\chi^4}{f_\pi^4}\left(\mathrm{Tr}\left[\hat{\mathcal{M}}-\hat{\mathcal{M}}^\dagger\right]\right)^2,
\end{aligned}
\tag{2.91}
$$

$$
\mathcal{L}_{(4)z} = iz_4\,\mathrm{Tr}\left[V_{\mu\nu}\hat{\alpha}_{\perp}^{\mu}\hat{\alpha}_{\perp}^{\nu}\right] + iz_5\mathrm{Tr}\left[V_{\mu\nu}\hat{\alpha}_{\parallel}^{\mu}\hat{\alpha}_{\parallel}^{\nu}\right],
\tag{2.92}
$$

with

$$\hat{\mathcal{M}} = \xi_L \mathcal{M} \xi_R^\dagger = 2B\xi_L \left(\mathcal{S} + i\mathcal{P}\right) \xi_R^\dagger. \tag{2.93}$$

This $\hat{\mathcal{M}}$ field takes into account the explicit chiral symmetry breaking through

$$\mathcal{S} = \begin{pmatrix} m_u & 0 & 0 \\ 0 & m_d & 0 \\ 0 & 0 & m_s \end{pmatrix}. \tag{2.94}$$

Finally, the anomalous parity Wess–Zumino–Witten (WZW) terms $\mathcal{L}_{\rm anom}$ are written as

$$\Gamma_{\rm WZW} = \int d^4x \mathcal{L}_{\rm anom} = \frac{N_c}{240\pi^2}\int_{M^5} {\rm Tr}[\alpha^5] + \frac{N_c}{16\pi^2}\int_{M^4}\sum_{i=1}^{3} c_i \mathcal{L}_i, \tag{2.95}$$

where $\alpha = \frac{1}{i}(\partial_\mu U)U^{-1}dx^\mu$, M^4 stands for the 4-dimensional Minkowski space and the second term of (2.95) called "homogeneous WZ term" that we will refer to as hWZ has three components

$$\mathcal{L}_1 = i\,{\rm Tr}\left[\hat{\alpha}_L^3\hat{\alpha}_R - \hat{\alpha}_R^3\hat{\alpha}_L\right], \tag{2.96a}$$

$$\mathcal{L}_2 = i\,{\rm Tr}\left[\hat{\alpha}_L\hat{\alpha}_R\hat{\alpha}_L\hat{\alpha}_R\right], \tag{2.96b}$$

$$\mathcal{L}_3 = {\rm Tr}\left[F_V\left(\hat{\alpha}_L\hat{\alpha}_R - \hat{\alpha}_R\hat{\alpha}_L\right)\right]. \tag{2.96c}$$

Here we have used the following 1- and 2-forms defined before

$$\begin{aligned}
\hat{\alpha}_{\rm L} &\equiv \frac{1}{i}\mathcal{D}\xi_{\rm L}\cdot\xi_{\rm L}^\dagger = \alpha_{\rm L} - V = \alpha_\parallel - \alpha_\perp - V = \hat{\alpha}_\parallel - \hat{\alpha}_\perp\ , \\
\hat{\alpha}_{\rm R} &\equiv \frac{1}{i}\mathcal{D}\xi_{\rm R}\cdot\xi_{\rm R}^\dagger = \alpha_{\rm R} - V = \alpha_\parallel + \alpha_\perp - V = \hat{\alpha}_\parallel + \hat{\alpha}_\perp\ , \\
\alpha_{\rm L,R} &= \frac{1}{i}d\xi_{\rm L,R}\cdot\xi_{\rm L,R}^\dagger, \\
F_{\rm V} &\equiv dV - iV^2.
\end{aligned} \tag{2.97}$$

2.6.3 *Baryons in HLS*

We next include baryons in HLS. In exploring dense matter, it is found to be more powerful to implement the robust features of topological structure gotten in HLS in the HLS Lagrangian with baryons explicitly included. And the calculation is also a lot easier and controllable. This will be treated in detail in Chapter 8.

We first consider two-flavor case in which only iso-doublet baryons, proton and neutron, are considered. The baryon field is

$$\Psi = \begin{pmatrix} p \\ n \end{pmatrix}. \tag{2.98}$$

After a proper redefinition, one can regard the baryon field transforming as an isodoublet under chiral transformation [Weinberg (2001)], i.e.,

$$\Psi \to \Psi' = h\Psi, \tag{2.99}$$

where $h \in SU(2)_V$. To couple pions to baryons, we write

$$U(x) = u^2(x). \tag{2.100}$$

Then pions interact with baryons in derivative coupling. Up to one derivative, the Lagrangian then is

$$\mathcal{L}_{\pi N}^{(1)} = \bar{\Psi}\left(i\not{D} - m_N + \frac{g_A}{2}\gamma^\mu\gamma_5 A_\mu\right)\Psi. \tag{2.101}$$

The covariant derivative in (2.101) is defined as

$$D_\mu\Psi = (\partial_\mu + \Gamma_\mu)\Psi, \tag{2.102}$$

with $\Gamma_\mu = (u^\dagger\partial_\mu u + u\partial_\mu u^\dagger)/2$ and A_μ as

$$A_\mu = i(u^\dagger\partial_\mu u - u\partial_\mu u^\dagger). \tag{2.103}$$

Extended to three flavors, the octet $\frac{1}{2}^+$ baryon field is

$$B = \sum_{a=1}^{8}\lambda_a B_a = \begin{pmatrix} \frac{1}{\sqrt{2}}\Sigma^0 + \frac{1}{\sqrt{6}}\Lambda & \Sigma^+ & p \\ \Sigma^- & -\frac{1}{\sqrt{2}}\Sigma^0 + \frac{1}{\sqrt{6}}\Lambda & n \\ \Xi^- & \Xi^0 & -\frac{2}{\sqrt{6}}\Lambda \end{pmatrix}. \tag{2.104}$$

This octet forms the adjoint representation of the $SU(3)_V$ subgroup of chiral symmetry, transforming as

$$B(x) \to VB(x)V^\dagger \tag{2.105}$$

with $V \in SU(3)_V$. Then the LO — up to one derivative — effective Lagrangian takes the form [Georgi (1984)]

$$\begin{aligned}\mathcal{L}_{\pi B}^{(1)} &= \mathrm{tr}\left[\bar{B}\left(i\not{D} - m_B\right)B\right] \\ &\quad - \frac{D}{2}\mathrm{tr}\left(\bar{B}\gamma^\mu\gamma_5\left\{A_\mu, B\right\}\right) - \frac{F}{2}\mathrm{tr}\left(\bar{B}\gamma^\mu\gamma_5\left[A_\mu, B\right]\right),\end{aligned} \tag{2.106}$$

where the covariant derivative is defined as

$$D_\mu B = \partial_\mu B + \left[\Gamma_\mu, B\right]. \tag{2.107}$$

The constants D and F determined by the semi-leptonic decays $B \to B' + e^- + \bar{v}_e$ at tree level are $D \approx 0.80, \quad F \approx 0.50$.

HLS can be straightforwardly implemented. For the two-flavor case, identifying h in (2.99) with HLS, i.e., $h \in [SU(2)]_{local}$, the Lagrangian is

$$\mathcal{L}_{\pi N}^{\rm HLS} = \bar{\Psi}\left(i\not{D} - m_N + \frac{g_A}{2}\gamma^\mu\gamma_5\hat{\alpha}_{\perp\mu} + \frac{g_V}{2}\gamma^\mu\hat{\alpha}_{\parallel\mu}\right)\Psi, \tag{2.108}$$

with the covariant derivative

$$D_\mu\Psi = \left(\partial_\mu - iV_\mu\right)\Psi. \tag{2.109}$$

For the three-flavor case, the Lagrangian is of the form

$$\begin{aligned}\mathcal{L}_{\pi B}^{(1)} &= {\rm tr}\left[\bar{B}\left(i\not{D} - m_B\right)B\right]\\ &\quad - g_{A_1}{\rm tr}\left(\bar{B}\gamma^\mu\gamma_5\left\{\hat{\alpha}_{\perp\mu}, B\right\}\right) - g_{A_2}{\rm tr}\left(\bar{B}\gamma^\mu\gamma_5\left[\hat{\alpha}_{\perp\mu}, B\right]\right)\\ &\quad - g_{V_1}{\rm tr}\left(\bar{B}\gamma^\mu\gamma_5\left\{\hat{\alpha}_{\parallel\mu}, B\right\}\right) - g_{V_2}{\rm tr}\left(\bar{B}\gamma^\mu\gamma_5\left[\hat{\alpha}_{\parallel\mu}, B\right]\right),\end{aligned} \tag{2.110}$$

with the covariant derivative

$$D_\mu B = \partial_\mu B + \left[V_\mu, B\right]. \tag{2.111}$$

For completeness, a short remark on the power counting of the baryon mass is in order here although it does not figure explicitly in what follows in this volume. Since the baryon mass is of order of Λ_χ in QCDLite, the power counting with it is not straightforward. There have been a wide variety of approaches to overcome this difficulty, one most straightforward being the heavy-baryon [Jenkins and Manohar (1990); Krause (1990); Bernard, Kaiser, Kambor and Meissner (1992); Scherer (2003)]. Schematically the additional power counting with baryons included can be summarized as

$$\begin{aligned}&\Psi \sim \mathcal{O}(p^0), \qquad D_\mu\Psi \sim \mathcal{O}(p^0), \qquad (i\not{D}\Psi - m) \sim \mathcal{O}(p),\\ &\{1, \gamma_\mu, \gamma_5\gamma_\mu, \sigma_{\mu\nu}\} \sim \mathcal{O}(p^0), \qquad \gamma_5 \sim \mathcal{O}(p).\end{aligned} \tag{2.112}$$

2.6.4 *Scalar meson: Dilaton*

Together with the vector mesons, a scalar meson of comparable mass plays a crucial role in nuclear physics. What is involved is a subtle symmetry in strong-interaction physics, namely scale invariance.

In Subsection 2.4.2 above, we discussed scale symmetry from the QCD proper. As noted there, the symmetry is broken both explicitly and spontaneously. Here we return to the essential features of scale symmetry in the framework of EFT. This is one of the core subjects in this volume, so will be covered more fully in Chapter 6. As explained in Chapter 5, scale symmetry, when viewed from effective theory, is hidden, like HLS, but in a much subtler sense.

The principal motivation for considering scale symmetry in nuclear physics is that it is a potential source for a scalar meson with a low mass that lends itself to treatment in terms of a local field. In a highly practical approach to nuclear many-body problems, a scalar degree of freedom with mass $\sim 600 - 700$ MeV is found to play an extremely important role, as for instance in relativistic mean field (RMF) approach of Walecka's linear model type [Serot and Walecka (1986)]. In nature, there is a scalar resonance at about that energy with a very broad width and it is listed in the particle data booklet as $f_0(500)$. There is an extremely long history full of controversies associated with this object [Pelaez (2015)]. It is not at all obvious as to how to associate this object with what enters in RMF models where a local field is invoked. An attractive option that we will adopt in this volume is that proposed by Crewther and Lewis (CT) [Crewther and Tunstall (2015)]: that the f_0 is the pseudo-NG boson of scale symmetry spontaneously broken, i.e., a dilaton. This issue is highly controversial, a certain aspect of which will be discussed in Chapter 6. Here we will simply take this as a working hypothesis.

Following [Crewther and Tunstall (2015)], we assume that the scalar f_0 can be incorporated on the same level as the octet pseudo-scalar NG bosons generically denoted as π and that the large width comes from radiative corrections. Processes in scalar channels can then be dominated by tree-order graphs involving the scalar, and high orders are then to be organized in a way closely analogous to the SχPT. However the scalar coupling is non-derivative, so there is a problem with power counting. We will argue that this problem can be suitably resolved and a systematic perturbative procedure can be developed. This will be discussed in Chapter 6. The two symmetries, chiral and scalar, are closely interlocked with the scalar and pseudo-scalar NG bosons intervening on the same footing and lead to the combined symmetry, dubbed as "scale-chiral symmetry."

2.6.4.1 *Scale-chiral symmetry*

The attractive feature of the CT approach [Crewther and Tunstall (2015)], particularly for nuclear physics, is that a systematic perturbative scheme, scale-chiral perturbation theory, can be set up extending standard chiral perturbation combining the two symmetries. The scale-chiral counting consists of the usual chiral counting given above for standard chiral Lagrangian as well as HLS plus the counting brought in by scale symmetry, leading to scale-chiral perturbation theory denoted $\chi\mathrm{PT}_\sigma$. The dilaton coupling does not involve derivatives. It is via $\Delta\alpha_s = |\alpha_s^{\mathrm{IR}} - \alpha_s|$ that the dilaton figures in the power counting, where α_s^{IR} is the value at the assumed IR fixed point at which the beta function of QCD gauge coupling is zero, $\beta(\alpha_s^{\mathrm{IR}}) = 0$,

$$\Delta\alpha_s \sim \mathcal{O}(p^2). \tag{2.113}$$

How this counting is arrived at is explained in Chapter 6.

There is an alternative scheme by Golterman and Shamir (GS) [Golterman and Shamir (2016)] that has a similar power counting with

$$\Delta n_f = (N_f^c - N_f)/N_c \sim \mathcal{O}(p^2) \tag{2.114}$$

where N_f^c is the would-be number of flavors at which a putative IR fixed point exists. To the leading order in the scale-chiral counting,[20] the CT and GS formalisms are found to give the same Lagrangian.

Here we briefly mention what has been done in the past in incorporating the dilaton in nuclear physics, e.g., [Brown and Rho (1991)]. We recall that one can rewrite the linear sigma model in the form given in (2.52),

$$\mathcal{L}_{\mathrm{L}\sigma\mathrm{M}} = \frac{1}{2}\partial_\mu\sigma\partial^\mu\sigma + \frac{1}{2}\sigma^2\partial_\mu U(x)\partial^\mu U^\dagger(x) - V_0\left(\sigma^2\right)$$

with σ a chiral singlet, not the fourth component of the chiral four-vector (and not the σ of HLS, the would-be NG boson to be higgsed to give mass to the vector mesons). Now if we assign scale dimension 1 to σ, then the first two terms are scale invariant. The potential V is of course not scale invariant, but if one can dial a parameter in V such that $\sigma \to f_\pi$, one gets the nonlinear σ model with no scale nvariance, that is, scale symmetry is shoved into the kinetic energy term. This illustrates subtleness or perhaps a conceptual ambiguity in dealing with scale symmetry.

As will be shown explicitly in Chapter 6, it is not difficult to write down the $\chi\mathrm{PT}_\sigma$ Lagrangian to the leading order, $\mathcal{O}(p^2)$. To do this, define the "conformal compensator" field that has mass dimension 1 and scale dimension 1,

$$\chi = f_\sigma e^{\sigma/f_\sigma}. \tag{2.115}$$

Under scale transformation $x \to \lambda^{-1}x$, χ transforms linearly

$$\chi(x) \to \lambda\chi(\lambda^{-1}x) \tag{2.116}$$

[20] Since we will exploit this order Lagrangian at various places, we will dub it "LOSS" whenever feasible. We will explain how this approximation is arrived at in $\chi\mathrm{PT}_\sigma$ in Chapter 6.

whereas the σ field transforms *nonlinearly* as

$$\sigma(x) \to \sigma(\lambda^{-1}x) + f_\sigma \ln \lambda. \tag{2.117}$$

As is well-known, this shift transformation is symptomatic of a NG boson, here the dilaton with its decay constant f_σ. Using the compensator field, the $\mathcal{O}(p^2)$ Lagrangian is of the form

$$\mathcal{L} = \mathcal{L}_{\rm M} + \mathcal{L}_{\rm B} + V(\chi), \tag{2.118}$$

$$\mathcal{L}_{\rm M} = \frac{f_\pi^2}{4}\left(\frac{\chi}{f_\sigma}\right)^2 {\rm Tr}\left(\partial_\mu U \partial^\mu U^\dagger\right) + \frac{1}{2}\partial_\mu\chi\partial^\mu\chi, \tag{2.119}$$

$$\mathcal{L}_{\rm B} = \bar{\Psi}\left(i\not{D} - \frac{\chi}{f_\sigma}m_N + \frac{g_A}{2}\gamma^\mu\gamma_5 A_\mu\right)\Psi \tag{2.120}$$

with the covariant derivative D_μ and A_μ defined in Section 2.6.3. Here $V(\chi)$ is a potential that is supposed to capture the scale symmetry breaking, explicit as well as spontaneous. At present the precise form of $V(\chi)$ for QCD, *if any*, is not known. Even whether or not it makes sense to talk about $V(\chi)$ is controversial.

The above controversy notwithstanding, the procedure to write the leading-order χPT$_\sigma$ Lagrangian is rather simple: Given a matter field Lagrangian with various scale dimensions, multiply or divide by the power needed of the scale-dimension-1 compensator fields to have the scale dimension of the Lagrangian density to be 4 and put all scale-symmetry breaking in the potential $V(\chi)$. The structure of $V(\chi)$ will of course crucially depend on the IR structure of the theory. Now how the form, i.e., LOSS, of the Lagrangian (2.118) is arrived at has an intricate and complex underpinning. It is in principle feasible to go beyond the leading order and this will be discussed in Chapter 6. At present, applications to nuclear physics and compact star matter that will be made in the following chapters can be made only at the leading order in scale-chiral expansion. To this order, however, the precise structure of the potential $V(\chi)$, it appears, is not indispensable.

2.7 Topological solitons

In standard nuclear physics approaches (SNPA), nucleons figure explicitly in the effective Lagrangian. Chapter 4 gives an overview of various approaches wherein the nucleon field as a local interpolating field is explicitly present. We have noted above that the nonlinear realization of chiral symmetry in mesonic Lagrangian naturally encodes a topology and the model that is built on that topology, skyrmion, first conceived by T. H. R. Skyrme [Skyrme (1962)], captures the large N_c structure of nucleons in QCD.

Highly surprising though it may appear, this solitonic structure of the nucleon turns out to play a powerful and predictive role in strongly-correlated interactions with other nucleons in medium, in particular, leading to an appropriate equation of state (EoS) of dense compact-star matter that can accommodate massive neutron stars without invoking explicit QCD degrees of freedom. Very little appreciated in nuclear physics community, this introduces a totally novel aspect to the effort to make a contact with QCD in going to the high-density matter.

The topology is carried by the pion field only, so for topological properties, it suffices to consider the original model employed by Skyrme , i.e., the "Skyrme model." The soliton is built off the Lagrangian containing two terms, one the two-derivative (current algebra) term and the other quartic derivative term called "Skryme term." For realistic nuclear dynamics involved in dense matter, the HLS fields and the dilaton field must also be included. Hence in confronting Nature, we will include all relevant mesonic fields in examining topological impact on dense matter. Here for the basic idea, we restrict our discussions to the Skyrme model to illustrate the topological aspect. In the next chapter, we shall connect, using the topology carried by the pion field, the topological soliton structure to the nucleon structure implied by QCD in terms of quarks and gluons. For this we shall resort to what is called "Cheshire Cat phenomenon."

2.7.1 *The Skyrme model*

The Lagrangian for the Skyrme model is

$$\mathcal{L}_{\text{Skyr}} = \frac{f_\pi^2}{4}\text{Tr}\left[\partial_\mu U \partial^\mu U^\dagger\right] + \frac{1}{32e^2}\text{Tr}\left\{\left[U^\dagger\partial_\mu U, U^\dagger\partial_\nu U\right]\left[U^\dagger\partial^\mu U, U^\dagger\partial^\nu U\right]\right\}, \tag{2.121}$$

where f_π is the pion decay constant as given before and e is a dimensionless parameter which could be connected to other physical observable, e.g., g_A. In (2.121), the first term is the current algebra term and the second term is the so-called Skyrme term without which the soliton will collapse.[21] Here, the soliton is of the classical, stable structure with finite energy in the nonlinear field theory. The solution, generically called skyrmion, with

[21] This quartic term turns out to be one of the L_4 terms in the SχPT Lagrangian that dominates the $\pi\pi$ scattering at low-energy [Diakonov and Petrov (2000)].

the winding number $1(B = 1)$ is identified as a fermion. The skyrmion baryon given by the Skyrme model will also be referred to as "skyrmion$_\pi$," indicating that only the pion field is involved.

The static energy of the system — skyrmion$_\pi$ mass — satisfies

$$\begin{aligned} E^{\text{Skyr}}_{\text{static}} &\geq \frac{f_\pi^2}{4} \int d^3x \left| \text{Tr}\left(\frac{1}{ef_\pi}\epsilon_{ijk}L_iL_jL_k\right)\right| \\ &\geq 0\,, \end{aligned} \tag{2.122}$$

where $L_i = U^\dagger \partial_i U$. This inequality means that the static energy is bounded from below, and that for the Skyrme model should be larger than or equal to zero. In terms of the topological charge B from Eq. (2.66), the above relation becomes

$$E^{\text{Skyr}}_{\text{static}} \geq 6\pi^2 \frac{f_\pi}{e} |B|\,, \tag{2.123}$$

which is known as the Bogomol'nyi bound. The lower limit will be saturated if L_i is self-dual,

$$L_i = \frac{1}{2ef_\pi}\epsilon_{ijk}L_iL_j\,, \tag{2.124}$$

but the Skyrme model does not satisfy it so the soliton energy should always be larger than the Bogomol'nyi bound. A soliton model constructed so that it satisfies the Bogomol'nyi bound gives what is called BPS skyrmion. Such a description turns out to give a surprisingly satisfactory model for binding energies of heavy nuclei. This matter will be discussed in Chapter 7.

From (2.121), we have the equation of motion (EoM) for the skyrmion

$$\frac{f_\pi^2}{2}(\partial_\mu L^\mu) - \frac{1}{8e^2}\partial_\mu\left[L_\nu, [L^\mu, L^\nu]\right] = 0. \tag{2.125}$$

This equation is a highly nonlinear equation which can be handled only in some special cases. Under the assumption of maximal symmetry, Skyrme proposed the hedgehog ansatz

$$U(\boldsymbol{x}) = \exp\left(i\boldsymbol{\tau}\cdot\hat{\boldsymbol{x}}F(r)\right) = \cos F(r) + i\boldsymbol{\tau}\cdot\hat{\boldsymbol{x}}\sin F(r) \tag{2.126}$$

with which the soliton mass is

$$M_{\text{Skyr}} = 4\pi\int drr^2\left[\frac{f_\pi^2}{2}\left(2\frac{\sin^2 F}{r^2} + F'^2\right) + \frac{1}{2e^2}\frac{\sin^2 F}{r^2}\left(\frac{\sin^2 F}{r^2} + 2F'^2\right)\right]. \tag{2.127}$$

Minimizing the soliton mass (2.127) one can obtain the EoM for the profile function $F(r)$ as

$$\begin{aligned} 0 &= \left(r^2 + \frac{2}{e^2f_\pi^2}\sin^2 F\right)F'' + 2rF' \\ &\quad + \frac{1}{e^2f_\pi^2}\sin 2F\,F'^2 - \sin 2F - \frac{\sin^2 F\sin 2F}{e^2f_\pi^2 r^2}. \end{aligned} \tag{2.128}$$

To obtain the baryon number-one solution, one has to impose the boundary conditions

$$F(r=0)=\pi, \qquad F(r\to\infty)=0. \tag{2.129}$$

It is convenient to make the coordinate transformation

$$r \to r'/(ef_\pi). \tag{2.130}$$

Then the skyrmion mass takes the form

$$M_{\text{Skyr}} = 4\pi\frac{f_\pi}{e}\int drr^2\frac{1}{2}\left[\left(2\frac{\sin^2 F}{r^2}+F'^2\right)+\frac{\sin^2 F}{r^2}\left(\frac{\sin^2 F}{r^2}+2F'^2\right)\right] \tag{2.131}$$

and the profile function $F(r)$ satisfies

$$\begin{aligned}\left(r^2+2\sin^2 F\right)F''+2rF'+\sin 2F\,F'^2&\\ -\sin 2F-\sin^2 F\sin 2F=0.&\end{aligned} \tag{2.132}$$

Since the solution $F(r)$ is independent of f_π and e, as one can see in (2.132), the soliton mass is given by $C\times\frac{f_\pi}{e}\propto\mathcal{O}(N_c)$ where C is a constant. The solution to the EoM (2.132) was first obtained independently in 1983 by [Jackson and Rho (1983); Adkins, Nappi and Witten (1983)].[22] With the numerical result obtained therefrom, the skyrmion$_\pi$ mass comes out to be

$$M_{\text{Skyr}}\simeq 74.58\times\left(\frac{f_\pi}{e}\right). \tag{2.133}$$

With the hedgehog ansatz for the profile function (2.126), the topological charge density can also be written down explicitly,

$$B^0(r)=\frac{1}{2\pi^2}\sin^2 F\frac{F'}{r^2}, \tag{2.134}$$

which leads to the topological charge

$$B=\frac{1}{\pi}\left(F(0)-F(\infty)\right)+\frac{1}{2\pi}\left(\sin 2F(\infty)-\sin 2F(0)\right). \tag{2.135}$$

Therefore, with the boundary conditions (2.129), the Skyrme model indeed describes the $B=1$ classical baryon physics.

What we have obtained thus far is still a classical object. It does not have quantum numbers. To endow skyrmion with definite spin and isospin

[22] It is surprising that the differential equation for the profile function F had remained unsolved until 1983. When one of the authors of [Jackson and Rho (1983)] discussed the importance of solving it at a seminar in early 1982, he was told by a mathematician in the audience that the solution did not exist!

quantum numbers, the system should be quantized. This quantization could be done by introducing the SU(2) collective variables through

$$U_c(\boldsymbol{x}) \to U(\boldsymbol{x}, t) = C(t) U_c(\boldsymbol{x}) C^\dagger(t), \tag{2.136}$$

where $C(t)$ is a unitary matrix satisfying $C(t)C(t)^\dagger = C(t)^\dagger C(t) = 1$ and the subscript c indicates that the field U is independent of time. Defining the angular velocity Ω_i corresponding to the collective coordinate rotation by

$$\frac{i}{2}\boldsymbol{\tau} \cdot \boldsymbol{\Omega} \equiv C^\dagger \partial_0 C, \tag{2.137}$$

one can express the rotational energy $E^{\rm Skyr}_{\rm rotation}$ in terms of the angular velocity as

$$E^{\rm Skyr}_{\rm rotation} = \mathcal{I}_{\rm Skyr} {\rm Tr}\left[\partial_0 C^\dagger(t) \partial_0 C(t)\right] = \frac{1}{2}\mathcal{I}_{\rm Skyr}\Omega^2, \tag{2.138}$$

with $\mathcal{I}_{\rm Skyr}$ being the moment of inertia of the soliton configuration with respect to the rotation (2.136) which can be expressed as a function of $F(r)$ as

$$\mathcal{I}_{\rm Skyr} = \frac{8\pi}{3}\int_0^\infty r^2 dr \sin^2 F \left\{ f_\pi^2 + \frac{1}{e^2}\left(F'^2 + \frac{\sin^2 F}{r^2} \right) \right\}. \tag{2.139}$$

This can be calculated by using the solution of EoM for $F(r)$

$$\mathcal{I}_{\rm Skyr} \simeq 38.11 \times \left(\frac{1}{e^3 f_\pi} \right). \tag{2.140}$$

One sees that $\mathcal{I}_{\rm Skyr} \sim O(N_c)$. The classical angular momentum of the skyrmion can be gotten by

$$\boldsymbol{J} = \frac{\partial E^{\rm Skyr}_{\rm rotation}}{\partial \boldsymbol{\Omega}} = \mathcal{I}_{\rm Skyr}\boldsymbol{\Omega}. \tag{2.141}$$

So the rotational energy is[23]

$$E^{\rm Skyr}_{\rm rotation} = \frac{\boldsymbol{J}^2}{2\mathcal{I}_{\rm Skyr}}. \tag{2.142}$$

Following the standard quantum mechanics, the quantized angular momentum is given by

$$\boldsymbol{J}^2 = j(j+1)\hbar^2, \tag{2.143}$$

[23]In Chapter 7, the rotational quantization in the iso-spin direction will be found to give a cusp structure, an important topological input, to the nuclear symmetry energy important in the EoS of dense baryonic matter.

where $j = 0, 1/2, 1, 3/2, 2, \cdots$. We will take hereon $\hbar = 1$. Then the baryon masses with $1/N_c$ corrections are given by

$$m_B = M_{\rm static}^{\rm Skyr} + \frac{j(j+1)}{2\mathcal{I}_{\rm Skyr}}. \tag{2.144}$$

Specifically, the nucleon $J = 1/2$ and Δ resonance $J = 3/2$ masses are

$$M_N = M_{\rm Skyr} + \frac{3}{8\mathcal{I}_{\rm Skyr}}, \quad M_\Delta = M_{\rm Skyr} + \frac{15}{8\mathcal{I}_{\rm Skyr}}. \tag{2.145}$$

The skyrmion$_\pi$ mass (2.133) can be roughly estimated given the values of the two constants that we have in the chiral limit. The constant e can be determined by fitting the axial current coupling contant $g_A \simeq 1.27$. It comes out to be $1/e \approx 0.21$. With $f_\pi \approx 93$ MeV, the skyrmion mass is $M_{\rm Skyr} \approx 1400$ MeV. With the $1/N_c$ correction $\lesssim 100$ MeV from (2.145), the predicted nucleon mass ~ 1500 MeV is much too big compared with the experimental 940 MeV. For this reason, in the literature, the two parameters in the model have been fiddled to fit the experimental mass. That required f_π to be $f_\pi \approx 58$ MeV. The justification for this quenched value of the pion decay constant is that it is an "effective" constant, renormalized from the vacuum value. We think that this is not a compelling argument.

The reason is simply that the above computation ignores that there can be a big correction from the Casimir energy, which is $\mathcal{O}(N_c^0)$. The Casimir problem, a notoriously difficult problem in general, is even more daunting in the skyrmion approach as it is aggravated by that the Lagrangian needed to compute the effect is not known at short distances (as we will see in later chapters). Rough though they are, estimates in nonlinear sigma model indicate that the Casimir correction can be as large as ~ -500 MeV (see [Nowak, Rho and Zahed (1996)] on this matter). This could not be simulated by effective constants.[24]

2.7.2 *Skyrmions on crystal for nuclear matter*

We see that in the large N_c limit at which the notion of skyrmions as baryons makes sense, the nucleon mass, that is the ground state of baryonic excitations, comes much too high. This presages the potential difficulty in using skyrmions in accessing nuclear and dense matter at the leading order. It is a well-known fact in nuclear physics that while it is the ground state of

[24]Schematically the N_c counting for the nucleon mass goes (in MeV) as

$$M = M_{N_c} + M_{N_c^0} + M_{1/N_c} \approx 1000 \approx 1400 - 500 + 100.$$

nuclear matter that is the most difficult object to get right, the excitations relative to the ground state are more easily controllable. This is because the scale involved for excitations is much smaller than the scale for the ground state. For instance in relativistic mean-field approach, there is a competition between two large numbers, namely the big attraction due to the scalar degree of freedom and the big repulsion due to the isoscalar vector repulsion, both $\sim 1/3$ of the nucleon mass, whereas the excitations within a nucleus or nuclear matter involve much smaller scale, some two orders of magnitude smaller. In accessing nuclei and nuclear matter in terms of the skyrmion description, however, the situation is immensely different. Large N_c arguments suggest that the binding energy of nuclei would also involve the scale of the strong interactions, namely, f_π, and indeed one finds that in the skyrmion prediction in the Skyrme model, the binding energy is big, increasing with the mass number. For example, it is $\sim$ 100 MeV for the alpha particle [Sutcliffe (2011)]. In Nature, the binding energy in nuclei is tiny, $\sim$ 16 MeV per particle in infinite nuclear matter. There are possible remedies to this problem which will be discussed in Chapter 7.

Here we will by-pass this difficult issue and address certain robust features supplied by the topological structure that can be suitably implemented in EFT anchored on scale-invariant-HLS. The basic idea that will be developed is that the baryonic matter can be simulated by putting the skyrmions onto a crystal lattice [Klebanov (1985)]. Taking the skyrmion crystal solution as "background classical fields",[25] we can then interpret the fluctuating fields on top of it as the corresponding mesons in dense baryonic matter. Following the procedure proposed in [Ma, Harada, Lee, Oh, Park and Rho (2014)], we write the minimum energy solution as $U_{(0)}$ and introduce the fluctuating fields as

$$U = u_{(0)}\check{U}u_{(0)}, \tag{2.146}$$

where $\check{U} = \exp(2i\tau_a\check{\pi}_a/f_\pi)$ stands for the corresponding fluctuating field and $u_{(0)}^2 = U_{(0)}$. By substituting the fields in Eq. (2.146) into the Skyrme model Lagrangian, one can obtain the Lagrangian for pion in medium. The medium modified hadron properties in skyrmion crystal matter will be discussed in Chapter 7.

Of greater interest is the approach to dense matter via skyrmion topology. The most illustrative case is the alpha particle with two neutrons and two protons.

[25]This topological object will be connected later — particularly in Chapters 8 and 9 — to the notion of "intrinsic density dependence (IDD)" of the QCD vacuum structure in the presence of medium.

It has been shown that the alpha particle, a tightly bound nucleus, can be described by a cubic lattice structure of half-skyrmions put at the eight vertices [Battye, Manton and Sutcliffe (2016); Manton (2017)]. This structure is found to have an enhanced cubic symmetry favored over the face-centered-cubic (FCC) crystal (see Fig. 2.2), which is the preferred symmetry at lower density [Kugler and Shtrikman (1988, 1989)]. The point of interest in this volume, exploited in Chapters 7-8, lies in that when put on FCC crystal lattice and squeezed, the skyrmions in the system get transformed to half-skyrmions in cubic crystal (CC) at a density above that of nuclear matter. The density at which this happens will be denoted from here on as $n_{1/2}$. This takes place at a density somewhat higher than the

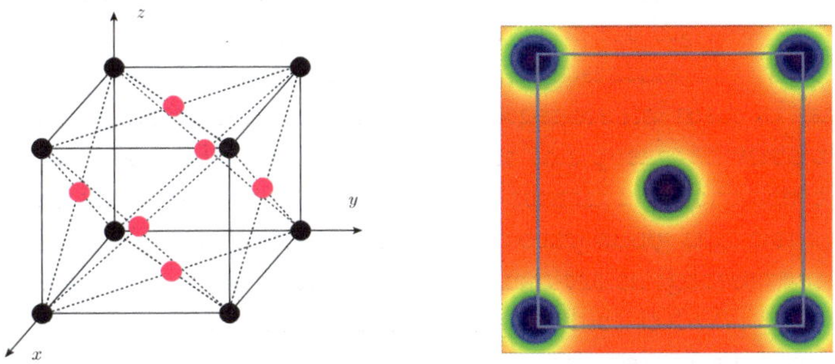

Fig. 2.2 Arrangement of the skyrmions on the FCC crystal lattice and baryon density distribution.

normal nuclear matter density, an estimate giving $n_{1/2} \sim 2n_0$. Here $n_{1/2}$ represents the skymion-to-half-skyrmion changeover density, referred to as half-skyrmion density. Now this transformation, being topological, is highly robust. It depends only on the topology of the system, so only on the pion field that carries topology whereas other degrees of freedom such as vector mesons and scalar (i.e., dilaton) affect only the numerical details.

The half-skyrmion structure in dense medium is quite similar to that present in the alpha particle with one difference. In the half-skyrmion state of denser matter, the chiral condensate proportional to the quark condensate $\langle\bar{q}q\rangle$ vanishes on the average in the unit cell while it should not in the alpha particle. The pion is still present in the half-skyrmion phase, so this means that the leading term in the chiral condensate vanishes, but the pion decay constant is not equal to zero. This feature, resembling the pseudogap phase in high T_c superconductivity [Mueller (2018); Zarembo

(2002)], indicates that the pion decay constant in the half-skyrmion phase is of higher order in the quark bilinears. The half-skyrmions are confined, so they are not propagating degrees of freedom, but they are quite fragile in low-density system like the alpha particle. The interplay between the cubic symmetry structure of half-skyrmions and the FCC structure of skyrmions could lead to the shell structure and small binding energy [Manton (2017)].

The mathematical complexity that blocks more fruitful and adroit exploitation of the skyrmion structure notwithstanding, this skyrmion-half-skyrmion topology change, considered to be robust, will play an important — and fascinating — role in what follows.

Chapter 3

Hadron-Quark Continuity

Effective field theories (EFTs) use hadronic languages to represent QCD. The "translation" between the two languages, EFT and QCD, cannot be necessarily unique. Here we discuss one possible translation that renders the connection visible that exploits the chiral bag model [Brown and Rho (1988)] that marries QCD dynamics and mesons via topology, specifically the skyrmion. First we establish the link for the baryon between the skyrmion and the MIT bag [Johnson (1975)] and then later extend the notion to dense baryonic matter. The basic idea to be developed can be phrased as a "trading-in" of QCD degrees of freedom for topology in going to density regimes that are inaccessible by perturbative QCD. We will see that this trading-in figures importantly in compact-star observables, e.g., the star sound velocity v_s.

3.1 Cheshire Cat

At some high energy/momentum scale, the basic degrees of freedom of QCD, quarks and gluons, will certainly become the relevant degrees of freedom for dynamics. In low-energy nuclear physics, however, quarks and gluons are invisible. Atoms consist of nuclei and electrons, nuclei consist of nucleons, protons and neutrons. They are visible. Now the nucleon is made up of quarks, but the question remains whether quark degrees of freedom will ever be "visible." One of the principal objectives of the modern nuclear physics is how one can go from the physics of nuclear matter that is known, both theoretically and experimentally in the vicinity of nuclear matter density n_0, to high density at which quark-gluon degrees of freedom are expected. One possible way — which is generalized widely in the later part of this volume — of addressing this problem is in terms of what's known as "chiral-bag model" [Brown and Rho (1988)] with its natural outcome, the

Cheshire Cat (CC) picture, which posits that one can go from the former to the latter in a smooth way with no change of phases, that is, with a smooth translation of languages. We identify in this volume this feature with the general notion of *quark-hadron continuity*. In this chapter we will illustrate this feature. In confronting compact-star matter (in Chapter 9), we exploit the intuition gained from this model to map what happens in the equation of state (EoS) in going from n_0 to densities relevant to the interior of compact stars, $\sim (5-7)n_0$. We will see that this continuity can be established *both* with and without quark-gluon degrees of freedom. In fact one can use "hadronic variables" with the help of soliton structure all the way from nuclear matter to hadron-quark crossover to color-flavor locked state.

3.2 Chiral bag model for Cheshire Cat phenomena

We first guess, using intuition, an effective Lagrangian for the CC structure. We will work with a toy model in (1+1) dimensions, where the description is simple, to gain an intuition and then extend, and make it more realistic, to (3+1) dimensions with certain reasonable approximations injected. To a certain extent the derivation is exact in the former, but it is at best approximate, apart from the baryon charge, in the latter. We will detail in what sense this is the case. For this, we consider a model of hadrons with three regions. (1) The interior region represented by the volume V — that we consider spherical for simplicity — in which quarks and gluons are "confined." (2) The region outside of the bag denoted by the volume $\overline{V}$ populated by mesons, i.e., pseudo-Nambu–Goldstone (pNG) bosons (pions, kaons), η', vector mesons (ρ, ω, ...). The boundary that delineates V and $\overline{V}$ is denoted ∂V. The two regions, inside and outside, will be connected by boundary conditions subject to the symmetries imposed by QCD. The boundary conditions serve the role of translator between the two "languages." We take the boundaries to be sharp for simplicity, but in nature, this may not be the case. A sort of smoothing should make it closer to nature. This is, however, a technical detail we shall not get into. In our discussions, we will do the smoothing in interpretation.

3.2.1 *Cheshire Cat as a gauge artifact*

The basic notion of CC phenomenon is that the "bag" implementing confinement in the sense of the MIT bag model [Johnson (1975)] is an artifact that has no strict physical meaning. A highly potent and elegant way to view this is to interpret the "bag" as a gauge artifact and treat it as a gauge degree of freedom [Damgaard, Nielsen and Sollacher (1992)]. Thus a description in terms of a particular "bag radius" that was thought to represent the size of confinement region in the MIT bag model [Johnson (1975)] can be interpreted as a particular gauge fixing. In this view, confinement size will therefore have no physical meaning. It could be taken as a modeling of "hadron-quark continuity" that implies, in principle, when formulated completely, physics should be strictly equivalent for *all* gauge choices ("bag radii") one makes. This is a drastically different way — even in this 21st century — of looking at the classical confinement embodied by the MIT bag. Although the idea can perhaps be exactly formulated only in (1+1) dimensions thanks to the exact bosonization technique, it is plausible that the same holds, albeit approximately, in four dimensions in low-energy regimes.

Consider massless fermions ψ coupled to some external vector $\mathcal{V}_\mu$ and axial-vector $\mathcal{A}_\mu$ fields, the generating functional of which (in Minkowski space) is[1]

$$Z[\mathcal{V},\mathcal{A}] = \int [d\psi][d\bar{\psi}]e^{iS}, \tag{3.1}$$
$$S = \int d^2x\bar{\psi}\gamma^\mu\left(i\partial_\mu + \mathcal{V}_\mu + \mathcal{A}_\mu\gamma_5\right)\psi.$$

The well-known massless Thirring model is gotten by adding a term $\sim \mathcal{V}^\mu\mathcal{V}_\mu$ and integrating over the external vector field. One can get the massive Thirring model if one adds scalar and pseudoscalar sources in addition. Thus the model is quite general. Now do the field redefinition while enlarging the space by introducing a fictitious (or redundant) field θ,

$$\psi(x) = e^{i\theta(x)\gamma_5}\chi(x). \tag{3.2}$$

Since this is just a change of symbols, the generating functional is unmodified:

$$Z[\mathcal{V},\mathcal{A}] = \int [d\chi][d\bar{\chi}]\, J\, e^{iS'}, \tag{3.3}$$
$$S' = \int d^2x\bar{\chi}\gamma^\mu\left(i\partial_\mu + \mathcal{V}_\mu + \mathcal{A}_\mu\gamma_5 - \partial_\mu\theta(x)\gamma_5\right)\chi\,,$$

[1] Our convention in (1+1) dimesnions is as follows. The metric is $g_{\mu\nu} = \mathrm{diag}(1,-1)$ with Lorentz indices $\mu,\nu = 0,1$ and the γ matrices are in Weyl representation, $\gamma_0 = \gamma^0 = \sigma_1$, $\gamma_1 = -\gamma^1 = -i\sigma_2$, $\gamma_5 = \gamma^5 = \sigma_3$ with the usual Pauli matrices σ_i.

where J is the Jacobian of the transformation given by

$$J = \exp\left\{ i \int d^2x \left(\frac{1}{2\pi}\partial_\mu\theta\partial^\mu\theta + \frac{1}{\pi}\epsilon^{\mu\nu}\mathcal{V}_\mu\partial_\nu\theta - \frac{1}{\pi}\mathcal{A}_\mu\partial^\mu\theta \right) \right\}. \quad (3.4)$$

This nontrivial Jacobian arises because the measure is not invariant under the local chiral transformation (3.2). The important thing to note is that since *by definition* (3.1) and (3.3) are the same, the physics should not depend upon $\theta(x)$. The latter is redundant. Therefore — modulo, an infinite constant which requires gauge fixing as described below — we may integrate over the field $\theta(x)$ in (3.3) without changing physics. Doing the integral in the path integral amounts to elevating the redundant $\theta(x)$ to a dynamical variable. Now if we rewrite (3.3) in terms of $\mathcal{L}'$ defined as

$$Z[\mathcal{V}, \mathcal{A}] = \int [d\chi][d\bar{\chi}] e^{i\int d^2x \mathcal{L}'}, \quad (3.5)$$

$$\mathcal{L}' = \bar{\chi}\gamma^\mu \left(i\partial_\mu + \mathcal{V}_\mu + \mathcal{A}_\mu\gamma_5 - \partial_\mu\theta\gamma_5 \right)\chi + \frac{1}{2\pi}\partial_\mu\theta\partial^\mu\theta + \frac{1}{\pi}\epsilon^{\mu\nu}\mathcal{V}_\mu\partial_\nu\theta - \frac{1}{\pi}\mathcal{A}_\mu\partial^\mu\theta, \quad (3.6)$$

we uncover a (hidden) local gauge symmetry, namely, that (3.6) is invariant under the transformation

$$\chi(x) \to e^{i\alpha(x)\gamma_5}\chi(x), \quad \bar{\chi}(x) \to \bar{\chi}(x)e^{i\alpha(x)\gamma_5}, \quad \theta(x) \to \theta(x) - \alpha(x) \quad (3.7)$$

provided of course the Jacobian is again taken into account. Thus by introducing a new dynamical field θ, we have gained a gauge symmetry at the expense of enlarging the space. This notion will be exploited later and throughout this volume in (emergent) hidden gauge symmetries that will play an important role in describing hadron phenomena both in normal nuclear matter and under extreme conditions.

Now in order to quantize the theory, we need to fix the gauge. To do this we pick a θ by the general gauge condition

$$\Phi(\theta) = 0. \quad (3.8)$$

Then following the standard text-book (Faddeev–Popov) method, we write

$$Z[\mathcal{V}, \mathcal{A}] = \int [d\chi][d\bar{\chi}][d\theta]\delta(\Phi[\theta])|\det\left(\frac{\delta\Phi}{\delta\theta}\right)|e^{i\int d^2x\mathcal{L}'}. \quad (3.9)$$

Note that if we choose $\Phi = \theta$, we recover the original generating functional (3.1) which describes everything in terms of fermions. We would like to choose the gauge condition relevant to the physics of the chiral bag we are dealing with. There the axial current, i.e., chiral symmetry, plays an essential role. A suitable choice is

$$\Phi[\theta, \chi, \bar{\chi}] = \Delta \int_{x_0}^{x} d\eta^\nu \bar{\chi}(\eta)\gamma_\nu\gamma_5\chi(\eta) + (1-\Delta)\frac{1}{\pi}\theta(x) = 0 \qquad (3.10)$$

which corresponds to partitioning the total axial current into the fraction Δ of the fermionic (quark) piece and the fraction $(1-\Delta)$ of the bosonic (pionic) piece. For the reason which will be clearer below, we call this "chiral bag gauge."

To compute the Faddeev–Popov determinant corresponding to (3.10), write the constraint

$$\begin{aligned}\delta(\Phi[\theta, \bar{\chi}, \chi]) &= \int [db] e^{i\int d^2x\, b\,\Phi} \\ &= \int [db] e^{i\int d^2x\ \mathcal{L}_{gf}} \end{aligned} \qquad (3.11)$$

defining the gauge-fixing Lagrangian $\mathcal{L}_{gf}$. Next do the (infinitesimal) gauge transformation (3.7) on (3.11). One gets

$$\delta_\alpha \mathcal{L}_{gf} = -\frac{1}{\pi}\{(1-\Delta) + \Delta\}\alpha \qquad (3.12)$$

where the first term in curly bracket comes from the shift in θ and the second term from the Fujikawa measure. Now the right-hand side of this equation is just $b\delta_\alpha\Phi$, from which follows

$$\det\left(\frac{\delta\Phi}{\delta\alpha}\right) = \det\left(-\frac{1}{\pi}\right). \qquad (3.13)$$

The CC structure is manifested in the choice of Δ. If one takes $\Delta = 0$, we have the pure fermion theory as one can see trivially. If one takes instead $\Delta = 1$, one obtains, after some calculation [Damgaard, Nielsen and Sollacher (1994)],

$$Z[\mathcal{V}, \mathcal{A}] = \text{const.} \times \int [d\theta] e^{i\int d^2x \mathcal{L}^\theta} \qquad (3.14)$$

with

$$\mathcal{L}^\theta = \frac{1}{2}\partial_\mu\theta\partial^\mu\theta - \frac{1}{\sqrt{\pi}}\epsilon^{\mu\nu}\mathcal{V}_\mu\partial_\nu\theta + \frac{1}{\sqrt{\pi}}\mathcal{A}_\mu\partial^\mu\theta. \qquad (3.15)$$

This is just the bosonized form of the fermion theory coupled to the vector fields. The CC statement is that the theory is identical for *any* arbitrary value of Δ.

One can go further and take Δ to be a local function. In this case, one can continuously change representation from one region of space-time to another, choosing different gauge fixing conditions in different space-time regions. This leads to a smooth CC bag. The original chiral bag model [Brown and Rho (1988)] corresponds to the gauge choice

$$\Delta(x) = \Theta(x - z) \tag{3.16}$$

with $z = \hat{r}R$. The usual bag boundary conditions employed in phenomenological studies arise after certain suitable averaging over part of the space, and, hence, are a specific realization of the CC scheme. Possible discontinuities or anomalies caused by sharp boundary conditions used in actual calculations should not be considered to be defects of the effective model. An interesting conclusion of this point of view is that one can in principle construct an *exact* CC model without knowing the exact bosonized version of the theory considered.

Since the Cheshire Cat phenomenon in (3+1) dimensions is in practice not exact, using the strategy of hidden gauge symmetry (here $U(1)$ associated with $\theta(x)$) provides certain flexibility at the expense of non-uniqueness that goes with the specific gauge choice. Depending upon what processes one studies, a certain gauge choice might be more efficient than others. Thus for long-wavelength processes, a gauge closer to $\Delta = 1$ (meson degrees of freedom) could be more efficient in capturing the strong interaction physics whereas processes sampling short distances would do better with gauge choices closer to $\Delta = 0$ (quark degrees of freedom).

3.2.2 *Baryon charge*

Having shown how the "chiral bag gauge" works, we now consider the baryon charge. The baryon charge should be conserved, so the question is how the baryon charge "leaks," when the bag is squeezed, out of the bag in which the baryon charge is made up of quarks into the meson sector in two dimensions where the Cheshire Cat Principle (CCP) can be formulated exactly. This gives us a general idea of the way the principle works in illustrating how elusive the notion of "confinement" is. Later we will make a leap to reality in four dimensions and mention how much one can take over from the (1+1) case. The baryon charge is topological and hence illustrates how the trading-in between quarks/gluons and topology precisely works out. We shall suggest that the result is generic and should not sensitively depend upon details of the models. Thus other models of similar structure containing the same symmetries, such as chiral quark soliton model, could be equally used. We will leave the construction of CCP or approximate CCP to the aficionados of their own favorite models.

We write the action in (1+1) dimensions (in Minkowski space) in the simplest form

$$S = S_V + S_{\overline{V}} + S_{\partial V} \tag{3.17}$$

where

$$S_V = \int_V d^2x \bar{\psi} i\gamma^\mu \partial_\mu \psi + \cdots , \tag{3.18}$$

$$S_{\overline{V}} = \int_{\overline{V}} d^2x \frac{1}{2}(\partial_\mu \phi)^2 + \cdots \tag{3.19}$$

where ϕ is the bosonic scalar field of odd intrinsic parity, and $S_{\partial V}$ is the boundary term which plays the role in translating the languages of the two regions. Here the ellipses stand for other terms such as interactions, fermion masses etc. consistent with the assumed symmetries. For simplicity without losing generality, we ignore the mass term. Unlike in the (3+1) dimensional case, one can write down the boundary term in a unique way subject to chiral symmetry, the discrete symmetries $\mathcal{P}$, $\mathcal{C}$ and $\mathcal{T}$, and the constraint that there be no decoupled boundary conditions, that is to say, that the boundary conditions involve only ψ and ϕ fields. It has the simple form

$$S_{\partial V} = \int_{\partial V} d\Sigma^\mu \left\{ \frac{1}{2} n_\mu \bar{\psi} e^{i\gamma_5 \phi/f} \psi \right\} \tag{3.20}$$

with the ϕ "decay constant" $f = 1/\sqrt{4\pi}$ and n_μ is a (outward-directed) unit four vector. This is unique in (1+1) dimensions, but in higher dimensions the boundary conditions are likely more complex and not unique.

At the classical level the action (3.17) gives rise to the bag confinement, namely that, inside the bag, the fermion (which we shall call "quark" from now on) obeys (dropping the mass terms)

$$i\gamma^\mu \partial_\mu \psi = 0 \tag{3.21}$$

while outside the bag the boson (which we shall call "pion") satisfies

$$\partial^2 \phi = 0 \tag{3.22}$$

subject to the boundary conditions on ∂V

$$i n^\mu \gamma_\mu \psi = -e^{i\gamma_5 \phi/f} \psi, \tag{3.23}$$

$$n^\mu \partial_\mu \phi = f^{-1} \bar{\psi} \left(\frac{1}{2} n^\mu \gamma_\mu \gamma_5 \right) \psi. \tag{3.24}$$

Equation (3.23) is the familiar "MIT confinement condition" with chiral symmetry implemented, which is simply a statement of the classical conservation of the vector current $\partial_\mu j^\mu = 0$ or $\bar{\psi} \frac{1}{2} n^\mu \gamma_\mu \psi = 0$ at the surface, while Eq. (3.24) is just the statement of the conserved axial vector current $\partial_\mu j_5^\mu = 0$.

These classical observations, however, are invalidated by quantum mechanical effects. In particular, whereas the axial current continues to be conserved, the vector current fails to do so due to "quantum anomaly." In field theory without boundaries, there is no anomaly in the vector current. With the bag, however, the vector anomaly occurs due to the "artificial" boundary we elected which acts as an induced local axial field.

There are several ways of seeing that something is amiss in the picture with the classical conservation law. The simplest is what is known as the "infinite hotel scenario" [Nadkarni, Nielsen and Zahed (1985); Nielsen and Wirzba (1988)]. To illustrate what happens, let us imagine that the quark is "confined" in the space $-\infty \leq r \leq R$ with a boundary at $r = R$. Now the vector current $j_\mu = \frac{1}{2}\bar{\psi}\gamma_\mu\psi$ is conserved inside the "bag"

$$\partial^\mu j_\mu = 0, \qquad r < R. \tag{3.25}$$

If one integrates this from $-\infty$ to R in r, one gets the time-rate change of the fermion (i.e., quark) charge

$$\dot{Q} \equiv \frac{d}{dt}Q = 2\int_{-\infty}^{R} dr\partial_0 j_0 = 2\int_{-\infty}^{R} dr\partial_1 j_1 = 2j_1(R) \tag{3.26}$$

which is just

$$\bar{\psi}n^\mu\gamma_\mu\psi, \qquad r = R. \tag{3.27}$$

This vanishes classically as we mentioned above. But quantum mechanically, this is not correct. The reason is that it is ill-defined locally in time. In other words, $\psi^\dagger(t)\psi(t+\epsilon)$ goes like ϵ^{-1} and so is singular as $\epsilon \to 0$. To circumvent this difficulty, we regulate the bilinear product by point-splitting in time

$$j_1(t) = \bar{\psi}(t)\frac{1}{2}\gamma_1\psi(t+\epsilon), \qquad \epsilon \to 0. \tag{3.28}$$

Now using the boundary condition

$$\begin{aligned} i\gamma_1\psi(t+\epsilon) &= e^{i\gamma_5\phi(t+\epsilon)/f}\psi(t+\epsilon) \\ &\approx e^{i\gamma_5\phi(t)/f}[1 + i\epsilon\gamma_5\dot{\phi}(t)/f]e^{\frac{1}{2f^2}[\phi(t),\phi(t+\epsilon)]}\psi(t+\epsilon), \; r = R \end{aligned} \tag{3.29}$$

and the commutation relation $[\phi(t), \phi(t+\epsilon)] = i \text{ sign}(\epsilon) = \pm i$, we obtain

$$j_1(t) = -\frac{i}{4f}\epsilon\dot{\phi}(t)\psi^\dagger(t)\psi(t+\epsilon) = \frac{1}{4\pi f}\dot{\phi}(t) + O(\epsilon), \qquad r = R \tag{3.30}$$

where we have used $\psi^\dagger(t)\psi(t+\epsilon) = \frac{i}{\pi\epsilon} + [\text{regular}]$, the boundary condition (3.23) and $f^2 = (4\pi)^{-1}$. Thus quarks can flow out or in if the pion field varies in time. But they cannot simply disappear into nowhere if we impose that the fermion number be conserved. To see what happens, rewriting (3.30) using the surface tangent $t^\mu = \epsilon^{\mu\nu} n_\nu$, we have

$$t \cdot \partial\phi = \frac{1}{2f}\bar{\psi} n \cdot \gamma \psi = \frac{1}{2f}\bar{\psi} t \cdot \gamma\gamma_5 \psi, \qquad r = R \tag{3.31}$$

where we have used the relation $\bar{\psi}\gamma_\mu\gamma_5\psi = \epsilon_{\mu\nu}\bar{\psi}\gamma^\nu\psi$ valid in two dimensions. Equation (3.31) together with (3.24) is nothing but the bosonization relation

$$\partial_\mu \phi = f^{-1}\bar{\psi}\left(\frac{1}{2}\gamma_\mu\gamma_5\right)\psi \tag{3.32}$$

at the point $r = R$ and time t. As is well known, this is a particular feature of (1+1) dimensional fields that makes one-to-one correspondence between fermions and bosons.

Equation (3.26) with (3.30) is the vector anomaly. What it says is that the vector charge, which in this case is equivalent to the fermion (quark) number inside the bag, is not conserved. Physically, what happens is that the amount of fermion number ΔQ corresponding to $\Delta t\dot{\phi}/\pi f$ is pushed into the Dirac sea at the bag boundary, and so is lost from inside and gets accumulated at the pion side of the bag wall. Since the fermion charge is to be conserved, this accumulated baryon charge must be carried by something residing in the meson sector, namely the pion field outside. For this to be realized, the pion field must support a soliton. In the present model, we find that one unit of fermion charge $Q = 1$ is partitioned as

$$Q = 1 = Q_V + Q_{\overline{V}} \tag{3.33}$$

with

$$Q_{\overline{V}} = \theta/\pi, \ \theta = \phi(R)/f. \tag{3.34}$$

We thus learn that the quark charge is partitioned into the bag and outside of the bag, without, however, any dependence of the total on the size or location of the bag boundary. This partition is exact. In the (1+1)-dimensional case, one can calculate other physical quantities such as the energy, response functions and, more generally, partition functions and show that the physics does not depend upon the presence of the bag. We could work with quarks alone, or pions alone or any mixture of the two. If one works with the quarks alone, we have to do a proper quantum treatment to obtain something which one can obtain in mean field with the pions alone. In some situations, the hybrid description is more economical than the pure ones. *The complete independence of the physics on the bag — size or location — is the precise statement of "Cheshire Cat Principle"* with the corresponding mechanism referred to as "Cheshire Cat mechanism."

We went into a great deal of detailed reasoning in arriving at the exact CCP in 2D. It is unwarranted for the objective of the volume. We do this because this will be the only luxury we can entertain and savor in this volume where calculations can *actually* be done explicitly and simply. In what follows dealing with the real world, we are unable to make such a clear-cut reasoning and calculations. In 4D, things are vastly more complicated. There is no known bosonization technique for four dimensions. So whatever is bosonized in 4D must be approximate. Furthermore, it cannot be captured in a simple form, involving most likely an infinite number of boson fields as we observe in holographic QCD (see Chapter 7). Nonetheless an approximate CCP does hold and works even quantitatively in certain processes. All topological quantities like the baryon charge do obey exact CCP. The baryon charge is partitioned in a similar way in (3+1) as in (1+1) dimensions as first shown for the magic chiral angle $\theta = \pi/2$ [Rho, Goldhaber and Brown (1983)] and then generalized for any angle [Goldstone and Jaffe (1983)]. In (3+1) dimensions, one gets $Q_{\bar{V}} = \frac{B}{\pi}(\theta - \sin\theta\cos\theta)$ for baryon number B. Non-topological quantities such as masses, static properties and also some non-static properties satisfy it only approximately, but rather well. They are described in various reviews and won't be repeated here — except for the celebrated case of the flavor-singlet axial charge (FSAC) of the proton described below.

It should be mentioned here that there is one particular partition of baryon number that is phenomenologically favored in the application of the chiral bag to nuclear physics. Given that the CCP is approximate in (3+1) dimensions, how to partition the space into two depends on the dynamics involved in the given process. It turns out that at low energy, nuclear processes are optimally described by the chiral bag with the baryon charge equally partitioned into inside and outside. The chiral angle at which this takes place, $\theta = \pi/2$, is referred to as "magic angle." This angle is often invoked for certain applications of the model.

3.2.3 *Wess–Zumino term*

The topology winding number is the same as the baryon number and this information is encoded in the Wess–Zumino term, i.e., the term that encodes anomalies [Witten (1983)]. We wish to know how the CC phenomenon reproduces the Wess–Zumino term. For 4D, the Wess–Zumino term will be in 5D with $N_f \geq 3$. We will work in $(1+1)$ dimensions, so the Wess–Zumino term will be in 3D with $N_f > 1$.

As before we have the two spaces given by V and $\overline{V}$. We start with the Wess–Zumino term defined in the space $\overline{V}$. Written in $D+1=3$ dimensions, it has the form

$$S^{WZ}_{\overline{V}} = \frac{N_c}{12\pi}\int_{\overline{V}\times[0,1]} d^3x\epsilon^{\mu\nu\alpha}\mathrm{Tr}(L_\mu L_\nu L_\alpha), \tag{3.35}$$

where $L_\mu = U^\dagger\partial_\mu U$ with $U = e^{2i\phi/f}$, $f = 1/\sqrt{4\pi}$, $\phi = \frac{T^a}{2}\phi^a$. The variation of this action as needed in the equation of motion gives

$$\begin{aligned}\delta S^{WZ}_{\overline{V}} &= 3\frac{N_c}{12\pi}\int_{\overline{V}\times[0,1]} d^2x\epsilon^{\mu\nu}\mathrm{Tr}(U^\dagger\delta U L_\mu L_\nu)\\ &\quad + 3\frac{N_c}{12\pi}\int_0^1 ds\int_{\overline{V}\times[0,1]} d\Sigma_\mu\epsilon^{\mu\nu\alpha}\mathrm{Tr}(U^\dagger\delta U L_\nu L_\alpha)\end{aligned} \tag{3.36}$$

where the Maurer–Cartan equation

$$\delta L_\mu = \partial_\mu(U^\dagger\delta U) + [L_\mu, U^\dagger\delta U] \tag{3.37}$$

is used. Now the surface term in (3.36) is unphysical and should be canceled away. How is this done? To do this fill the bag $\boldsymbol{B} = V\times S_1$ with dynamical quarks, chirally coupled to the bag surface $\partial\boldsymbol{B}$ [Rho and Zahed (1986)]. The equation of motion for the bagged quarks in V, working in Euclidean space, is (from (3.17))

$$\gamma\cdot\partial\psi(x,t) = 0, \quad (e^{i\gamma_5\theta(\beta,t)/f} - \gamma\cdot n)\psi(\beta,t) = 0 \tag{3.38}$$

with the boundary at $x=\beta$. Here θ is time-dependent chiral angle. To pin down "induced" Wess–Zumino term in the fermion vacuum inside the bag, time-dependent treatment with the time-dependent bounary condition is needed. The general solution to (3.38) is

$$\psi(x,t) = e^{-i[\frac{1}{2}(1+\gamma_5)\theta_-/f + \frac{1}{2}(1-\gamma_5)\theta_+/f]}\psi_0(x,t) \tag{3.39}$$

with the boundary condition

$$\theta_-(\beta - t) - \theta_+(\beta + t) = 0 \tag{3.40}$$

where $\theta_\pm = \theta_\pm(x\pm t)$ and ψ_0 is the MIT bag wavefunction. Making a homotopy extension $\phi(x,t)\to\phi(x,t,s)$ such that $\phi(x,t,s=1)=\phi(x,t)$ and $\phi(x,t,s=0)=0$, exploiting the multiple reflection analysis of the bagged quark propagator [Goldstone and Jaffe (1983)] and limiting to the lowest derivative, one arrives at an action of the form

$$\sim \int_0^1 ds \int_V dxdt \mathrm{Tr}(\partial_s\theta\partial_t\theta\partial_x\theta) + \cdots \tag{3.41}$$

where the ellipsis stands for higher derivatives. Imposing chiral invariance and restoring the constant factor, one gets

$$S_V^{WZ} = \frac{N_c}{12\pi}\int_{V\times[0,1]} d^3x\epsilon^{\mu\nu\alpha}\mathrm{Tr}(L_\mu L_\nu L_\alpha) \tag{3.42}$$

which is the bag complement to the outside Wess–Zumino term (3.35). Here the Wess–Zumino term inside of the bag is given in terms of L_μ, which is justified because it is a topological term arising from the vacuum fluctuation induced inside the bag by the boundary, as is the case with the baryon current. Topological objects satisfy the CC principle.

3.2.4 *Color anomaly*

In the MIT bag model, not only the baryon charge but also the color is confined inside the bag. We have seen above that the baryon charge leaks out of the bag once the pion is coupled to the bag surface as is required by chiral symmetry. It turns out that the same is true with the color. The color also leaks due to the coupling to the η' meson which in the limit $N_c \to \infty$ would be a NG boson like the pion is in the chiral limit. How the leaking occurs and yet the color symmetry is not violated goes in a way analogous to the baryon charge and also to the Wess–Zumino term [Nielsen, Rho, Wirzba and Zahed (1991)].

The MIT bag boundary conditions for the color electromagnetic field are

$$\hat{\boldsymbol{n}}\cdot\boldsymbol{E}^a = 0,\ \hat{\boldsymbol{n}}\times\boldsymbol{B}^a = 0 \tag{3.43}$$

where $\boldsymbol{n}$ is the outward normal to the bag surface as defined before. The indices a, b, c *etc.* denote color. These conditions show that the color electric fields E_μ^a point along the surface while the color magnetic fields B_μ^a point perpendicular to the surface. Since the color electric field must point parallel to the surface, color cannot go across the boundary and hence will be confined. Now suppose that the η' field couples to the fermion as required by the $U_A(1)$ chiral symmetry, i.e., in Minkowski space,

$$\left(i\gamma\cdot\boldsymbol{n} + e^{i\gamma_5\eta'(\beta)/f_0}\right)\psi(\beta) = 0 \tag{3.44}$$

where f_0 is the η' decay constant and β is the point on the boundary surface Σ of the bag. Then it turns out that the color confinement gets spoiled at the quantum level: In the presence of the η' field at the boundary, a color electric field normal to the surface is induced in the form of $\frac{\eta'}{f_0}\hat{\boldsymbol{n}}\cdot\boldsymbol{B}^a$. Thus color will be pulled out of the bag at a rate proportional to $\frac{\eta'}{f_0}\hat{\boldsymbol{n}}\cdot\boldsymbol{B}^a$.

Consider the color charge inside V

$$Q^a = \int_V d^3x J_0^a \tag{3.45}$$

where

$$J_\mu^a = g_s\bar{\psi}\frac{\lambda^a}{2}\psi - g_s f^{abc}(\boldsymbol{A}^b \times \boldsymbol{B}^c). \tag{3.46}$$

Here and throughout this volume, we use g_s for the color (QCD) gauge coupling (to be distinguished from flavor hidden gauge coupling g or g_V — with $V = \rho, \omega$). At the classical level both (3.43) and (3.44) render color charge conserved, that is, $\frac{d}{dt}Q = 0$. However at the quantum level, it diverges and requires regularization. This leads to anomaly in color charge, in a way similar to the anomaly in baryon charge. Using the technique used for the baryon-charge leakage and the Wess–Zumino term, after detailed calculations, one arrives at [Nielsen, Rho, Wirzba and Zahed (1991)]

$$\frac{\delta}{\delta\eta'(x,t)}Q^a(t) = \frac{g_s^2}{8\pi^2}\int_{\partial V} d\beta\hat{\boldsymbol{n}}\cdot\boldsymbol{B}^a(\beta,t). \tag{3.47}$$

This is the leakage of color charge violating color gauge invariance. Since the charge generated at quantum level (i.e., anomaly) has to be conserved, a surface term has to be added to the Lagrangian to "kill" this charge accumulated on the surface. This counter term will have the form

$$S_{CT} = -\frac{g_s^2 N_F}{8\pi^2}\int_{\partial V} d\beta A_0^a\hat{\boldsymbol{n}}\cdot\boldsymbol{B}^a\eta'/f_0. \tag{3.48}$$

Below we will incorporate this term in the three-flavor chiral Lagrangian.

3.2.5 *Cheshire Cat and color-flavor locking*

At asymptotic high densities, the three light quarks, u, d and s which can be taken to be massless, pair into diquarks with color and flavor locked forming the color-flavor-locked (CFL) phase [Alford, Rajagopal and Wilczek (1999)] and the CFL phase exhibits a spectrum with matching quantum numbers to that of the zero density phase [Schäfer and Wilczek (1999)]. We show that this is a manifestation of the CC phenomenon at superhigh density.

In the framework developed in this volume, the CFL phase does not actually figure in the description of compact stars, but it is highly relevant to the notion developed in this volume, namely, the notion of hadron-quark continuity from zero density to superhigh density. We illustrate this with a baryon in the CFL phase as described as skyrmion, that we call "superqualiton" [Hong, Rho and Zahed (1999)]. The superqualitons carry quantum numbers $B = Y = (1 \bmod 2)/3$ and $S = 1/2$.

We start with a brief review of the CFL for the essence of the idea [Alford, Rajagopal and Wilczek (1999)].

At very high density, quarks with opposite Fermi momenta tend to pair with color and flavor locked and get condensed. The pertinent condensate can be taken in the $(\bar{3}, \bar{3})$ color-flavor representation as

$$\left\langle q^{ia}_{L\alpha} q^{jb}_{L\beta} \right\rangle = -\left\langle q^{ia}_{R\alpha} q^{jb}_{R\beta} \right\rangle = \kappa\, \epsilon^{ij} \epsilon^{abI} \epsilon_{\alpha\beta I}\,. \tag{3.49}$$

Here κ is some constant, i, j are $SL(2,C)$ indices, a, b are color indices, and α, β are flavor indices. For finite κ, both global color $SU(3)_c$ and flavor $SU(3)_{f,L,R}$ symmetries are broken. The flavor-color locking in (3.49) implies spontaneous breaking through the color-flavor diagonal. Specifically: $SU(3)_c \times (SU(3)_L \times SU(3)_R) \to SU(3)_{c+L+R}$, with the emergence of 8 pseudoscalar Nambu–Goldstone (NG) bosons, denoted below as Π, together with 8 scalar NG bosons Σ to be "eaten up" by gluons through the Higgs mechanism. There is an extra NB boson associated with $U(1)_B \to Z_2$, but it has no relevance to our discussion. So we will ignore it. In an exact analogy to the octet pions at low density, we introduce the chiral fields[2]

$$U = \xi_L^\dagger \xi_R = e^{2i\Pi/F_\pi}, \tag{3.51}$$

$$\xi_{L,R} = e^{\mp i\Pi/F_\pi} e^{i\Sigma/F_\sigma} \tag{3.52}$$

where ξ_L is a map from space-time to $M_L = SU(3)_c \times SU(3)_L / SU(3)_{c+R}$ to describe the excitations of the right-handed diquark condensate and likewise with $L \leftrightarrow R$. Under $SU(3)_c \times SU(3)_L \times SU(3)_R$ transformation by unitary matrices (g_c, g_L, g_R), ξ_L transforms as $\xi_L \mapsto g_c^* \xi_L g_L^\dagger$ and ξ_R transforms as $\xi_R \mapsto g_c^* U_R g_R^\dagger$.

The effective Lagrangian for the CFL phase at asymptotic densities follows by integrating out the 'hard' quark modes at the edge of the Fermi

[2]In terms of quark fields, one can write [Hong, Rho and Zahed (1999)], for instance,

$$\xi_{L\alpha a}(x) = \lim_{y \to x} \frac{|x-y|^{\gamma_m}}{\kappa} \epsilon^{ij} \epsilon_{abc} \epsilon_{\alpha\beta\gamma} q^{b\beta}_{Li}(-\vec{v}_F, x) q^{c\gamma}_{Lj}(\vec{v}_F, y) \tag{3.50}$$

where γ_m is the anomalous dimension of the diquark field of order α_s and $q(\vec{v}_F, x)$ denotes the quark field with momentum close to a Fermi momentum $\mu\vec{v}_F$.

surface. The effective Lagrangian for $U_{L,R}$ is a standard non-linear sigma model in D=4 dimensions and should include the interaction of NB bosons with colored but "screened" gluons G. The $SU(3)_c$ current of NB bosons in the CFL phase consists of two terms,

$$J_{cL}{}^{A\mu} = \frac{i}{2}F^2 \text{tr}\ U_L^\dagger T^A \partial^\mu U_L + \frac{1}{24\pi^2}\epsilon^{\mu\nu\rho\sigma} \text{tr}\ T^A U_L^\dagger \partial_\nu U_L U_L^\dagger \partial_\rho U_L U_L^\dagger \partial_\sigma U_L, \tag{3.53}$$

where the first term is the Noether current and the second one is from the Wess–Zumino term. Expanding in powers of derivative, the effective Lagrangian for the (colored) NB bosons is then

$$\mathcal{L} = +\frac{F_T^2}{4}\ \text{tr}(\partial_0 U_L \partial_0 U_L^\dagger) - \frac{F_S^2}{4}\ \text{tr}(\partial_i U_L \partial_i U_L^\dagger) + g_s G \cdot J_{cL} + n_L \mathcal{L}_{WZW} \\ + (L \to R) + \mathcal{O}\left(\frac{\partial^4}{(4\pi\Delta)^4}\right). \tag{3.54}$$

The coefficient $n_{L,R}$ of the Wess–Zumino terms is fixed to 1 by color-flavor anomalies. Note that in the presence of a chemical potential, Lorentz symmetry is broken down to $O(3)$. In carrying out the derivative expansion, the (superconducting) scale $4\pi\Delta$ is taken to be large. The temporal and spatial decay constants $F_{T,S}$ can be fixed by the 'hard' modes at the Fermi surface.

As one can see from the U field defined in (3.51), there is the same redundancy as with hidden local symmetry (HLS) at low density and hence a local gauge symmetry. (Unlike in the case of low-energy gauge fields, however, here the local symmetry is explicit, i.e., color symmetry, which is *not* hidden.) One can therefore generalize Lagrangian (3.54) to a local gauge invariant form in analogy to HLS.

Much like the effective Lagrangian for QCD at low density (giving rise to skyrmions), the low-energy effective Lagrangian in the CFL phase admits a stable (static) soliton solution, with a winding number given by the homotopy $\Pi_3(M) = Z$. It has a baryon number 1/3 and resembles the qualiton introduced by Kaplan [Kaplan (1991)] for the constituent quark. Unlike the qualiton in the vacuum which was found unstable, at super high density, the superqualiton is stabilized by the balance between the kinetic energy (attractive force) and the Coulomb energy (repulsive force).[3] The soliton quantum numbers are determined by the Wess–Zumino term upon

[3]There can be additional higher derivative terms that may significantly contribute to the stabilization of the qualiton. Differently to the qualiton in the vacuum, one would also have to consider terms that are specific to the Fermi surface.

quantization as in low-energy QCD. The usual collective quantization then gives a spin-half particle transforming under the fundamental representation of both the flavor group and the color group, which leads to a massive left-handed or right-handed quark in the CFL phase. The details of the picture are given in [Hong, Rho and Zahed (1999)] that we do not go into here. The point stressed here is that the HLS structure of the low-density sigma model and the color gauge symmetric sigma model of high density are continuously connected by the skyrmion description, as a manifestation of a CC phenomenon. We will see this phenomenon manifested in the EoS of massive compact stars in Chapter 9. As a further support for this continuity, it is possible to show that the pion decay constant f_π continuously evolves from low density to high density, say, even up to the CFL density as suggested by Song *et al.* [Song and Baym (2017)].

3.3 The Cheshire Cat facing Nature

3.3.1 *Three-flavor chiral bag Lagrangian*

We are now ready to explicitly write down the Lagrangian that is applicable for three-flavor chiral bag. The Lagrangian will have the color-symmetry broken at the classical level but consistent at the quantum level. Usually anomalies occur at quantum level with a Lagrangian that is anomaly free. Here the situation is opposite. In QCD, color is confined and the gauge symmetry is unbroken. Thus the chiral Lagrangian written down below makes sense only when quantum corrections are taken into account. We will show how this comes about with the flavor-singlet axial current for nucleons.

As done before we separate the space into inside the bag V (taken to be spherical), the outside $\overline{V}$ and the surface ∂V. Generalizing the formulation given above for (1+1) dimension to (3+1) dimensions, the total action for for $SU(N_f)_L \times SU(N_f)_R$ (for $N_f = 3$) (in Minkowski space) is

$$S = S_V + S_{\overline{V}} + S_{\partial V}, \tag{3.55}$$

$$\begin{aligned}
S_V &= \int_V d^4x \left(\bar{\psi} i \not{D} \psi - \frac{1}{2g_s^2} \mathrm{tr}\, G_{\mu\nu} G^{\mu\nu}\right) + \cdots \\
S_{\overline{V}} &= \frac{f^2}{4} \int_{\overline{V}} d^4x \left(\mathrm{Tr}\ \partial_\mu U^\dagger \partial^\mu U + \frac{1}{4N_f} m_{\eta'}^2 (\mathrm{Tr}\ln U - \mathrm{Tr}\ln U^\dagger)^2\right) \\
&\quad + S_{WZ} + \cdots \\
S_{\partial V} &= \frac{1}{2} \int_{\partial V} d\Sigma^\mu \left\{ (n_\mu \bar{\psi} U^{\gamma_5} \psi) + i \frac{g_s^2}{16\pi^2} K_{5\mu} (\mathrm{Tr}\ \ln U^\dagger - \mathrm{Tr}\ \ln U) \right\}.
\end{aligned} \tag{3.56}$$

Here U is a unitary $U(3)$ matrix, ψ is the quark field $\psi^T = (u\ d\ s)$, G_μ the octet gluon field, $G_{\mu\nu}$ the gluon field tensor, g_s the "color" gauge coupling[4] and K_5^μ is the (properly regularized) Chern–Simons current

$$K_5^\mu = \epsilon^{\mu\nu\alpha\beta}\left(G_\nu^a G_{\alpha\beta}^a - \frac{2}{3} g_s f^{abc} G_\nu^a G_\alpha^b G_\beta^c\right), \tag{3.57}$$

the chiral field U is

$$U = e^{i\eta'/f_0} e^{2i\pi/f} \tag{3.58}$$

and U^{γ_5} is

$$U^{\gamma_5} = e^{i\eta'\gamma_5/f_0} e^{2i\pi\gamma_5/f}. \tag{3.59}$$

Note that the Wess–Zumino term S_{WZ} is given in the volume $\overline{V}$. Its complement in the space V should be given by quantum effects in the same way as the baryon charge as described above for (1+1) dimensions. The CC allows to replace it by the same form defined in the volume V as in the case of (1+1) dimensions. We have used loosely the decay constants f_0 and f which will be specified more precisely later. The action (3.55) contains full hadron dynamics for baryons at low-energy ($E \ll \Lambda_\chi$ where Λ_χ is the chiral symmetry scale) including anomaly structure. Physics in 4D is not fully captured by analogy to 2D, so there could of course be many caveats. Let us simply describe here what each term represents, with focus on the terms that we will consider later for an illustrative example, the so-called "proton spin."

(1) The action S_V is the QCD action in quark-gluon variables with the ellipses standing for the quark mass term, assumed to hold within the bag volume. The quarks ψ and gluons G_μ are confined in the volume V.
(2) The action in the complementary volume $\overline{V}$ consists of hadron fields, here, the octet pseudoscalars π, the first term being the current algebra term, the second term the η' mass term representing the broken $U(1)_A$ symmetry, i.e. $U(1)_A$ anomaly, S_{WZ} the five-dimensional Wess–Zumino term present for the number of flavors $N_f \geq 3$, and the ellipses stand for higher derivatives, pseudoscalar mass terms and terms involving vector and other massive fields.
(3) The surface action $S_{\partial V}$ is subtler, so requires more detailed explanation. The first term is the usual MIT bag boundary term suitably made

[4]We use "tr" for color trace and "Tr" for flavor trace. We use the normalization for the group generators $\mathrm{Tr}(t^a t^b) = \frac{1}{2}\delta^{ab}$ and likewise for the color.

chirally symmetric with the π field. It confines, classically, outside of the bag not only the octet pion field but also the singlet field η' field associated with $U(1)_A$. As described the time variation of this field on the surface generates an anomaly which makes the color charge carried by the quarks and gluons leak outside. This leakage is caused by a quantum anomaly effect as described above for (1+1) dimensions. Now the raison d'être of the second term is to prevent this leakage since the leakage produces the violation of the color gauge invariance.

3.3.2 *Flavor-singlet axial charge* $a^0 \equiv g_A^0$

We illustrate how an *approximate* Cheshire Cat phenomenon works in one of the most subtle phenomena in nucleon structure, namely, the notorious case of what has been wrongly termed the "proton spin crisis." This problem has been studied extensively not only in the context of hadronic models but also from the point of view of QCD proper. For a review and references, see Bass [Bass (2005)]. The objective of the present discussion is to illustrate the subtlety involved in the way the CC *can* manifest in a problem of this sort.

The flavor-signet axial charge (FSAC), involving the chiral singlet η' and hence the chiral anomaly, is a very unique case where the notion of the CC and its subtle communication between the language of quarks and gluons and that of the hadrons are manifested. How this comes about is quite illustrative of the working of the CC in nature, so we will devote a great deal more space to this matter. For other matters discussed in what follows, we will be much less detailed. Those who are persuaded that the CC notion makes a great sense and are not interested in the details of how it works out could simply skip this part and jump to the result given in Fig. 3.1.

Here we derive the flavor-singlet axial current from the chiral bag model Lagrangian (3.55). We write the axial current A_μ ($\equiv J_{5\mu}$) in the chiral bag model as a sum of two terms, one from the interior of the bag and the other from the outside contributed by the meson field η' (how to account for the NG pion fields is known; they will be taken into account for the baryon charge leakage)

$$A^\mu = A_B^\mu \Theta_B + A_M^\mu \Theta_M. \tag{3.60}$$

For notational simplicity, we will omit the flavor (singlet) index in the current. We shall use the short-hand notations $\Theta_B = \theta(R - r)$ and $\Theta_M =$

$\theta(r-R)$ — assuming a spherical bag configuration — with R the radius of the bag. We interpret the $U_A(1)$ anomaly as given in this model by[5]

$$\partial_\mu A^\mu = \frac{\alpha_s N_f}{2\pi} \sum_a \vec{E}^a \cdot \vec{B}^a \Theta_B + f m_\eta^2 \eta \Theta_M \tag{3.61}$$

where $\alpha_s = g_s^2/4\pi$, $\vec{E}$ and $\vec{B}$ are, respectively, color electric and magnetic fields and f is the η' decay constant. This result follows from that the color anomaly induced inside the bag — quantum effect — is cancelled by the boundary term in the classical action (3.56). We are assuming here that in the nonperturbative sector outside of the bag, the only relevant $U_A(1)$ degree of freedom for the problem at hand is the massive η' field.[6] This allows us to write

$$A_M^\mu = A_\eta^\mu = f\partial^\mu \eta \tag{3.62}$$

with the divergence

$$\partial_\mu A_\eta^\mu = f m_\eta^2 \eta. \tag{3.63}$$

It turns out to be more convenient to write the current as

$$A^\mu = A_{B_Q}^\mu + A_{B_G}^\mu + A_\eta^\mu \tag{3.64}$$

such that

$$\partial_\mu (A_{B_Q}^\mu + A_\eta^\mu) = f m_\eta^2 \eta \Theta_M, \tag{3.65}$$

$$\partial_\mu A_{B_G}^\mu = \frac{\alpha_s N_f}{2\pi} \sum_a \vec{E}^a \cdot \vec{B}^a \Theta_B. \tag{3.66}$$

The subindices Q and G imply that these currents are written in terms of quark and gluon fields respectively. In writing (3.65), the up and down quark masses are ignored. Since we are dealing with an interacting theory, there is no unique way to separate the different contributions from the gluon, quark and η components. In particular, the separation we adopt, (3.65) and (3.66), is neither unique nor gauge invariant, although the sum is without ambiguity. This separation, however, is found to lead to a natural partition of the contributions in the framework of the bag description for the confinement mechanism that is used here.

Since the calculation involved is straightforward, we will merely sketch the steps involved.

---

[5]In what follows, we drop prime on η'.

[6]The pionic field figues for the baryon charge as described above.

- **The quark current $A^{\mu}_{B_Q}$**

The quark current is given by

$$A^{\mu}_{B_Q} = \bar{\Psi}\gamma^{\mu}\gamma_5\Psi \tag{3.67}$$

where Ψ represents the *bagged* (MIT) quark field. Therefore the quark current contribution to the FSAC is given by

$$a^0_{B_Q} = \langle p| \int_B d^3r \bar{\Psi}\gamma_3\gamma_5\Psi|p\rangle. \tag{3.68}$$

The calculation of this matrix element in the chiral bag model involves the baryon charge leakage described above. The leakage produces an R dependence.

- **The meson current A^{μ}_{η}**

Due to the coupling of the quark and η fields at the surface, we can simply write the η contribution in terms of the quark contribution,

$$a^0_{\eta} = \frac{1+y_{\eta}}{2(1+y_{\eta})+y^2_{\eta}} \langle p| \int_B d^3r \bar{\Psi}\gamma_3\gamma_5\Psi|p\rangle \tag{3.69}$$

where $y_{\eta} = m_{\eta}R$. Since the η field has no topological structure, its contribution also vanishes in the skyrmion ($R=0$) limit. Due to the baryon charge leakage, however, this contribution increases slowly as the bag increases. We may summarize the analysis of these two contributions by stating that no trace of the Cheshire Cat phenomenon is apparent from the "matter" contribution ($a^0_{B_Q} + a^0_{\eta}$). As will be seen in the figure below, there is a sensitive dependence on R. Thus if an approximate CCP were to emerge, the only possibility would be that the gluons play the crucial role!

Let us now look at the gluon contribution. The gluon current is split into two pieces

$$A^{\mu}_{B_G} = A^{\mu}_{G,stat} + A^{\mu}_{G,vac}\,. \tag{3.70}$$

The first term arises from the quark and η sources, while the latter is associated with the properties of the vacuum of the model. One might worry that this contribution could not be split into these two terms without double counting. However this worry is unfounded. Technically, it is easy to check it by noticing that the former acts on the quark Fock space and the latter on the gluon vacuum. Thus, one can interpret the former as a one-gluon exchange correction to the quantity. One can also show this intuitively by making the analogy to the condensate expansion in QCD, where the perturbative terms and the vacuum condensates enter additively to the lowest order.

- **The gluon static current** $A^{\mu}_{G,stat}$

Let us first describe the static term.

We ignore the η coupling for the time being. Afterwards, the η contribution can be added. The boundary conditions for the gluon field would correspond to the original MIT ones. The quark current is the source term that remains in the equations of motion after performing a perturbative expansion in the QCD coupling constant, i.e. the quark color current

$$g_s \bar{\Psi}_0 \gamma_\mu \lambda^a \Psi_0 \tag{3.71}$$

where the Ψ_0 fields represent the lowest cavity modes. In this lowest mode approximation, the color electric and magnetic fields are given by

$$\vec{E}^a = g_s \frac{\lambda^a}{4\pi} \frac{\hat{r}}{r^2} \rho(r) \tag{3.72}$$

$$\vec{B}^a = g_s \frac{\lambda^a}{4\pi} \left(\frac{\mu(r)}{r^3} (3\hat{r}\vec{\sigma} \cdot \hat{r} - \vec{\sigma}) + \left(\frac{\mu(R)}{R^3} + 2M(r) \right) \vec{\sigma} \right) \tag{3.73}$$

where ρ is related to the quark density ρ' as[7]

$$\rho(r, \Gamma) = \int_\Gamma^r ds \rho'(s)$$

and μ, M to the vector current density

$$\mu(r) = \int_0^r ds \mu'(s),$$

$$M(r) = \int_r^R ds \frac{\mu'(s)}{s^3}.$$

The lower limit Γ is taken to be zero in the MIT bag model — in which case the boundary condition is satisfied only *globally*, that is, after averaging — and $\Gamma = R$ in the so called *monopole solution* — in which case, the boundary condition is satisfied *locally*. We take the latter since consistency with the CCP condition rules out the MIT condition.

We now proceed to introduce the η field. We perform the same calculation with, however, the color boundary conditions of (3.55) taken into

[7]Note that the quark density that figures here is associated with the color charge, *not* with the quark number (or rather the baryon charge) that leaks due to the hedgehog pion.

account. In the approximation of keeping the lowest non-trivial term, the boundary conditions take the form

$$\hat{r} \cdot \vec{E}^a_{stat} = \frac{N_f g_s^2}{8\pi^2 f} \hat{r} \cdot \vec{B}^a_g \eta(R) \tag{3.74}$$

$$\hat{r} \times \vec{B}^a_{stat} = -\frac{N_f g_s^2}{8\pi^2 f} \hat{r} \times \vec{E}^a_g \eta(R). \tag{3.75}$$

Here $\vec{E}^a_g$ and $\vec{B}^a_g$ are the lowest order fields given by (3.72) and (3.73) and $\eta(R)$ is the meson field at the boundary. The η field is given by

$$\eta(\vec{r}) = -\frac{g_{NN\eta}}{4\pi M} \vec{S} \cdot \hat{r} \frac{1 + m_\eta r}{r^2} e^{-m_\eta r} \tag{3.76}$$

where the coupling constant is determined from the surface conditions.

Note that the magnetic field is not affected by the new boundary conditions, since $\vec{E}^a_g$ points into the radial direction. The effect on the electric field is just a change in the charge, i.e.

$$\rho_{stat}(r) = \rho(r, \Gamma) + \rho_\eta(R) \tag{3.77}$$

where

$$\rho_\eta(R) = \frac{N_f g_s^2}{64\pi^3 M} \frac{g_{NN\eta}}{f} (1 + y_\eta) e^{-y_\eta}. \tag{3.78}$$

The contribution to the FSAC arising from these fields is determined from the expectation value of the anomaly

$$a^0_{G,stat} = \langle p| - \frac{N_f \alpha_s}{\pi} \int_B d^3 r x_3 \vec{E}^a_{stat} \cdot \vec{B}^a_{stat} |p\rangle. \tag{3.79}$$

One finds that including the η contribution in $\rho_{stat}(r)$ brings a non-negligible modification to the FSAC but does not modify the result qualitatively.

- **The gluon Casimir current** $A^\mu_{G,vac}$

Up to this point, the FSAC is zero for $R = 0$ and non-zero for $R \neq 0$. This is in principle a violation of the CCP although the magnitude of the violation is small. We now show that it is the vacuum contribution through Casimir effects that the CCP is approximately restored. Being a Casimir problem, it is the most uncertain part of this calculation. What's given therefore is subject to uncertainty. However the qualitative feature of CCP does come out well.

The quantity to calculate is the gluon vacuum contribution to the flavor-singlet axial current of the proton, which can be gotten by evaluating the expectation value

$$\langle 0_B| - \frac{N_f \alpha_s}{\pi} \int_V d^3 r x_3 (\vec{E}^a \cdot \vec{B}^a) |0_B\rangle \tag{3.80}$$

where $|0_B\rangle$ denotes the vacuum inside the bag. To calculate this, we invoke at this point *the CCP which states that at low energy, hadronic phenomena do not discriminate between QCD degrees of freedom (quarks and gluons) on the one hand and meson degrees of freedom (pions, etas,...) on the other, provided that all necessary quantum effects (e.g., quantum anomalies) are properly taken into account.* If we consider the limit where the η excitation is a long wavelength oscillation of zero frequency, the CCP asserts that it does not matter whether we choose to describe the η, in the interior of the infinitesimal bag, in terms of quarks and gluons or in terms of mesonic degrees of freedom. This statement, together with the color boundary conditions, leads to an extremely simple and useful *local* formula

$$\vec{E}^a \cdot \vec{B}^a \approx -\frac{N_f g_s^2}{8\pi^2} \frac{\eta}{f} \frac{1}{2} G^2, \tag{3.81}$$

where only the term up to the first order in η is retained in the right-hand side. Here we adapt this formula to the chiral bag model. This means that the couplings are to be understood as the average bag couplings and the gluon fields are to be expressed in the cavity vacuum through a mode expansion. That the surface boundary condition can be interpreted as a local operator is a rather strong CCP assumption which while justifiable for small bag radius, can only be validated *à posteriori* by the consistency of the result. This procedure is the substitute to the condensates in the conventional discussion.

Substituting Eq. (3.81) into Eq. (4.45) we obtain

$$\begin{aligned} &\langle 0_B| - \frac{N_f \alpha_s}{\pi} \int_V d^3 r x_3 (\vec{E}^a \cdot \vec{B}^a) |0_B\rangle \\ &\approx \left(-\frac{N_f \alpha_s}{\pi}\right) \left(-\frac{N_f g_s^2}{8\pi^2}\right) \frac{y(R)}{f_0} \langle p|S_3|p\rangle (N_c^2 - 1) \\ &\qquad \sum_n \int_V d^3 r (\vec{B}_n^* \cdot \vec{B}_n - \vec{E}_n^* \cdot \vec{E}_n) x_3 \hat{x}_3, \end{aligned} \tag{3.82}$$

where we have used the fact that η has a structure of $(\vec{S} \cdot \hat{r})y(R)$ with $\vec{S}$ being the spin operator. Since we are interested only in the first order perturbation, the field operator can be expanded by using MIT bag eigenmodes (the zeroth order solution). Thus, the summation runs over all the classical MIT bag eigenmodes. The factor $(N_c^2 - 1)$ comes from the sum over the abelianized gluons. It is worth emphasizing that the calculation involves no unknown parameters. The results are summarized in Fig. 3.1.

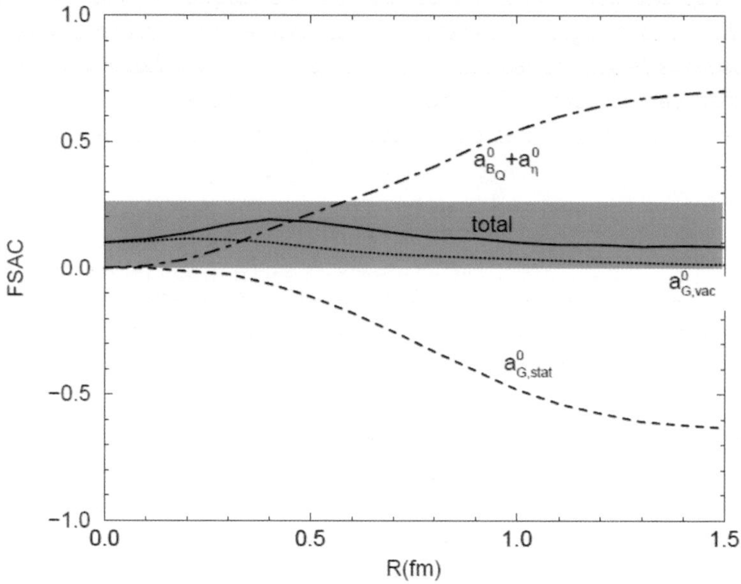

Fig. 3.1 Various contributions to the flavor-singlet axial charge of the proton as a function of bag radius R and comparison with the experiment: (a) quark plus η (or "matter") contribution ($a^0_{B_Q} + a^0_\eta$), (b) the contribution of the static gluons due to quark source ($a^0_{G,stat}$), (c) the gluon Casimir contribution ($a^0_{G,vac}$). The "total" is the sum of (a), (b) and (c). The shaded area corresponds to the range admitted by experiments.

An important point to re-stress here is that in the way formulated, neither the matter contribution (i.e., quarks) nor the gluon contribution to a^0, both of which are (color) gauge-non-invariant *and* CCP-violating, is physical. Only the total which is color gauge invariant and Cheshire Cat gauge invariant is physical and CCP-preserving. In this connection, an interesting question is how the individual part of the chiral singlet axial charge figures in the proton spin.

3.3.2.1 *Proton spin?*

The quantity that figures in the "proton-spin crisis" is the first moment of the spin-dependent structure function $g_1(x, Q^2)$ [Bass (2005)]

$$\Gamma_1^p \equiv \int_0^1 g_1^p(x)dx \tag{3.83}$$

where the superscript p stands for proton which can be expressed in terms of the matrix elements of the octet axial currents

$$J_{5\mu}^i = \bar{\psi}\gamma_\mu\gamma_5\frac{\lambda_i}{2}\psi \quad i = 1, 2....8 \tag{3.84}$$

where $\psi^T = (\psi_u, \psi_d, \psi_s)$ and the flavor-singlet axial current

$$J_{5\mu}^0 = \bar{\psi}\gamma_\mu\gamma_5\psi\,. \tag{3.85}$$

The matrix elements that enter are

$$\langle p', S|J_{5\mu}^i|p, S\rangle = \frac{1}{2}a^i S_\mu \quad \text{for} \quad i = 3, 8 \tag{3.86}$$

and

$$\langle p', S|J_{5\mu}^0|p, S\rangle = a^0 S_\mu\,. \tag{3.87}$$

The a^i's in principle are form factors dependent on $Q^2 = (p' - p)^2$. Assuming that the octet currents are conserved, one takes a^3 and a^8 to be constant. The singlet current is not conserved (i.e., anomaly), so it should depend on Q^2. To see how these matrix elements are related to the experimentally measured quantity Γ_1, one resorts to a renormalization scheme. In the $\overline{MS}$ scheme used frequently, at leading twist valid for $Q^2 \gg m_N^2$, one gets [Leader (2016)]

$$\Gamma_1^p(Q^2) = \frac{1}{12}\left[\left(a_3 + \frac{1}{3}a_8\right)\Delta C_{NS} + \frac{4}{3}a_0(Q^2)\Delta C_S\right], \tag{3.88}$$

where the $\Delta C_{S,NS}$ are the known singlet and non-singlet Wilson coefficients. Since $a_{3,8}$ are known from beta decay, measuring the left-hand side of (3.88) for given Q^2 then gives a_0 for the given Q^2. What was found in the experiment was $a_0 \sim 0$. This small value was "predicted" in the Cheshire Cat model described above. What was heralded as "proton spin crisis" is that in the naive parton model,

$$a_0 = 2S_z^p = 1 \tag{3.89}$$

where S_z^p is the proton spin ($= 1/2$). It is now understood that this interpretation is incorrect [Leader (2016)]. The correct interpretation of (3.88) in terms of the proton spin is

$$\frac{1}{2} = \sum_{x=q,G} (\langle S_z^x \rangle + \langle L_z^x \rangle) = \frac{1}{2} a_0 + \Delta G + \langle L^q \rangle + \langle L^Q \rangle \tag{3.90}$$

where q and G stand respectively for quark and gluon and S and L stand respectively for canonical spin and orbital angular momentum operators, the matrix elements of which should be evaluated in a particular gauge., i.e., $A^+ = 0$ or $A^0 = 0$ [Leader (2016)]. Note that all quantities in (3.90) are individually Q^2 dependent but the sum is not. Also the spin matrix element which is not gauge-invariant by itself gets merged into a_0. It should be noted that what was computed above for the flavor-singlet axial coupling in Section 3.3.2 corresponds to $a_0 = a_0(Q^2 = 0)$ which is well-defined for the weak coupling constant but extremely subtle for its role for the proton spin.

3.3.3 *Nucleon form factors*

The FSAC discussed above probes the long wavelength kinematics and the Cheshire Cat mechanism applies to it more or less accurately even though it is not topological. The situation is different when the kinematics involves shorter length scales.

For this, we consider the electromagnetic form factor in the chiral bag description of the nucleon at momentum transfers $q^2 \neq 0$. The CCP suggests that as q^2 increases, higher energy/momentum scale will intervene, so either the QCD degrees of freedom inside the bag become more important or more massive degrees of freedom outside of the bag will have to be taken into account. The latter is most efficiently captured by vector mesons introduced as hidden local gauge bosons discussed in Chapter 5 [Bando, Kugo and Yamawaki (1988); Harada and Yamawaki (2003)]. According to the CC, the more higher energy/momentum degrees of freedom are included in the exterior to the bag, the smaller the bag can be.

3.3.3.1 *Chiral bag at the magic angle: Role of lowest vector mesons*

As a simplified case, consider the photon coupling to the nucleon given by the chiral bag with the exterior populated by the pion and the lowest-lying vector mesons ρ and ω. For processes involving relatively low momentum transfers $q^2 \lesssim 1/2\ \mathrm{GeV}^2$, these three degrees of freedom will suffice with an adroit choice of the bag radius. It turns out for nuclear processes at low momentum scales that the optimal partition of interactions is at the magic angle $\theta_{\text{magic}} = \pi/2$. At this angle, the baryon charge is partitioned equally into the interior and exterior of the bag. Needless to say, in CC, there is nothing "magic" about this angle: Depending on the kinematics involved, it could be anywhere near that angle within the spirit of CC. Let us pick it for convenience and see how it works.

Let us first recall a well-known fact in low-energy hadronic physics that while the vector meson dominance (VMD) works well for meson EM form factors, it famously fails, as has been well known since a long time, for the nucleon form factor. Why the VMD fails for the nucleon form factor has been a long-standing puzzle, pointing to a fundamental difference in structure between the baryons and mesons, which seems at odds with the MIT bag model.

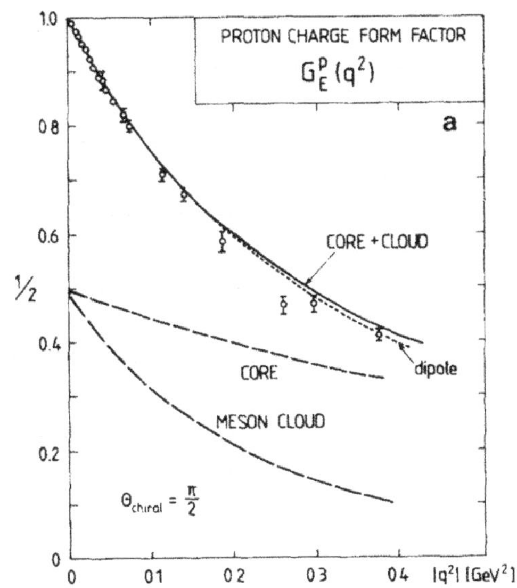

Fig. 3.2 Proton charge form factor vs. momentum transfer q^2 in the chiral bag model at the magic chiral angle $\theta_{\text{magic}} = \pi/2$.

With the exterior sector described with the hidden local symmetric effective Lagrangian, the photon couples to the nucleon in the exterior of the chiral bag via VMD with the coefficient $a = 2$ in the KSRF relation for the vector-meson mass, $m_V^2 = ag^2 f_\pi^2$. At the magic angle, the baryon charge is partitioned into half and half, so there will be the "core" contribution from the bag and the "cloud" contribution from the vector mesons accordingly partitioned. Given in Fig. 3.2 is the proton charge form factor calculated at the magic angle which captures the essential feature of the Cheshire Cat mechanism [Brown, Rho and Weise (1986)]. At $q^2 = 0$, for the magic angle, the charge is partitioned equally into the quark-gluon sector and the mesonic sector. But being controlled by topology, the charge distribution should not depend on the chiral angle. However for the given chiral angle, as the momentum transfer is increased, the quark-gluon contribution will become more important. This feature is confirmed by the calculation.

3.3.3.2 *Role of the infinite-tower vector mesons*

Within the given framework, the more massive degrees of freedom are taken into account in the exterior sector, the smaller the contribution from the bag is expected to become. This expectation is beautifully confirmed in holographic QCD models with the infinite tower of hidden local symmetric vector mesons, both iso-vector and iso-scalar [Hong, Rho, Yee and Yi (2008); Hashimoto, Sakai and Sugimoto (2008)]. There the bag can be shrunk to zero, replaced entirely by the infinite tower of vector mesons and full vector dominance is recovered. For the pion form factors, while the infinite tower does indeed bring improvement to the VMD over and above what is already obtained from the lowest-lying ρ meson, it amounts to a mere fine-tuning. In contrast, the effect in the nucleon form factors is qualitatively important and crucial. This phenomenon can be understood from the point of view of holography in 5D holographic QCD mentioned in Chapter 7 and extensively reviewed in [Rho and Zahed (2016)]. Briefly stated, the reasoning goes as follows [Nielsen and Zahed (2016)].

In holographic QCD, specifically in the model by Sakai and Sugimoto [Sakai and Sugimoto (2005)] which has the right chiral symmetry properties of QCD, the baryon is a flavor instanton in the holographic (5th) dimension, denoted Z, which is warped by gravity. For large Z, the warped instanton configuration is unknown. However for large 't Hooft constant defined by $\lambda = g_s^2 N_c$, the instanton configuration is forced to size $Z \sim 1/\sqrt{\lambda}$. In the large N_c and large λ limit, the baryon is the holonomy of the instanton configuration along the "conformal" direction Z. It has been shown in semiclassical quantization [Nielsen and Zahed (2016)] that a zero-size skyrmion emerges as a bosonized form of a heavy-quark sitting still in the conformal direction. The core skyrmion is dressed with the infinite tower of vector mesons coupling in the holographic direction a distance Z_c away from the core. Z_c divides in the Z direction a core dominated by an instanton and a cloud of infinite number of vector mesons. The CC phenomenon in this description is that physics should be independent of where Z_c — the "smile" of the Cheshire Cat — is located. In the EM form factors, the VMD in this description materializes with the photon coupling to the "core," i.e., the instanton, vanishing, and replaced by the infinite tower of hidden local vector mesons, which mocks up the quark bag needed in the magic angle description given above.

3.3.4 *Compact stars*

Properly addressing how the CCP could apply to compact stars requires background details on compact star physics, e.g., the equation of state (EoS) required from the nuclear physics side in the Tolman–Oppenheimer–Volkov (TOV) equation. This matter therefore can be more appropriately — and will be — treated in Chapter 9. Here we briefly mention the essence of the issues involved. As we will see later in the book — and more specifically in Chapter 9, the inner core of a massive neutron star could support densities as high as $(5-7)$ times the normal nuclear matter density n_0. Thus the EoS will probe a wide range of strong interactions as the density increases.

Up to the matter density n_0, the standard nuclear physics (hadronic) approach (SNPA), established with precision since many years — e.g., chiral perturbation expansion among others — works. It may be extrapolated, perhaps to $\sim 2n_0$, with help from experiments at RIB-type accelerators. However as one increases the density beyond $n \sim (2-3)n_0$, baryons are expected to start overlapping and quarks will start percolating with correlated quark interactions intervening. The SNPA will surely break down then. Ultimately we will encounter the color-flavor-locked quark matter at some large density, say, $n > (10-100)n_0$. How this can come about is discussed in Chapter 9 in terms of the "3-window" phenomenological model of [Kojo, Powell, Song and Baym (2016)]. It is argued there that this multi-window scenario can be given a Cheshire Cat description, with the skyrmion-half-skyrmion topology change treated in Chapter 7 supplying a dual description to the crucial intermediate window $n \sim (2-4)n_0$. As argued there, the smooth cross-over from the hadron phase to the strong-coupled quark-gluon phase, appropriately described in terms of the quarkyonic matter [McLerran and Pisarski (2007)], can be given a dual picture in terms of the skyrmion-to-half-skyrmion phase change. This dual picture is shown to correctly account for the massive $\sim$ 2-solar mass stars observed in nature with additional novel predictions on star properties, such as the sound velocity and the possible emergence of hidden symmetries of QCD in a phase that cannot be continuously connected to lower-density baryonic matter. The hadronic description of two windows in the three-window scenario is properly captured by topology-guided description. The third window involving the color-flavor-locked superconducting matter does not intervene in this CC picture of massive stars, but it can be brought in, as mentioned above, via "superqualiton" description with generalized HLS at super-high density.

Chapter 4

Effective Field Theories for Nuclear Interactions

4.1 Web of effective field theories

Although an effective field theory (EFT) with one unified Lagrangian is the principal tool in our exposition, a web of theories with a variety of effective fields will be invoked in going from normal nuclear matter to high density compact-star matter. They may look different at different scales but what is important to recognize is that they are *intrinsically* related to each other via the unified framework developed in this volume. Which is the relevant effective field theory for given observables depends on the energy/momentum scale probed. In the principal spirit of this volume, one suitably defined EFT is to cover the whole range provided the "sliding vacuum structure" of the matter is appropriately taken into account.

Starting from the highest scale that concerns us, we have an effective theory defined at the chiral scale $\Lambda_\chi \sim 4\pi f_\pi \sim 1$ GeV. At this scale the chiral Lagrangian can be matched to QCD via correlators of the EFT and QCD. In our approach this is the only place where QCD makes its *direct* input. The next important scale for nuclear physics is the scale at which nuclear forces are defined. From the chiral scale down to this nuclear-force scale figure, typically, massive vector and scalar mesons, in addition to pions and of course nucleons. Nuclear interactions intervening in nuclear structure take place on top of the Fermi sea characterized by the Fermi momentum k_F or density n, involving Fermi-liquid fixed points — described precisely below — in the vicinity of normal nuclear matter density $n_0 \approx 0.16$ fm^{-3}. Here the pertinent EFT is no longer the starting chiral Lagrangian but the Landau Fermi-liquid EFT Lagrangian anchored on Wilsonian renormalization group (RG) approach, with certain symmetries not directly linked to chiral symmetry. How to arrive at Fermi liquid directly from QCD is not known.[1] Although there is no rigorous proof, there is a compelling argument that chiral Lagrangian suitably matched to QCD at the matching scale, treated at the mean field level — along the paradigm of Walecka relativistic mean-field (RMF) theory — captures the Fermi-liquid fixed-point theory, as suggested by, among others, Matsui [Matsui (1981)]. This provides, as will be discussed in this volume, a logical (albeit implicit) link between the Fermi-liquid structure and QCD. This link is described below. It also makes a connection to what is known as "energy density functional (EDF) theory" in analogy to the celebrated "density functional (DF) theory" in atomic and chemical physics. Going further down in scale, in the long wavelength limit, one arrives at pionless effective field theory ($\not{\pi}$ EFT) resulting from integrating out, not just the massive vector and scalar mesons, but also the pions.[2]

In this chapter, we construct a web of effective field theories, in the spirit of the "Folk Theorem" (FT), which may appear at first glance disparate and unrelated. The point of importance in our theme is that just as the Lagrangian that we start with at the chiral scale is not "derived" from — but "matched" to — QCD proper, the chain of effective theories we will employ need not be *directly* connected to each other or may not necessarily have a vestige of the fundamental symmetries possessed by QCD proper. One extreme case, apparently far from the spirit of EFT for QCD à la FT, is the BPS model for nuclear binding energy, which naively appears disconnected from the basic premise of EFT in nuclear physics. This model will be introduced and briefly discussed in Chapter 7. Here we simply mention that this model, while using a triplet of pseudo-scalar field possessing $SU(2)_V$ symmetry that carries topology, has no obvious connection with the fundamental symmetries of QCD. Most of all, it encodes no soft-pion theorems that must be satisfied for low-energy pion-nuclear interactions, hence the pseudoscalar field is not the pion of non-linear sigma model. It has, however, much of the symmetries, properly speaking emergent, of many-nucleon systems attributed to heavy nuclei not directly connected to QCD. As discussed in Chapter 7, one can however say that there is an as yet un-clarified link to the FT for QCD, given that the BPS model shares certain characteristics of nuclear structure with the gravity-gauge

[1] An initial discussion going via χPT is sketched in [Rho (1991)].

[2] It is not clear to the authors how $\not{\pi}$ EFT can address soft-pion-dominated processes, such as the "chiral filter" mechanism important for long-wavelength axial charge (first-forbidden) transitions discussed above in Chapter 2 which gets a factor of ~ 2 enhancement in the matrix element by soft-pion effects.

dual model of instanton baryons in the large N_c and large 't Hooft limit, which is regarded to be dual to QCD, though with an ultraviolet completion which may not be the same as the QCD relevant to nuclear dynamics. How the BPS nuclear model is reconcilable with the current algebras encoded in non-linear sigma model is currently unknown, but it will be too premature to dismiss it. This point, one of the key observations made in this volume — that different density regimes may have a chain of different effective field theories — has implications on some of the important issues raised in the literature, such as for instance imposing certain "constraints" in the equation of state (EoS) determined at normal matter density n_0, for the equation of state at higher densities relevant to compact star matter starting from nuclear matter density. This matter will be addressed in later chapters.

4.2 EFT strategy in dense baryonic matter

As stated in Chapter 2, the impressive success with EFT in nuclear physics as is being witnessed in the nuclear physics community, in particular precision calculations of electroweak transitions in nuclei [Kubodera and Rho (2011); Rho (2008)], can be taken as a convincing "folk proof" of Weinberg's FT. The processes for which the FT fares extremely well are in the domain of nuclear physics that have been extensively studied, both theoretically and experimentally, since many years. There the systematic formulation of the relevant field theory, namely, generic χPT, be that either standard (S)χPT or generalized (G)χPT, improves on the long-standing standard nuclear physics approach (SNPA) that has exploited highly sophisticated phenomenological potentials that have been developed since many years. What χPT brings in as a powerful new input is the systematic procedure for making higher-order corrections, reorganizing correction terms in consistency with the symmetries of the basic theory, QCD. This is the premise of the FT. This point is recognized and stated as [Weinberg (1997)]: "The use of effective quantum field theories has been extended more recently to nuclear physics, where although nucleons are not soft they never get far from their mass shell, and for that reason can be also treated by similar methods as the soft pions. Nuclear physicists have adopted this point of view, and I gather that they are happy about using this new language because it allows one to show in a fairly convincing way that what they've been doing all along (using two-body potentials only, including one-pion exchange and a hard core) is the correct first step in a consistent approximation scheme." This could be considered as a "Corollary" to the FT.

Now the question arises as to whether the FT can be exploited also to explore the regime for which no experiments are currently available, i.e., dense matter, and how to go about it? This constitutes making genuine predictions and as such would require certain concepts that go beyond the SχPT approach. In this chapter we discuss this matter.

4.2.1 *Renormalization decimations*

QCD, although complete by itself, cannot access the highly nonperturbative regime at low energy and high density relevant to compact-star matter, the system we are primarily interested in. The only tool available at present is quantum EFT. Since QCD can be trusted at high energy/momentum scale (e-scale for short), to do nuclear physics one can then set up the EFT Lagrangian at high e-scale and then decimate, defined precisely below, via RG to the appropriate low scale with density dependence suitably taken into account. There are two issues involved in doing this. One is the e-scale involved and the other the vacuum property affected by the medium, i.e., density. How to account for both in the framework of EFT is described in this chapter in general terms. How to apply the technique to specific problems in nuclear physics and in compact-star physics will be given in later chapters, Chapters 8 and 9.

First consider the problem of decimation in the e-scale. Suppose that the Lagrangian is given at a scale Λ_M. What it can be will be defined precisely below.

The starting element is the Lagrangian defined at a suitable scale, say, Λ_M, at which the EFT is defined with as *complete* a set of relevant degrees of freedom as feasible. Now what the relevant degrees of freedom are depends on one's intuition and expertise and could in some situations be the source of uncertainty. In the case we are considering, the pertinent Lagrangian is the one consisting of relevant hadronic degrees of freedom, the correlators of which are matched to the known theory, QCD, through its correlators. We identify Λ_M as the "matching scale," the scale at which EFT is optimally matched to QCD. Whether that matching scale can be precisely located is an open question. A guess work, guided by intuition, is involved here. At the matching scale, one calculates the appropriate correlators in the EFT sector at *tree order* and in the QCD sector in Wilsonian operator product expansion (OPE) [Harada and Yamawaki (2003)]. Done this way, the "bare" parameters of the EFT Lagrangian will carry information on nonperturbative characteristics of QCD, typically given in terms of various condensates, such as vacuum expectation values of bilinear quark operators $\bar{q}q$, gluon field tensor $G_{\mu\nu}G^{\mu\nu}$ and cross terms, as the vacuum changes. The quantum theory is then to decimate the generating functional from Λ_M to the scale at which low-energy physics is done. In doing this, there may be several steps involved. Instead of decimating all the way at one go, one may have multiple decimations. In the present case, we will essentially be dealing with two decimations. This is the double-decimation introduced in [Brown and Rho (2004)].

Following the notation of Polchinski [Polchinski (1992)], we consider a quantum field theory with a characteristic scale, say, $\tilde{\Lambda}$. For low-energy nuclear processes, we are interested in physics at a much lower scale $E \ll \tilde{\Lambda}$. For considering multiple scales, one can set up a series of such scales. To illustrate the idea, consider a system delineated by only one e-scale. We pick the matching scale Λ_M slightly below $\tilde{\Lambda}$ and divide the fields in the path integral into high and low components

$$\phi = \phi_{\text{high}} + \phi_{\text{low}}. \tag{4.1}$$

How to do the separation to "high" and "low" components depends on what system is being considered. In the system considered below, the separation is made at the energy scale up to which experimental information is available. Let it be denoted as Λ. The idea is that "high" components are not relevant for dynamics involving "low" components, so one integrates out the "high" components from the generating functional

$$\mathcal{Z} = \int [\mathcal{D}\phi_{\text{high}}][\mathcal{D}\phi_{\text{low}}]e^{iS[\phi_{\text{high}},\phi_{\text{light}}]} = \int [\mathcal{D}\phi_{\text{low}}]e^{iS_\Lambda[\phi_{\text{light}}]}. \tag{4.2}$$

The action $S_\Lambda[\phi_{\text{light}}]$, called "effective action" or "Wilsonian action," encodes physics above the cutoff Λ as

$$e^{iS_\Lambda[\phi_{\text{low}}]} = \int [\mathcal{D}\phi_{\text{high}}]e^{iS[\phi_{\text{high}},\phi_{\text{low}}]}. \tag{4.3}$$

Given $S_\Lambda[\phi_{\text{light}}]$ obtained from integrating out above the cutoff Λ, one can then do the functional integration over the low field ϕ_{low}. One can partition this decimation over low into further scales as one wishes. In fact, in our treatment described below, there will be another cutoff $\Lambda' < \Lambda$. Thus it will involve "double decimations", the first decimation from Λ to Λ' and the second from Λ' to zero. There are several ways to do these decimations. One way, a standard approach in the spirit of the standard chiral perturbation theory, currently popular in the field, is to expand $S_\Lambda[\phi_{\rm light}]$ (likewise for Λ' etc.) in terms of local operators $\mathcal{O}_i$

$$S_\Lambda = \int d^D x \sum_i c_i \mathcal{O}_i. \tag{4.4}$$

In nuclear physics as in condensed matter physics, one is breaking Lorentz invariance by integrating out the modes in energy $\omega > \Lambda$, so the operators $\mathcal{O}$ are nonlocal in time (i.e. energy-dependent) [Polchinski (1992)]. One then does the dimensional analysis to determine whether the operators involved are "relevant," "marginal" or "irrelevant." The fate of the matter considered, stable or unstable, will then depend on which sorts of operators are involved.

4.2.2 *Fermi-liquid fixed point*

In applying the above strategy to nuclei and nuclear matter, it is found to be convenient to resort to the double-decimation (DD) procedure, the first that goes from a characteristic energy scale to $\Lambda' \sim k_F$ above the Fermi surface where k_F is the Fermi momentum characterizing the Fermi sea or density n of nuclear matter and the second for the quantum fluctuations from Λ' to the Fermi surface. To start with, one has to determine what the "bare Lagrangian" defined at the characteristic energy is. For nuclear dynamics, it is the chiral scale $\Lambda_\chi \sim 4\pi f_\pi \sim 1$ GeV that is supposedly to delineate the hadronic sector to which EFT is applicable from the quark-gluon sector for which QCD is the pertinent theory. In practice, it is taken just above the vector-meson mass scale m_ρ which is 770 MeV in the vacuum.[3] From the point of view of RG, the most important aspect in nuclear physics is that nuclear matter is at a Fermi-liquid fixed point. So the first thing to do is to establish that one has a Fermi-liquid theory at the mean field level. For this we resort to two approaches: One is RMF and the other "V_{lowk}" approach.

Given a Lagrangian defined at a scale $\overline{\Lambda}$, one can do a mean-field approximation in the presence of Fermi surface(s). As will be specified below, in the presence of a Fermi surface, the relevant scale is the momentum scale relative to the Fermi momentum k_F, $\overline{\Lambda} = \Lambda - k_F$ where Λ is the cutoff scale for the first decimation. An appropriate Lagrangian that we will focus on

[3]As described in Chapter 8, the mass scale associated with the ρ meson in nuclear medium is subtle in that it depends on the baryonic matter density involved. In fact in dense medium it can become comparable to the pion mass, so how to deal with such excitation becomes highly non-trivial.

later is the effective Lagrangian, bsHLS, that incorporates baryon fields to scale-invariant hidden local symmetry (HLS) sHLS described in detail in Chapter 8. A linearized version of bsHLS gives essentially an improved version of Walecka linear mean-field model [Serot and Walecka (1997)]. In the literature, the mean-field approach with various complications (such as multi-dimensional fields) is applied with fair success to a wide range of nuclear processes including compact-star matter. Why this approach that relies on the mean-field nature works is, as mentioned above, because it corresponds essentially to Landau Fermi-liquid fixed-point theory [Matsui (1981)]. In the classification given above, this can be considered as "single-decimation" RG. One can go, as mentioned, beyond the mean-field approximation via what's known as V_{lowk}RG approach, precisely defined below. In the framework we are developing, this requires "intrinsic density dependence" (IDD for short) in the bare parameters of EFT Lagrangian input by QCD at the matching scale Λ_M between EFT and QCD. This is a concept that underlies the basic strategy of our EFT and will be defined precisely in Chapter 8. In this chapter, without being specific, it could be simply taken as a given density dependence of the EFT Lagrangian inherited from QCD as the vacuum slides with density.

4.2.3 V_{lowk}RG *approach*

To go beyond the mean field, we are interested in the "double-decimation" procedure utilizing the V_{lowk} [Bogner, Kuo and Schwenk (2003)]. We would like our theory not to depend on what's going on for the energy scale $> \Lambda$ where short-distance uncertainty lies. In doing the decimation, one exploits precise experimental information available up to E_{lab}. In two-nucleon systems, accurate phase shifts and deuteron properties are available up to $E_{lab} \sim 350$ MeV.[4] That suggests Λ to be

$$\Lambda \sim 2.1 \text{ fm}^{-1}. \tag{4.5}$$

We shall take that as the cutoff for the first decimation. In nuclear processes, it is more efficient to use the Weinberg scheme to first calculate the nuclear potential V as two-particle irreducible graphs and treat the reducible graphs in Schrödinger equation or Lippmann–Schwinger equation. The potential V can be taken from high-order chiral perturbation theory or alternatively — and equally well — from sophisticated phenomenological potentials fitted to the scale commensurate with $E_{lab} \sim 350$ MeV. It should

[4]This is valid for two-nucleon systems. For many-body systems n-body forces for $n > 2$ need to be taken into account.

not matter which, given that the short-distance part of the potential is integrated out, with its effects subsumed in the coefficients of the "low" field Lagrangian subject to the RG invariance (given below). Now the physical amplitude to go from $|I\rangle$ to $|F\rangle$ we want to compute is of the form

$$\langle F|\mathcal{A}|I\rangle = \langle F|V|I\rangle + \sum_{n=0}^{\infty} \frac{\langle F|V|n\rangle\langle n|V|I\rangle}{E_I - E_n} + \mathcal{O}(V^3) \,. \tag{4.6}$$

Rewrite

$$\langle F|\mathcal{A}|I\rangle = \langle F|V_{lowk}|I\rangle + \sum_{n=0}^{\Lambda} \frac{\langle F|V_{lowk}|n\rangle\langle n|V_{lowk}|I\rangle}{E_I - E_n} + \mathcal{O}(V_{lowk}^3) \tag{4.7}$$

which *formally* defines V_{lowk}. That physics should not depend on where the cutoff is set

$$\frac{d}{d\Lambda}\langle F|\mathcal{A}|I\rangle = 0 \tag{4.8}$$

gives the RG equation

$$\frac{d}{d\Lambda}V_{lowk} = \beta([V_{lowk}], \Lambda). \tag{4.9}$$

How to solve for the V_{lowk} is the rest of the story. It involves doing renormalization with suitable counter terms included.

While the procedure to calculate V_{lowk} for comparison with nature is involved, requiring numerous simplifications and approximations, how it works out in nuclear processes is nicely illustrated by a toy model given in [Bogner, Kuo and Schwenk (2003)]. We reproduce the details here.

The "bare" potential, taken to be the potential with which RG decimations are to be made, is picked in the separable form

$$\begin{aligned} V_{\rm bare}(k', k) &= L(k')\, R(k) \\ &= \frac{g}{(k'^2 + m^2)^n}\left(-\frac{g}{(k^2+m^2)^n} + \frac{\alpha\, g\,(k^2+m^2)^n}{(k^2+\eta^2\, m^2)^{2n}}\right). \end{aligned} \tag{4.10}$$

This potential has five parameters, g, m, n, α and η. The separable form allows an exact calculation of the two-nucleon scattering, so the RG strategy can be worked out analytically. The potential contains an attractive part and a short-range part, repulsive with $\eta > 1$. It turns out that by choosing the strength of the repulsion, for given η and other parameters, one can actually fit the ^{1}S$_0$ phase shift data rather reasonably. Thus this model is actually realistic enough for the illustration.

For separable potentials, the Lippmann–Schwinger equation can be resummed explicitly. For the bare potential (4.10), the T matrix is given by

$$T_{\rm bare}(k', k; k^2) = \frac{L(k')\, R(k)}{1 + \frac{2}{\pi}\mathcal{P}\int\limits_0^\infty \frac{L(p)\, R(p)}{p^2 - k^2} p^2 dp}. \tag{4.11}$$

Here $\eta = 2$ is chosen to approximately reflect the ratio of the repulsive to attractive meson contributions in terms of meson exchanges, which is $m_\omega/2m_\pi$. The remaining parameters, g, m and α, are then fit to reproduce the effective range expansion,

$$T_{\rm bare}(k, k; k^2) = \frac{1}{\frac{1}{a_s} - \frac{1}{2} r_0\, k^2}, \tag{4.12}$$

and the observed change in sign of the experimental phase shifts $T_{\rm bare}(\kappa, \kappa; \kappa^2) = 0$, say, at $\kappa = 1.79\,{\rm fm}^{-1}$ for a given power n.

Given the number of parameters, one can change the parameters of the bare potential while assuring the same T matrix for low-energy scattering. One picks the potentials that contain different short distance components without affecting the scattering at low energy. The next thing to do is to calculate the T matrix using V_{lowk} that is obtained by decimating the high-energy component subject to the RG invariance of the the T matrix. For separable interactions, the RG equation, e.g., for the diagonal components of the low momentum interaction, can be obtained in a closed form,

$$\frac{d}{d\Lambda} V_{\rm lowk}(k, k; \Lambda) = \frac{2/\pi}{1 - (k/\Lambda)^2} \frac{V_{\rm lowk}(k, k; \Lambda)\, V_{\rm lowk}(\Lambda, \Lambda; \Lambda)}{1 + \frac{2}{\pi}\mathcal{P}\int\limits_0^\Lambda \frac{V_{\rm low\ k}(p, p; \Lambda)}{p^2 - \Lambda^2} p^2 dp} \tag{4.13}$$

where $\mathcal{P}$ stands for the principal value integral. The initial condition is

$$V_{\rm lowk}(k, k; \Lambda_0) = V_{\rm bare}(k, k), \tag{4.14}$$

at an initial scale Λ_0 where the high momentum modes are to be negligible,

$$V_{\rm bare}(\Lambda_0, \Lambda_0) \approx 0. \tag{4.15}$$

Now the RG equation is used to evolve the bare potentials of different short-distance properties to low momenta, thus removing the model-dependent high momentum parts. The solution of the Lippmann–Schwinger equation given in terms of V_{lowk} — in place of V_{bare} — is

$$T_{\rm lowk}(k, k; k^2) = \frac{V_{\rm lowk}(k, k; \Lambda)}{1 + \frac{2}{\pi}\mathcal{P}\int\limits_0^\Lambda \frac{V_{\rm lowk}(p, p; \Lambda)}{p^2 - k^2} p^2 dp} \tag{4.16}$$

and the phase shift is

$$\tan\delta = -k\, T_{\text{lowk}}(k,k;k^2). \tag{4.17}$$

In [Bogner, Kuo and Schwenk (2003)], the separable RG equation is solved with the cutoff set at $\Lambda = 2.1\,\text{fm}^{-1}$, which corresponds to a laboratory energy scale up to which the realistic nucleon-nucleon interactions are constrained by experiment, $E_{\text{lab}} \sim 350\,\text{MeV}$.

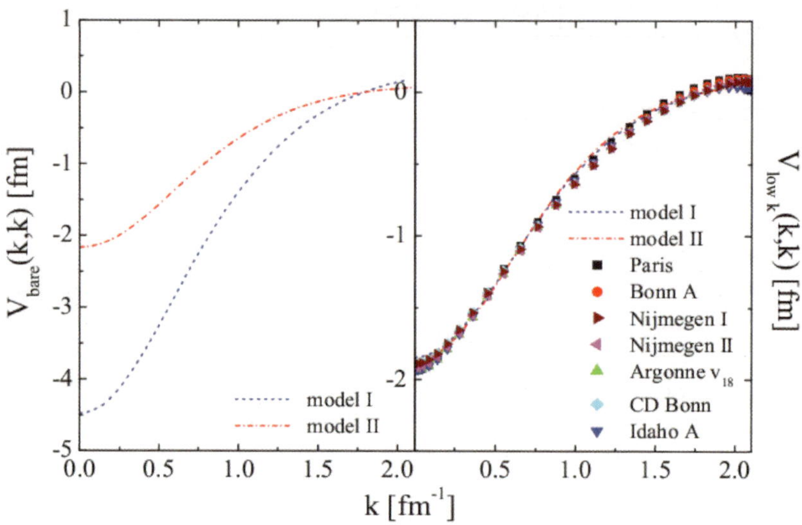

Fig. 4.1 The diagonal momentum-space matrix elements of the two V_{bare}'s for models I and II of [Bogner, Kuo and Schwenk (2003)] and $V_{\text{low k}}$ derived from them. As a comparison, those derived from the high precision sophisticated nucleon-nucleon interactions are also shown.

In Fig. 4.1 are shown the resulting V_{lowk}'s from two different V_{bare}'s that fit the $^1S_0$ phase shift reproduced from [Bogner, Kuo and Schwenk (2003)]. Although the bare separable models I and II shown on the left panel differ substantially, the V_{lowk}'s agree very well for the given cutoff. It is intriguing that they also agree with the V_{lowk}'s gotten from the high precision nucleon-nucleon interactions. This observation is telling us that for the processes taking place at $E \ll \Lambda$, the short-distance ambiguities do not affect the dynamics.[5]

[5]That a toy model with a separable potential, when treated subject to RG invariance with the cutoff in consistency with data works on par with "sophisticated" potentials

For later applications, it is worth clarifying the essential points of the strategy. Diagrammatically, $V_{\rm lowk}$ sums all ladders with bare potential vertices and intermediate momenta $p > \Lambda$. The result is nonlocal in energy given that the decimation is done in momentum space. Subsequently, the energy dependence is removed by further many-body techniques, e.g., the folded diagram technique described below. For the two-nucleon problem treated above, the nuclear model space effective action methods are equivalent to solving an RG equation obtained by requiring the RG invariance $\frac{d}{d\Lambda}T(k', k; k^2) = 0$ with the various bare potentials as large Λ conditions. For a suitable cutoff, all model potentials flow to a unique low-momentum interaction $V_{\rm lowk}$. That this convergence takes place precisely is the proof for the validity of the method, offering the possibility of applying to dense baryonic matter.

4.2.4 *Going beyond the V_{lowk}RG?*

Although highly successful, an important question that requires to be addressed for further progress — and also interpreting some of the new results obtained in Chapter 9 as will be seen — is how to go beyond the approximations involved in the V_{lowk}RG formulation. Since the V_{lowk}RG procedure adopted in this volume is at best an approximate RG scheme, this question would involve examining what sort of terms are left out in the procedure. As stressed, our approach adheres to the notion that the matter we are dealing with is a Landau Fermi-liquid both at low density near n_0 and, as will be argued later, at higher density relevant to massive compact stars, $\sim (5-7)n_0$. We should however recognize that this notion could be inapplicable at high density for several reasons. The first could be that the effective Lagrangian with a possible topology change that we exploit crucially in dense matter may be incorrect. The second could be that certain important fluctuations are missing from the V_{lowk}RG strategy. This question remains to be answered with *bs*HLS Lagrangian with the IDDs incorporated. There is, however, a study addressed to the role of quantum fluctuations over the mean field with a model Lagrangian that corresponds to the Walecka linear mean-field Lagrangian [Serot and Walecka (1986)] discussed below in Section 4.5.2 in the context of EDF approaches with the

as well as high-order chiral perturbation expansion is arguing for that the standard nuclear physics approach to nuclear potentials is as good as high N^{n}LO chiral potentials, confirming the Weinberg statement that "nuclear physicists have known what they have been doing." In other words, there is no big deal in the claim that high-order χPT approach is doing something more fundamental.

pions restored in hidden local symmetric way. It resorts to a nonperturbative method that is known as "functional renormalization group (FRG)." In principle, given a Lagrangian, fundamental or effective (with a cutoff), one can get an "exact" energy density of the system. However the standard chiral Lagrangian anchored on non-linear sigma model (discussed below) is not suitable for practical calculations in FRG as will be explained below.

What is available in the literature up to date [Drews and Weise (2017)] is an analysis using the action $\Gamma = \int d^4x\mathcal{L}$ with

$$\mathcal{L} = \bar{\psi}\big(i\gamma^\mu\partial_\mu\psi + \frac{1}{2}\partial^\mu\sigma\partial_\mu\sigma + \frac{1}{2}\partial_\mu\boldsymbol{\pi}\cdot\partial^\mu\boldsymbol{\pi}\big)\psi - \bar{\psi}\Big[g(\sigma + i\gamma_5\boldsymbol{\tau}\cdot\boldsymbol{\pi}) + \gamma_0(g_v\omega_0 + g_t\rho_{3,0}\tau^3)\Big]\psi + \mathcal{U} \tag{4.18}$$

with

$$\mathcal{U} = \frac{1}{2}(\sigma^2 + \boldsymbol{\pi}^2) - m_\pi^2 f_\pi(\sigma - f_\pi) - \frac{1}{2}m_V^2(\omega_0^2 + \rho_{3,0}^2). \tag{4.19}$$

This is the Gell-Man–Lèvy (GML) $SU(2)\times SU(2)$ linear σ coupled to the vector fields ρ and ω in hidden gauge symmetry with the following assumptions. The vector mesons are massive so that the kinetic energy terms can be suppressed and contribute only through the mean fields of their time components. The vector couplings are taken unequal $g_v \neq g_t$, so the flavor $U(2)$ symmetry is broken even though their masses are taken equal, $m_\rho = m_\omega = m_V$. Note that the scalar mass is taken $m_\sigma \sim m_V$.

In nature, chiral symmetry is realized nonlinearly in the matter-free vacuum, which corresponds to taking $m_\sigma \to \infty$ and $m_\rho \sim m_\omega$. So (4.18) is not correct for the low-energy theorems, i.e., soft-pion theorems, KSRF relations etc. However in linear Walecka RMF model, it is required that $m_\sigma \sim m_V$ so as to have the intricate balance between attraction and repulsion in nuclear matter. More importantly the scalar in Walecka model cannot be the fourth component of the chiral four-vector of the linear sigma model. Although it is an effective field, it should be predominantly — if not entirely — a chiral singlet. Later in the volume it will be seen that the GML linear sigma model structure does however arise at high density as what will be called "dilaton-limit fixed point." This phenomenon will turn out to play a crucial role in our development.

The FRG analysis consists of first computing the mean-field contribution and then computing fluctuations nonperturbatively, with the parameters suitably fixed by fits to experimentally known properties. The mean-field contribution is essentially what one obtains in standard RMF approaches listed below in Section 4.5.2. There is one basic strategy in

RMF approaches that differs from the mean-field approach in the present FRG calculation. The RMF approaches purport to capture Landau Fermi-liquid fixed-point theory, with corrections to be suppressed by $\bar{\Lambda}/k_F$ where $\bar{\Lambda}$ — that decreases as density increases — is the decimation cutoff relative to the Fermi momentum. In this FRG, the mean-field result is defined with respect to what's included in the fluctuations computed nonperturbatively and hence need not be tuned to the Landau Fermi-liquid fixed-point structure. This would suggest that the mean-field result in the FRG calculation should not be directly compared to the mean field in RMF theory.

The numerical results of [Drews and Weise (2017)] illustrate the principal points mentioned above. In Fig. 4.2 is given E_n/A for pure neutron matter computed in FRG. It is notable that the various calculations, i.e.,

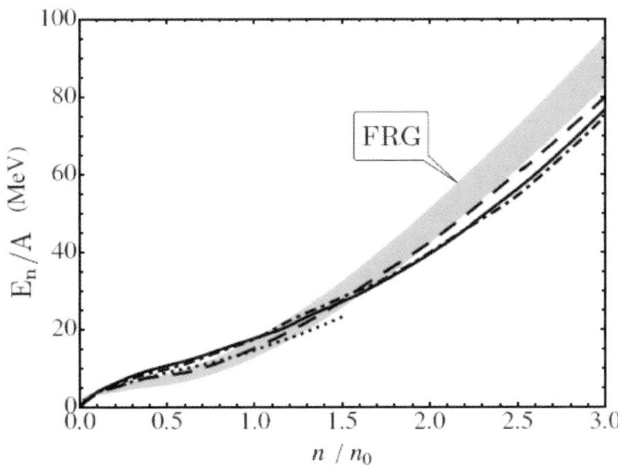

Fig. 4.2 Energy per particle for pure neutron matter in FRG for the range of symmetry energies $29 \leq E_{sym}/\text{MeV} \leq 33$ compared with the chiral perturbation EoS (full line), the quantum Monte Carlo calculation with (a) "realistic" potential (dashed line) and(b) chiral potentials (dotted line) and the Akmal–Pandharipande–Ravenhall EoS (dashed-dotted line).

standard high-order χPT (described below), standard nuclear physics approach, quantum Monte Carlo based on sophisticated phenomenological potentials, all give more or less the same results. The FRG result, somewhat sensitively dependent on the symmetry energy E_{sym} important in compact stars, indicates quantitatively somewhat more repulsion than the other methods. Note that the standard chiral perturbation approach fares fairly well with the phenomenological calculations in SNPA and is close to the FRG result including the chiral order parameter σ/f_π.

An observation most relevant to what we will find in our development is the property of the chiral order parameter at increasing density. In Fig. 4.3 is plotted the in-medium pion decay constant, the chiral order parameter, vs. density. What is the most noteworthy — and extremely

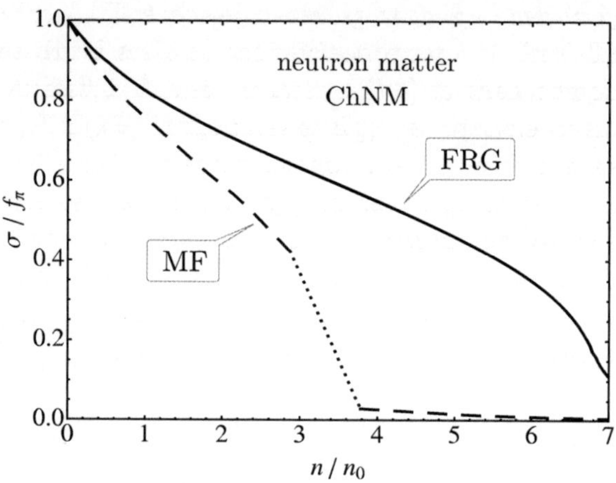

Fig. 4.3 Chiral order parameter vs. density. MF stands for mean field in the FRG calculation of [Drews and Weise (2017)].

significant for the results given in Chapter 9 — is that the fluctuations on top of the mean fields have strong effects to slow down the dropping of the chiral order parameter. In fact it will be seen in Chapter 9 that in the V_{lowk}RG approach, the chiral condensate, which will be related to the dilaton condensate, stops "running" downward at a density $\sim (2-3)n_0$ and stays constant or "crawls," possibly up to the density at which the dilaton limit point is approached.

It is difficult to pinpoint the precise mechanism that makes this difference. A likely explanation is that with HLS properly implemented, the vector manifestation associated with the ρ-meson mass going toward zero — explained in detail in Chapter 8 — manifests as density increases. In addition, the scalar that enters in bsHLS Lagrangian is a dilaton associated with scale invariance and transmutes to the scalar of the linear sigma model with increasing density. These two phenomena are *not* encoded in the Lagrangian (4.18) where the vector manifestation is excluded *at initio*. This is the principal feature that differentiates our thesis from the conventional one.

4.3 Chiral perturbation theory for nuclear matter

Let us now look at symmetric ($N = Z$) nuclear matter in standard chiral perturbation theory (SχPT) involving only the doublet nucleons and the triplet of pions. This discussion will illustrate the basic difference in strategy between SχPT and $V_{\rm lowk}$RG. The effective chiral Lagrangian taken can be considered as the one in which all heavier degrees of freedom in bsHLS Lagrangian are integrated out, giving the non-linear sigma model Lagrangian with baryons incorporated. The "bare" parameters of the Lagrangian are supposed to be in the matter-free space, that is, no IDDs. The density involved in compact stars is at higher scale than that for which SχPT is arguably justified. So the question then would be whether chiral perturbation theory involving only pions and nucleons, *without* IDDs, unlike bsHLS that incorporates all degrees of freedom appropriate for an effective theory defined at a matching scale to QCD, could access the density regime relevant to compact-star densities that are expected to be considerably higher than that of ordinary nuclear matter. We will first see how it works up to the density relevant to nuclear matter. It will then be extrapolated in Chapter 9 to higher densities going beyond n_0 to the regime relevant to compact stars.

In applying chiral perturbation approaches to nuclei and nuclear matter, one first posits that there is a Fermi sea. Starting from a chiral Lagrangian that is to capture low-energy properties of hadrons, if one were to follow a bona-fide EFT strategy, one would have to establish first the presence of Fermi sea. This would require an approach that would involve a quantum critical phenomenon. The early attempt to derive a Fermi sea of a sort consisted of looking for a nontopological soliton of an effective chiral Lagrangian in the form of chiral liquid [Lynn (1993)]. There has been little success in this endeavor. The current practice is that one simply posits that there is a Fermi surface characterized by the Fermi momentum k_F and the non-relativistic nucleon with bare mass m_N propagates in the Fermi sea with the propagator

$$G(p_0,\vec{p}) = \frac{i}{p_0 - \vec{p}^2/2m_N + i\epsilon} - 2\pi\delta(p_0 - \vec{p}^2/2m_N)\,\theta(k_F - |\vec{p}\,|)\,. \quad (4.20)$$

The first term is the free-space propagator, and the second accounts for the filled Fermi seas. This clearly posits that there is an additional scale associated with the Fermi momentum to what is in free space. This means that the chiral perturbation series has an additional "small" expansion parameter, the Fermi momentum k_F, in addition to the usual ones in the vacuum $p \sim \partial$ and m_π. The effective chiral Lagrangian with which one does in-medium chiral perturbation calculations is supposed to have symmetries of QCD in matter-free vacuum, apart from the spontaneously broken Lorentz symmetry, with the "bare" parameters of the Lagrangian determined at the scale $\bar{\Lambda} \sim m_N$. The power counting with k_F then is $k_F/\bar{\hat{\Lambda}}$ where $\bar{\hat{\Lambda}}$ is typically the chiral scale or the nucleon mass ~ 1 GeV. The energy per particle for symmetric nuclear matter $\bar{E}(k_F)$ is then computed in power series in k_F/m_N. $\bar{E}(k_F)$ is the total energy per particle given by the closed nucleon loops with the nucleon propagator (4.20) with the free-space value subtracted. The net result with appropriate regularization is then an expansion in the powers of k_F. In [Kaiser, Fritsch and Weise (2002)], up to $O(k_F^5)$ terms involving one- and two-pion exchanges were computed. Typical loop diagrams arising from one-pion and two-pion exchanges going up to terms of $O(k_F^4)$ are given in Fig. 4.4.[6] Irreducible two-pion exchange

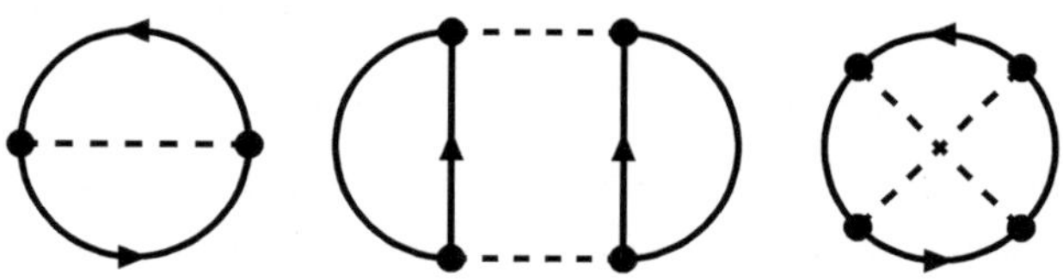

Fig. 4.4 One-pion-exchange Fock diagram (left) and iterated one-pion exchange direct (or Hartree) term (middle) and exchange (or Fock) term (right) contributing to the energy density $n\bar{E}(k_f)$ of nuclear matter.

graphs give terms of $O(k_F^5)$. Certain baryon resonances, in particular, the $\Delta(1232)$, replacing the nucleon intermediate nucleon particle line, can be included in the irreducible graphs if those resonances are not included as the relevant degrees of freedom. In fact it has been known since a long time that some of those terms with the Δ included in two-pion exchange graphs generate attraction resembling what is simulated by the exchange of a scalar meson of mass comparable to the lowest-lying vector mesons. The graphs can be calculated analytically and given in terms of g_A, f_π, m_N and m_π and the Fermi momentum k_F. One assumes that the parameters are

[6]The Hartree term with the pion field is absent since we are concerned with the symmetric matter.

given by experiments independent of the cutoff $\bar{\Lambda}$. The cutoff is a physical quantity setting the scale for the EFT. In principle physical quantities should of course be independent of the cutoff chosen, this according to the renormalization-group invariance.

It turns out that the result can be roughly captured by a simple formula consisting of three terms [Kaiser, Fritsch and Weise (2002)]

$$\bar{E}(k_F) = \frac{3k_F^2}{10m_N} - \alpha\frac{k_F^3}{m_N^2} + \beta\frac{k_F^4}{m_N^3}\,. \tag{4.21}$$

The first term is the kinetic energy term which is known, and the second and third terms represent, respectively, overall attraction and repulsion. Both effects are to be simulated at high order in SχPT involving the exchanges of pions, qualitatively reproduced in the mean field approach by the mean fields of scalar meson and vector meson of mass $\sim 600 - 700$ MeV. It turns out that a suitable choice of the parameters α and β gives a fairly good description of nuclear matter properties. For instance with $\alpha = 5.27$ and $\beta = 12.22$, obtained by fitting the saturation properties, $n_0 \approx 0.16$ fm^{-1} and binding energy ≈ 16 MeV, one gets the compression modulus $K \approx 240$ MeV. It is claimed to work up to density $n \approx 6n_0$.

At equilibrium, the matter density is $k_F \sim 2m_\pi$. A typical mass scale intervening in the chiral expansion, that is, the decimation cutoff, is that of the vector meson mass, say, $m_V \sim 6m_\pi$, so $k_F/m_V \sim 1/3$. This could be considered small enough so that the expansion in k_F may be justified for nuclear matter. One should however recognize that the simple perturbative approach as given above captures more or less correct description of matter at normal nuclear matter density is not a big deal. In fact there are a large number of different approaches to nuclear matter, purely phenomenological or more sophisticated, that more or less give correctly nuclear matter properties. Thus Walecka-type and more involved RMF or energy-density functional (EDF) models all do just as well. It is in going to higher densities, say, $n \gtrsim (4 - 7)n_0$ relevant to compact-stars, however, that various models vary widely unconstrained by anything. Given the paucity of data and lacking model-independent theory, the situation remains dismal. One extreme example, discussed in Chapter 9, is the nuclear symmetry energy E_{sym}: Beyond n_0, it varies wildly, totally unconstrained.

In an EFT with Fermi surface, the appropriate description is Landau Fermi-liquid theory [Polchinski (1992); Shankar (1994)]. In Landau Fermi-liquid theory, the momentum integration is to the Fermi surface, not to zero and the expansion is in $\overline{\Lambda}/k_F \equiv 1/N$ where $\overline{\Lambda} = \Lambda - k_F$ [Shankar

(1994)]. It is effectively an $1/N$ expansion, so the higher k_F the more reliable the quasiparticle description becomes. In terms of RMF Lagrangian approaches, this is equivalent to that mean-field approximation gets better higher the density. This suggests the k_F that figures in a chiral perturbation scheme will play a basically different role in Landau Fermi-liquid fixed-point theory. The unanswerable question is how far in density the Landau Fermi-liquid structure itself continues to hold. It is shown in Chapter 7 that there can be a topology change in the description of baryonic matter at $n \gtrsim 2n_0$ when looked at in terms of skyrmions, which would imply that the standard Fermi-liquid description could break down at that density. However this interpretation is misleading as will be seen in Chapter 9. Above the topology change density, there seems to emerge a "pseudo-conformal" structure with a conformal-like sound velocity $v_s^2 \approx 1/3$ in dense star matter which indicates a Landau Fermi-liquid fixed-point state.

4.4 Fermi-liquid fixed-point theory

4.4.1 *Relativistic mean field theory as a single-decimation fixed-point theory*

4.4.1.1 *From chiral Lagrangian to Fermi-liquid fixed-point theory*

As mentioned above, RMF approach can be interpreted as something equivalent to Fermi-liquid fixed-point theory. One may develop this point of view starting with a chiral Lagrangian described above in GχPT, i.e., in Section 4.3. Let us think of the chiral Lagrangian gotten from bsHLS with the massive mesons integrated out and expressed in terms of multi-nucleon fields. As first suggested in [Rho (1991)], one may take the mean field of the resulting Lagrangian and consider the procedure as giving the Fermi-liquid interactions [Matsui (1981)]. The resulting Landau–Migdal interaction is of the form

$$V_{LM} = (F_0 + G_0 \boldsymbol{\sigma}_1 \cdot \boldsymbol{\sigma}_2 + F_0' \boldsymbol{\tau}_1 \cdot \boldsymbol{\tau}_2 + G_0' \boldsymbol{\sigma}_1 \cdot \boldsymbol{\sigma}_2 \boldsymbol{\tau}_1 \cdot \boldsymbol{\tau}_2)\delta(\boldsymbol{r}) + V_{\rm NL} \,, \tag{4.22}$$

where $V_{\rm NL}$ contains possible nonlocal terms such as pionic tensor. Here F's and G's are density-dependent constants coming from IDDs (Lorentz-invariant) and also "induced density effects" (later defined as "induced density dependence" — which is non-Lorentz-invariant — or $\mathrm{DD}_{\rm induced}$ for short in contrast to IDD) including short-range multi-body terms.[7] In going

[7] They can contain, for instance, 3-body effects, dependent on density, that can be brought in as "irrelevant" in the RG sense. This means that they could contain more

into the RG strategy discussed below, the RMF can be considered as a single-decimation RG procedure [Song, Min and Rho (1998)]. The double-decimation procedure with V_{lowk} that we will develop below and further in Chapter 8 goes beyond the fixed point with higher-order correlations included, adapted to a quantitative confrontation with nature, in the second decimation.

4.4.1.2 *Simplified model*

For the purpose of this discussion, let us take the Lagrangian appropriate for symmetric nuclear matter and good enough for the essential argument, which comes, greatly simplified, from the scale-invariant hidden local symmetric Lagrangian (6.86) derived in Chapter 6,

$$\mathcal{L} = \bar{\psi}[\gamma_\mu(i\partial^\mu + g_v^*\omega^\mu) - m_N^* + h^*\phi]\psi$$
$$+ \frac{1}{2}[(\partial\phi)^2 - m_s^{*2}\phi^2] - \frac{1}{4}F_{\mu\nu}^2 + \frac{1}{2}m_\omega^{*2}\omega_\mu^2 + \cdots \tag{4.23}$$

where ψ is the nucleon field, ω_μ the isoscalar vector field with $F_{\mu\nu}$ being its field strength tensor, ϕ an isoscalar scalar field and the masses with asterisk are BR-scaling as first introduced in [Brown and Rho (1991)]. This scaling in the density regime $n \lesssim 2n_0$ (as will be clarified in Chapter 7) is essentially the sum of the IDD and the induced density dependence $\mathrm{DD}_{\mathrm{induced}}$ described precisely below. This is in a form identical to Walecka's linear mean field model and follows from the *bs*HLS Lagrangian with the pion and ρ field dropped and expanded to the lowest power in fields. The only difference is the density dependence involving BR scaling.

There are three issues involved here with baryonic matter. One is the Lorentz symmetry, the next is the thermodynamic consistency and the last is the relation between the mean-field treatment of *bs*HLS Lagrangian and Landau(–Migdal) Fermi-liquid theory.

In medium with Lorenrz symmetry broken down to $O(3)$, the coupling constant for the time component of covariant four-vector differs from that of the space component. The IDD is a Lorentz scalar, so that at the matching scale, the density dependence is a Lorentz invariant quantity. In doing decimations with the "bare" Lagrangian, it usually picks up contributions

than just the (Lorentz-invariant) IDDs, as will be discussed in Chapter 8, specifically in connection with the anomalous orbital gyromagnetic ratio δg_l. Unless otherwise specified, in discussing nuclear observables, we will understand by IDDs generically including induced effects as well.

that differentiate the time and space components. For instance, the zero-range three-body interactions that are of the scale commensurate with the cutoff scale brings a density-dependence not related to the IDD and this contribution belonging to the class of $\mathrm{DD_{induced}}$ will figure in the Brown–Rho scaling that enters in applications. We will denote the sum of IDD and $\mathrm{DD_{induced}}$ as IDD* whenever the distinction from the Lorentz-invariant IDD is called for. In the discussion that follows below, we will be concerned with the IDD proper, not necessarily BR scaling, so there will be no distinction between the time and space components in the coupling constants.

4.4.1.3 *Thermodynamic consistency*

If one takes the density n as a c-number, then it is easy to see that the conservation of the energy-momentum tensor $\theta_{\mu\nu}$ is not preserved. This has to do with the fact that the pressure of the system involves derivative with respect to density on the energy density $\mathcal{E}$ gotten from the Lagrangian and hence there will be terms involving the derivative on the masses and coupling constants that are not present in the energy momentum tensor gotten from the Lagrangian (4.23) by Noether theorem. This problem is known as "thermodynamic inconsistency."

How should one take the density dependence so that thermodynamic consistency is maintained?

To address this question, let us consider the Lagrangian (4.23) defined at a scale at which we can do mean-field approximation. This means that we assume the bare Lagrangian is decimated from a cutoff scale $\Lambda \sim m_V$. We will focus on ground-state properties and hence as in density-functional theory for atomic/molecular and chemical physics (as explained below), we put weight on baryon density as the primary degree of freedom. Thus up to nuclear matter density n_0 with which we are concerned, we will have the bare parameters of the effective Lagrangian defined at Λ. The local density function $n(\boldsymbol{r})$ is then $\bar{\psi}\gamma_0\psi(\boldsymbol{r})$ where ψ will be a quasi-particle nucleon field,

$$n(\boldsymbol{r}) = \psi^\dagger\psi(\boldsymbol{r}). \tag{4.24}$$

One can take the mean-field theory to be equivalent to Landau Fermi-liquid fixed-point theory [Matsui (1981)].

In writing the equations of motion (EoM) for the meson and baryon fields, we consider the bare parameters to be functional of $n(\boldsymbol{r})$. The EoM for the meson fields are simply

$$(\partial^\mu\partial_\mu + m_s^{*^2})\phi = h^*\bar{\psi}\psi \tag{4.25}$$
$$\partial_\nu F^{\nu\mu} + m_\omega^{*\,2}\omega^\mu = g_v^*\bar{\psi}\gamma^\mu\psi. \tag{4.26}$$

The EoM for the nucleon field is more involved

$$[i\gamma^\mu(\partial_\mu - ig_v^*\omega_\mu - iu_\mu\check{\Sigma}) - m_N^* + h^*\phi]\psi = 0 \tag{4.27}$$

with

$$\begin{aligned}\check{\Sigma} &= \frac{\partial\mathcal{L}}{\partial\check{n}}\\ &= m_\omega^*\omega^2\frac{\partial m_\omega^*}{\partial\check{n}} - m_s^*\phi^2\frac{\partial m_s^*}{\partial\check{n}} + \bar{\psi}\omega^\mu\gamma_\mu\psi\frac{\partial g_v^*}{\partial\check{n}}\\ &\quad - \bar{\psi}\psi\frac{\partial m_N^*}{\partial\check{n}} + \bar{\psi}\psi\phi\frac{\partial h^*}{\partial\check{n}}\end{aligned} \tag{4.28}$$

where we have defined

$$\check{n}u^\mu \equiv \bar{\psi}\gamma^\mu\psi \tag{4.29}$$

with unit fluid 4-velocity $u^\mu = \frac{1}{\sqrt{1-\vec{v}^2}}(1,\vec{v}) = \frac{1}{\sqrt{n^2-\vec{j}^2}}(n,\vec{j})$ with the baryon current density $\vec{j} = \langle\bar{\psi}\vec{\gamma}\psi\rangle$ and the baryon number density $n = \langle\psi^\dagger\psi\rangle = \sum_i n_i$. The conserved canonical energy-momentum tensor then is

$$\begin{aligned}T^{\mu\nu} &= i\bar{\psi}\gamma^\mu\partial^\nu\psi + \partial^\mu\phi\partial^\nu\phi - \partial^\mu\omega_\lambda\partial^\nu\omega^\lambda\\ &\quad -\frac{1}{2}[(\partial\phi)^2 - m_s^{*2}\phi^2 - (\partial\omega)^2 + m_\omega^{*2}\omega^2 - 2\check{\Sigma}\bar{\psi}\gamma\cdot u\psi]g^{\mu\nu}.\end{aligned} \tag{4.30}$$

Let us now compute thermodynamic quantities from (4.30) using the mean-field approximation which amounts to taking

$$\begin{aligned}\psi = \frac{1}{\sqrt{V}}\sum_i a_i\sqrt{\frac{E_{\kappa_i}+m_L^*}{2E_{\kappa_i}}}\begin{pmatrix}\chi\\ \frac{\vec{\sigma}\cdot\vec{\kappa}_i}{E_{\kappa_i}+m_L^*}\chi\end{pmatrix}&\\ \exp\left(i\vec{\kappa}_i\cdot\vec{x} - i(g_v^*\omega_0 - u_0\Sigma + E_i)t\right)&\end{aligned} \tag{4.31}$$

$$h^*\phi = C_h^2\langle\bar{\psi}\psi\rangle = C_h^2\sum_i n_i\frac{m_L^*}{\sqrt{\vec{\kappa}_i^2 + m_L^{*2}}} \tag{4.32}$$

$$g_v^*\omega_0 = C_v^2 n = C_v^2\sum_i n_i \tag{4.33}$$

$$g_v^*\vec{\omega} = C_v^2\vec{j} = C_v^2\sum_i n_i\frac{\vec{\kappa}_i}{\sqrt{\vec{\kappa}_i^2 + m_L^{*2}}} \tag{4.34}$$

where a_i is the annihilation operator of the nucleon i, with $n_i = \langle a_i^\dagger a_i \rangle$, and $\Sigma = \langle \check{\Sigma} \rangle$, $E_{\kappa_i} = \sqrt{\vec{\kappa}_i^2 + m_L^{*2}}$ with $m_L^* \equiv m_N^* - h^*\phi$, χ is the spinor and $\vec{\sigma}$ is the Pauli matrix. We have defined

$$C_v(\check{n}) \equiv \frac{g_v^*(\check{n})}{m_\omega^*(\check{n})}, \quad C_h(\check{n}) \equiv \frac{h^*}{m_s^*(\check{n})}, \quad \tilde{C}_h(\check{n}) \equiv \frac{1}{C_h(\check{n})} \tag{4.35}$$

$$\vec{\kappa} \equiv \vec{k} - C_v^2 \vec{j} + \vec{u}\Sigma. \tag{4.36}$$

In this approximation, the energy density is

$$\begin{aligned}\mathcal{E} &= \langle T^{00} \rangle \\ &= \langle i\bar{\psi}\gamma^0\partial^0\psi + \frac{1}{2}m_s^{*2}\phi^2 - \frac{1}{2}m_\omega^{*2}\omega^2 + \check{\Sigma}\bar{\psi}\gamma^\mu u_\mu \psi \rangle \\ &= \frac{1}{2}C_v^2(n^2 + \vec{j}^2) + \frac{1}{2}\tilde{C}_h^2(m_L^* - m_N^*)^2 + \sum_l n_l \sqrt{\vec{\kappa}_l^2 + m_L^{*2}} \\ &\quad - \Sigma \vec{u}\cdot\vec{j}.\end{aligned} \tag{4.37}$$

Note that the Σ-dependent terms cancel out in the comoving frame ($\vec{v} = 0$), so that the resulting energy density is identical to what one would obtain from the Lagrangian in the mean field with the density-dependent parameters *taken as field-independent quantities.*

Given the energy density (4.37), the pressure can be calculated by (at $T = 0$)

$$\begin{aligned}p &= -\frac{\partial E}{\partial V} = n^2 \frac{\partial \mathcal{E}/n}{n\partial n} = \mu n - \mathcal{E} \\ &= \frac{1}{2}C_v^2 n^2 - \Sigma_0 n - \frac{1}{2}\tilde{C}_h^2(m_L^* - m_N^*)^2 \\ &\quad - \frac{\gamma}{2\pi^2}\left(E_F\left(\frac{m_L^{*2}}{8}k_F - \frac{1}{12}k_F^3\right) - \frac{m_L^{*4}}{8}\ln[(k_F + E_F)/m_L^*]\right)\end{aligned} \tag{4.38}$$

where μ is the chemical potential, the first derivative of the energy density with respect to n in the comoving frame ($\vec{v} = 0$):

$$\mu \equiv \frac{\partial}{\partial n}\mathcal{E}|_{\vec{v}=0} = C_v^2 n + E_F - \Sigma_0 \tag{4.39}$$

with $E_F = \sqrt{k_F^2 + m_L^{*2}}$ and $\Sigma_0 = \langle \check{\Sigma} \rangle_{\vec{v}=0}$. To check that this is consistent, we calculate the pressure from the energy-momentum tensor (4.30) in the mean field at $T = 0$:

$$\begin{aligned} p_t &\equiv \frac{1}{3}\langle T_{ii}\rangle_{\vec{v}=0} \\ &= \frac{1}{3}\langle i\bar{\psi}\gamma^i\partial^i\psi - \frac{1}{2}(m_\omega^{*2}\omega^2 - m_s^{*2}\phi^2 - 2\check{\Sigma}\psi^\dagger\psi)g^{ii}\rangle_{\vec{v}=0} \\ &= \frac{1}{2}C_v^2 n^2 - \frac{1}{2}\tilde{C}_h^2(m_L^* - m_N^*)^2 - \Sigma_0 n \\ &\quad - \frac{\gamma}{2\pi^2}\left(E_F\left(\frac{m_L^{*2}}{8}k_F - \frac{1}{12}k_F^3\right) - \frac{m_L^{*4}}{8}\ln[(k_F+E_F)/m_L^*]\right). \end{aligned} \quad (4.40)$$

This agrees with (4.38). Thus our EoS conserves energy and momentum.

4.4.1.4 *Quasiparticle interactions*

The equivalence of the RMF written down above to Landau Fermi-liquid theory entitles one to write down the quasiparticle interactions.[8] Consider for illustration the quasiparticle interaction f_{ij}. Other components of quasiparticle interactions can be gotten in a similar way.

In terms of the single quasiparticle energy $\epsilon_i = \frac{\partial \mathcal{E}}{\partial n_i}$, we get

$$\begin{aligned} f_{ij} &= \frac{\partial \varepsilon_i}{\partial n_j}|_{\vec{j}=\vec{v}=0} \\ &= C_v^2 + 4C_v n\frac{\partial C_v}{\partial n} + \frac{m_L^*}{E_i}\frac{\partial m_L^*}{\partial n_j} + n^2\left[\left(\frac{\partial C_v}{\partial n}\right)^2 + C_v\frac{\partial^2 C_v}{\partial n^2}\right] \\ &\quad + (m_L^* - m_N^*)^2\left[\left(\frac{\partial \tilde{C}_h}{\partial n}\right)^2 + \tilde{C}_h\frac{\partial^2\tilde{C}_h}{\partial n^2}\right] \\ &\quad + 2\tilde{C}_h\frac{\partial\tilde{C}_h}{\partial n}(m_L^* - m_N^*)\frac{\partial}{\partial n_j}(m_L^* - m_N^*) \\ &\quad - 2\tilde{C}_h\frac{\partial\tilde{C}_h}{\partial n}(m_L^* - m_N^*)\frac{\partial m_N^*}{\partial n} - \tilde{C}_h^2\frac{\partial m_N^*}{\partial n}\frac{\partial}{\partial n_j}(m_L^* - m_N^*) \\ &\quad - \tilde{C}_h^2(m_L^* - m_N^*)\frac{n^2 m_N^*}{\partial n^2} - \left(C_v^2 - \frac{\Sigma_0}{n}\right)\frac{\vec{k}_i}{E_i}\cdot\frac{\partial\vec{j}}{\partial n_j} \end{aligned} \quad (4.41)$$

with $E_i = \sqrt{\vec{k}_i^2 + m_L^{*2}}$. Note that C_v, C_h, and m_N^* are functions of $\langle\check{n}\rangle = u_0 n - \vec{u}\cdot\vec{j}$ in the mean field approximation.

[8] An important point we will return to later in connection with the origin of proton mass is that the quasiparticle nucleon can be thought of as a bound state of N_c constituent quarks (hereon called "quasiquarks") in the constituent quark model. One can also think of the mesons as quasi-quark and quasi-antiquark bound states. This picture is to hold in medium up to the density $n_{1/2} \sim 2n_0$. However if the topology change involved there is real, then this picture could break down for $n > n_{1/2}$.

Writing in the standard way

$$f_l = (2l+1)\int \frac{d\Omega}{4\pi} P_l\left(\frac{\vec{k}_i\cdot\vec{k}_j}{k_F^2}\right) f_{ij}(|\vec{k}_i| = |\vec{k}_j| = k_F), \tag{4.42}$$

we see that the last term in (4.41) contributes to f_1 and the sum of the rest at the Fermi surface (i.e. $|\vec{k}_j| = k_F$) to f_0. So

$$\begin{aligned}
F_0 &\equiv \frac{\gamma k_F E_F}{2\pi^2} f_0 = \frac{3E_F}{k_F} n f_0 \\
&= \frac{3E_F}{k_F} n[C_v^2 + 4C_v n \frac{\partial C_v}{\partial n} + \frac{m_L^*}{E_F}\frac{\partial m_L^*}{\partial n_j} + n^2\left[\left(\frac{\partial C_v}{\partial n}\right)^2 + C_v \frac{\partial^2 C_v}{\partial n^2}\right] \\
&\quad + (m_L^* - m_N^*)^2\left[\left(\frac{\partial \tilde{C}_h}{\partial n}\right)^2 + \tilde{C}_h \frac{\partial^2 \tilde{C}_h}{\partial n^2}\right] \\
&\quad + 2\tilde{C}_h \frac{\partial \tilde{C}_h}{\partial n}(m_L^* - m_N^*)\frac{\partial}{\partial n_j}(m_L^* - m_N^*) \\
&\quad -2\tilde{C}_h \frac{\partial \tilde{C}_h}{\partial n}(m_L^* - m_N^*)\frac{\partial m_N^*}{\partial n} - \tilde{C}_h^2 \frac{\partial m_N^*}{\partial n}\frac{\partial}{\partial n_j}(m_L^* - m_N^*) \\
&\quad - \tilde{C}_h^2 (m_L^* - m_N^*)\frac{\partial^2 m_N^*}{\partial n^2}
\end{aligned} \tag{4.43}$$

and

$$F_1 \equiv \frac{\gamma k_F E_F}{2\pi^2} f_1 = -\frac{3(C_v^2 - \frac{\Sigma_0}{n})n}{E_F + (C_v^2 - \frac{\Sigma_0}{n})n}. \tag{4.44}$$

We now check some well-known properties of Fermi-liquid theory. First we look at the compression modulus:

$$\begin{aligned}
K &\equiv 9n \frac{\partial^2 \mathcal{E}(\boldsymbol{j}=0)}{\partial n^2} \\
&= \frac{3k_F^2}{E_F} + 9n\left[C_v^2 + 4C_v n \frac{\partial C_v}{\partial n} + \frac{m_L^*}{E_F}\frac{\partial m_L^*}{\partial n} + n^2\left\{\left(\frac{\partial C_v}{\partial n}\right)^2 + C_v \frac{\partial^2 C_v}{\partial n^2}\right\}\right. \\
&\quad + (m_L^* - m_N^*)^2\left\{\left(\frac{\partial \tilde{C}_h}{\partial n}\right)^2 + \tilde{C}_h \frac{\partial^2 \tilde{C}_h}{\partial n^2}\right\} \\
&\quad + 2\tilde{C}_h \frac{\partial \tilde{C}_h}{\partial n}(m_L^* - m_N^*)\frac{\partial}{\partial n}(m_L^* - m_N^*) \\
&\quad - 2\tilde{C}_h \frac{\partial \tilde{C}_h}{\partial n}(m_L^* - m_N^*)\frac{\partial m_N^*}{\partial n} - \tilde{C}_h^2 \frac{\partial m_N^*}{\partial\, n}\frac{\partial}{\partial n}(m_L^* - m_N^*) \\
&\quad \left. - \tilde{C}_h^2 (m_L^* - m_N^*)\frac{\partial^2 m_N^*}{\partial n^2}\right].
\end{aligned} \tag{4.45}$$

Comparing (4.43) and (4.45), we verify that our model satisfies the relativistic Landau Fermi-liquid formula for the compression modulus [Baym and Chin (1976)];

$$K = \frac{3k_F^2}{E_F}(1 + F_0). \tag{4.46}$$

Next we look at the first sound velocity c_1

$$c_1^2 \equiv \frac{\partial p}{\partial \mathcal{E}} = \frac{\partial}{\partial \mathcal{E}}\left(n^2 \frac{\partial \mathcal{E}/n}{\partial n}\right) = \frac{\partial n}{\partial \mathcal{E}}\frac{\partial}{\partial n}(\mu n - \mathcal{E}) = \frac{K}{9\mu}. \tag{4.47}$$

From (4.46), we have that

$$c_1 = v_F\sqrt{\frac{E_F}{3\mu}(1 + F_0)} \tag{4.48}$$

which comes out correct [Baym and Chin (1976)].

Finally Landau effective mass. Baym and Chin have shown [Baym and Chin (1976)] that the relativistic Landau liquid satisfies the mass relation

$$k_F\left(\frac{\partial k_i}{\partial \varepsilon_i}\right)_{k=k_F,\vec{v}=0} = \mu(1 + F_1/3). \tag{4.49}$$

In our model $k_F\left(\frac{\partial k_i}{\partial \varepsilon_i}\right)_{k=k_F,\vec{v}=0} = E_F$. One can see from equations (4.39) and (4.44) that (4.49) is satisfied exactly in our model.

4.4.1.5 *Test of the model*

Given the link between the chiral Lagrangians with IDDs and Fermi-liquid fixed-point theory, one should be able to make a connection between the parameters that enter into such nuclear static properties as (1) the δg_l, referred to in the literature as the "exchange-current" contribution to the orbital gyromagnetic ratio, (2) the effective Gamow–Teller coupling constant g_A^* and the effective mass of the vector mesons ω and ρ. In this subsection, we shall show that this is indeed possible [Friman and Rho (1996); Li, Ma and Rho (2018)]. The results illustrate the subtle interplay, pointed out previously, between the role played by intrinsic scale-chiral symmetry of QCD and the one by Yukawa meson exchange mechanism in EW currents in many-body systems. It is worth pointing out, largely ignored in the nuclear community, which solves long-standing puzzles in nuclear physics. It will be seen that the key element that is intimately related to the Landau parameter F_1 is the universal scaling factor Φ, to be distinguished from what takes place in nuclear interactions in the decimation from the cutoff to the Fermi-liquid scale. To clarify this point, consider the Landau effective mass of the nucleon m_L^* given in terms of the Landau parameter F_1:

$$\frac{m_L^*}{m_N} = 1 + \frac{F_1}{3} = \left(1 - \frac{\tilde{F}_1}{3}\right)^{-1}, \tag{4.50}$$

where $\tilde{F}_1 = (m_N/m_L^*)F_1$ and m_L^* is the Landau quasiparticle mass. Including the pion contribution, we have a short-ranged term and a long-ranged term

$$\tilde{F}_1 = \tilde{F}_1^\omega + \tilde{F}_1^\pi \tag{4.51}$$

where

$$\tilde{F}_1^\omega = \frac{m_N}{m_L^*} F_1^\omega = -C_\omega^2 \frac{2k_F^3}{\pi^2 m_N^\sigma}, \tag{4.52}$$

$$\tilde{F}_1^\pi = -3\frac{m_N}{k_F}\frac{d}{dp}\Sigma_\pi(p)|_{p=k_F}, \tag{4.53}$$

where the superscript denotes the relevant meson exchanged, Σ_π is the nucleon self-energy (Fock term) involving one-pion exchange[9] and

$$m_N^\sigma := m_N \Phi \tag{4.54}$$

the BR-scaled nucleon mass in the absence of pions. It follows from the quasiparticle velocity at the Fermi surface [Friman and Rho (1996)]

$$\frac{d}{dp}\epsilon(p)|_{p=k_F} = \frac{k_F}{m_L^*} = \frac{k_F}{m_N^\sigma} + \frac{d}{dp}\Sigma_\pi(p)|_{p=k_F} \tag{4.55}$$

given by the BR-scaled Lagrangian together with Eqs. (4.50) and (4.51) that the ω contribution to the Landau parameter F_1 is governed only by the factor Φ:

$$\tilde{F}_1^\omega = 3(1 - \Phi^{-1}). \tag{4.56}$$

This is the key relation that links the nucleon scaling present in mean field theories to the scaling of the vector mesons in medium derived via chiral symmetry plus scale anomaly. Understanding of this relation would be crucial if one wanted to have a unified description based on an effective chiral Lagrangian.

[9] As noted above and elaborated further in the volume in discussing Landau Fermi-liquid, the concept of large $N = k_F/\bar{\Lambda}$ where $\bar{\Lambda}$ is the renormalization-group cutoff relative to k_F is invoked. The nonlocal Fock term, a loop correction, is in principle $1/N$ down, but for the reason explained later, it plays an important — and perhaps indispensable — role in the context of the soft-pion theorem as is seen here.

The scaling factor $\Phi(n)$ is known from the QCD sum rule calculation for the in-medium mass of the ρ meson at $\rho = \rho_0$ [Jin and Leinweber (1986)]

$$\Phi(n_0) = 0.78 \pm 0.08. \tag{4.57}$$

Since the contribution from the pion exchange — which represents the most important long-ranged degree of freedom not encoded in the IDDs — is fixed by chiral symmetry for a given density, i.e. at $n = n_0$,

$$\begin{aligned} \frac{1}{3}\tilde{F}_1^\pi &= -\frac{3f_{\pi NN}^2 m_N}{8\pi^2 k_F}\left[\frac{m_\pi^2 + 2k_F^2}{2k_F^2}\ln\frac{m_\pi^2 + 4k_F^2}{m_\pi^2} - 2\right] \\ &= -0.153 \end{aligned} \tag{4.58}$$

with $f_{\pi NN} = g_{\pi NN}\frac{m_\pi}{2m_N}$, the Landau mass for the nucleon is entirely given once we assume the ω mass scales with the dilaton condensate:

$$\begin{aligned} \frac{m_L^*}{m_N} &= \Phi\left(1 + \frac{1}{3}F_1^\pi\right) \\ &= \left(\Phi^{-1} - \frac{1}{3}\tilde{F}_1^\pi\right)^{-1} \\ &= (1/0.78 + 0.153)^{-1} = 0.69 \end{aligned} \tag{4.59}$$

which should be identified with the nucleon effective mass determined by QCD sum rule at $\rho = \rho_0$[Furnstahl, Jin and Leinweber (1996)]

$$\frac{m_N^*}{m_N} = 0.69 \pm {}^{0.06}_{0.14}. \tag{4.60}$$

- **Anomalous orbital gyromagnetic ratio δg_l^p**

One of the strongest supports for the reasoning given above comes from the role that the Φ factor plays in δg_l, the exchange-current correction to the orbital gyromagnetic ratio of nuclei. The response to a slowly-varying electromagnetic field of an odd nucleon with momentum $\vec{p}$ added to a closed Fermi sea can, in Landau theory, be represented by the current

$$\vec{J} = \frac{\vec{p}}{m_N}\left(\frac{1+\tau_3}{2} + \frac{1}{6}\frac{F_1' - F_1}{1 + F_1/3}\tau_3\right) \tag{4.61}$$

where m_N is the nucleon mass in medium-free space. The long-wavelength limit of the current is not unique. The physically relevant one corresponds to the limit $q \to 0, \omega \to 0$ with $q/\omega \to 0$, where (ω, q) is the four-momentum transfer. The current (4.61) defines the gyromagnetic ratio

$$g_l = \frac{1+\tau_3}{2} + \delta g_l \tag{4.62}$$

where

$$\delta g_l = \frac{1}{6}\frac{F_1' - F_1}{1 + F_1/3}\tau_3 = \frac{1}{6}(\tilde{F}_1' - \tilde{F}_1)\tau_3. \tag{4.63}$$

That there is no anomalous isoscalar gyromagnetic ratio is well known. It is the charge conservation that requires it. It is however interesting how that comes about in this model. When the Lagrangian (4.23) is suitably gauged $U(1)_{\rm EM}$, the iso-scalar single-particle orbital current is $\vec{p}/m_N^*$ and it contributes to the iso-scalar gyromagnetic ratio

$$(g_l^{(0)})_1 = \frac{m_N}{m_N^*}. \tag{4.64}$$

This is not the whole story. Otherwise it will be at odds with charge conservation. There is an exchange current contribution to it coming from the ω meson exchange (via ω-tadpole)

$$(g_l^{(0)})_\omega = \frac{1}{2}\left(-\frac{m_N}{m_N^*} + 1\right). \tag{4.65}$$

Summing (4.64) and (4.65), one recovers the correct iso-scalar orbital gyromagnetic ratio $g_l^{(0)} = 1/2$.

Turning to the iso-vector orbital current, the expression (4.63) is recovered simply if one calculates the exchange of an ω and a ρ with BR scaling masses together with a pion exchange. The result obtained in [Friman and Rho (1996)] is

$$\delta g_l = \frac{4}{9}\left[\Phi^{-1} - 1 - \frac{1}{2}\tilde{F}_1^\pi\right]\tau_3. \tag{4.66}$$

This result is highly nontrivial in that (1) the ω vector-meson contribution restores the single-particle moment defined in terms of the free-space mass m_N, not of the BR scaled mass,[10] as required by Ward identities and (2) the correction occurs *only* in the isovector part. With (4.57) and (4.58) in (4.66), the numerical value for δg_l at nuclear matter density comes out to be

$$\delta g_l = 0.227\tau_3 \tag{4.67}$$

which agrees *exactly* with the experimental value obtained from giant dipole resonances in heavy nuclei[Nolte, Baumann and Schumacher (1986)]

$$\delta g_l^p = 0.23 \pm 0.03. \tag{4.68}$$

[10]This is reminiscent of the Kohn theorem for the cyclotron frequency of an electron in the metal in a magnetic field where the free-space mass of the electron, not the Landau mass, enters in the formula for the frequency.

- **Quenching of g_A in nuclei**

Another striking prediction of the model is the quenched g_A in nuclei which is intimately linked to δg_l. There is a long-standing puzzle from 1970's (see for review, [Wilkinson (1978)]) to today (reviewed, e.g., in [Ichimura, Sakai and Wakasa (2006); Engel and Menèndez (2017)] that the effective Gamow–Teller coupling constant g_A in nuclei is[11]

$$g_A^* \approx 1. \tag{4.69}$$

This is what is globally found in Gamow–Teller transitions in simple (small configuration space) shell model [Buck and Perez (1983); Caurier, Poves and Zuker (1995); Radha *et al.* (1997)], giant Gamow–Teller resonances [Sasano *et al.* (2011)], double beta decays [Engel and Menèndez (2017)] etc.

This puzzle can be resolved extremely simply in the model that resolves the δg_l puzzle.

For this purpose, we need the result of how scale invariance is implemented in the description discussed in detail in Chapter 6. There under scale transformation, g_A is found not to couple to the conformal compensator field χ which is related to the scalar field ϕ in the model (4.23).[12] This means that g_A does not scale by the dilaton condensate Φ. This is also given in the Skyrme model [Brown and Rho (1991); Rho (1991)]. Using the large N_c argument near the Fermi-liquid fixed point corresponding to the $\tilde{N} \to \infty$ limit (where $\tilde{N} = k_F/\bar{\Lambda}$ with $\bar{\Lambda}$ being the cutoff relative to the Fermi surface), we get

$$\frac{m_L^*}{m_N} \approx \Phi(n)\sqrt{\frac{g_A^L}{g_A}}. \tag{4.70}$$

It follows from this that

$$\frac{g_A^L}{g_A} \approx \left(1 - \frac{1}{3}\Phi\tilde{F}_1^\pi\right)^{-2}. \tag{4.71}$$

We note that here g_A^L is a Fermi-liquid fixed-point quantity as the Landau mass m_L^* is.

Remarkably, what determines g_A^L are the quantities that give the exchange current correction to δg_l^p. At nuclear matter density, (4.71) gives,

[11]In simple shell model, $g_A^* = 1.0 \pm 0.1$ covers the range involved in light nuclear systems. For the reason that will be explained in Chapter 6, the relation (4.69) is more appropriate.

[12]The scalar ϕ corresponds to the dilaton field σ in Chapter 6. In (4.23) we are using a generic scalar field, not necessarily the dilaton field coming from the specific implementation of scale invariance adopted by [Crewther and Tunstall (2015)].

for $g_A \approx 1.27$,

$$g_A^L \simeq 1.0. \tag{4.72}$$

We can now identify this Fermi-liquid fixed-point quantity with the effective constant (4.69). It is important to note that as defined (4.72) is a Fermi-liquid fixed-point quantity in the theory. It means that it is a quantity for which the beta function is zero, that is, loop corrections are suppressed by $1/\tilde{N} \to 0$. Thus the long-wavelength Gamow–Teller matrix element is given by one-quasi-particle-one-quasi-hole excitation from the ground state. This then gives the effective g_A operative in simple shell-model calculations

$$g_A^* = g_A^L. \tag{4.73}$$

This provides an elegant explanation for the observation of the quenched g_A in nuclei. Although (4.71) was computed for nuclear matter, the density dependence in Φ — decreasing with density — and that in $\tilde{F}_1^\pi$ — increasing with density — compensate each other nearly completely so the product is independent of density within the range of nuclei involved. g_A^L at $n = n_0/2$ differs from that at $n = n_0$ by less than 2%. Therefore the quenching (4.72) applies to both light and heavy nuclei.

To understand the above result, one should recognize that the Fermi-liquid fixed-point quantity subsumes *all* higher-order correlations. Thus if one were to do a microscopic many-body correlation calculations in full configuration space commensurate with the realistic nuclear force, what enters for g_A should be the unquenched $g_A = 1.27$, not the g_A^*. This is because of two important reasons. First there is no intrinsic density dependence due to the fact that g_A is invariant under scale-symmetry transformation as mentioned above and explained in detail in Chapter 6, and secondly the chiral filter mechanism strongly suppresses exchange current contributions to the single-particle Gamow–Teller operator [Kubodera, Delorme and Rho (1978); Rho (1991, 2008)]. A recent *ab initio* quantum Monte Carlo calculations in $A = 6 - 10$ nuclei by Pastore et al [Pastore *et al.* (2018)] gave a strong support to this reasoning: To within a few percent, the Gamow–Teller transitions can be explained with the unquenched g_A provided full correlations are taken into account.

A few additional remarks are in order here.

(1) Within the model adopted in this volume, the effective g_A is universally close to 1 but its precise value cannot have a profound meaning. For instance it cannot be interpreted, as was done sometimes in the literature, as a precursor to the chiral restoration at high density of spontaneously broken chiral symmetry. As it stands it's more of an accident.

(2) In *ab initio* microscopic calculations of the type of [Pastore *et al.* (2018)], high-order correlations should be particularly sensitive to the nuclear tensor force. Now the nuclear tensor force, treated in the framework of HLS, is subject to the IDD-sensitive cancelation between the π tensor and ρ tensor as density increases toward $\sim 2n_0$. This interplay between the two tensor force components, explaining the C14 dating Gamow–Teller matrix element [Holt, Brown, Kuo, Holt and Machleidt (2007)] and playing a crucial role in the EoS of compact stars (see Chapters 8 and 9), will tend to reduce the contribution from the correlations involving energy scales of order $\sim 200 - 300$ MeV strongly coupled to the Gamow–Teller operator. This would most likely leave some quenching unexplained for heavier nuclei with higher density in microscopic calculations restricted to nucleon configuration space. That would necessitate the intervention of non-nucleonic degrees of freedom, in particular, Δ-hole configurations. They can couple to the Gamow–Teller state at the energy scale where the nuclear tensor force is effective, namely, at the scale of $\sim 200 - 300$ MeV. The Δ-hole configurations could figure more importantly for heavier nuclei. Here a tensor-force structure in the Δ-N interactions, involving a cancellation between the π-tensor and ρ-tensor, similar to what's found in the NN interactions should be carefully taken into account.

(3) The link between the Landau parameter that figures in the Fermi-liquid structure of nuclear matter and the IDD scaling that figures in an effective chiral Lagrangian matched to QCD at a matching scale Λ_M supplies a stringent consistency check of the theory. The Landau parameter that represents a fixed-point quasiparticle interaction carries the IDD and hence the two are compounded in physical observables. The two effects are mutually constrained by various symmetry considerations — such as Ward identities — and this requires judicious choice of effective Lagrangians and quantum treatment thereof. This warns against the common practice of mixing in ad hoc, unconstrained, different models and approximations as done quite often in certain nuclear circles.

4.4.2 *Wilsonian renormalization and V_{lowk}*

Let us start with the *bs*HLS Lagrangian matched to QCD at Λ_M, which is at or slightly lower than the chiral scale. We first bring the action down to the scale at which the experimental data are available, say, $E_{lab} \sim 350$ MeV or $\Lambda \sim 2.1$ fm^{-1}. This means that we will be integrating out all meson fields with masses greater than ~ 350 MeV, i.e., $\rho(770)$, $a_1(1260)$, ω (782)

and $\phi(\sim 600)$. The resulting action will consist of multi-baryon fields and pion field. In a simplified form equivalent to the model (4.23), it has the form

$$\mathcal{L} = \bar{\psi}(i\gamma_\mu\partial^\mu - m_N^*)\psi + \frac{C_S^{*2}}{f^2}(\bar{\psi}\psi\bar{\psi}\psi) - \frac{C_V^{*2}}{f^2}(\psi^\dagger\psi\psi^\dagger\psi) + \cdots \tag{4.74}$$

where the IDDs — supplemented by $\mathrm{DD}_{\mathrm{induced}}$ — are indicated by $*$. The ellipsis contains other terms that depend on isospin. In principle the pion field cannot be integrated out for the processes involved in compact stars, but left out here just to compare with the simplified Lagrangian (4.23) considered above for symmetric nuclear matter at the mean-field level. The pion of course can contribute if one goes to the next corrections, namely, the Fock term as was found for δg_l. As discussed in Chapter 2 and extensively in [Rho (2008)], even for certain near zero-energy processes, such as axial-charge in nuclei, the pionic component cannot be ignored. It in fact makes an $\mathcal{O}(1)$ — relative to the leading order — correction enhancing the first-bidden beta decay transition matrix element by a factor of 2. Also for the δg_l treated above, the pionic contribution which comes in as a Fock term is found to be crucial for exchange currents in nuclei. We will discuss how to account for these effects below.

In nuclear matter, treated in the mean field, the Lagrangian (4.74) should give, with suitable choice of the parameters C, the same result as (4.23) in the mean field [Gelmini and Ritzi (1995)]. It would correspond to Landau Fermi-liquid in the sense specified above. How such a Lagrangian could yield Fermi-liquid fixed-point property can be understood in the following way. One can imagine that in the decimation down to the energy scale of nuclear processes, the Fermi sea is formed and the Lagrangian (4.74) transforms to a Fermi-liquid fixed-point theory with an effective action — built on Fermi surface — of the form [Shankar (1994); Polchinski (1992); Chen, Fröhlich and Seifert (1999)]

$$S = \int \bar{\psi}[i\omega - v^*k]\psi \frac{dk d\Omega d\omega}{(2\pi)^4} + \frac{1}{2!2!}\int u\bar{\psi}\bar{\psi}\psi\psi \tag{4.75}$$

where $v^* = k_F/m_L^*$ with m_L^* being the Landau quasiparticle mass. The Landau mass is to be at the fixed point (which can be assured by putting a counter term). Here u is a generic four-Fermi interaction representing all channels of quantum numbers. For symmetric nuclear matter, the leading terms are of the form of (4.74). RG analysis leads to the observation that the quartic interaction is marginal to leading order with the corrections suppressed as $\bar{\Lambda}/k_F \to 0$. This is precisely the Landau (or, more properly in nuclear physics, Landau–Migdal) parameter $F(z)$ defined above.

Now how to go from bsHLS to (4.74) and then to the fixed-point action of the form (4.75) is the main task here. One approach presented in this volume is via $V_{\rm lowk}$ [Holt, Brown, Holt and Kuo (2007)]. This approach has the advantage of being applicable to both finite nuclear systems and infinite dense matter.

We next discuss how to go from $V_{\rm lowk}$ to the fixed-point interactions F. For this procedure, we return to the integrating-out of the high momentum component of a half-on-shell T matrix in the momentum space (corresponding to Eq. (4.7))

$$T(p', p; p^2) = V(p', p) + \frac{2}{\pi}\mathcal{P}\int_0^\infty \frac{V(p', q)T(q, p; p^2)}{p^2 - q^2} q^2 dq. \tag{4.76}$$

Integrating out above the cutoff Λ, define the low-momentum T matrix as

$$T_{\rm lowk}(p', p; p^2) = V_{\rm lowk}(p', p) + \frac{2}{\pi}\mathcal{P}\int_0^\Lambda \frac{V_{\rm lowk}(p', q)T_{\rm lowk}(q, p; p^2)}{p^2 - q^2} q^2 dq. \tag{4.77}$$

Requiring that $T_{lowk}(p', p; p^2) = T(p', p; p^2)$ for $p', p < \Lambda$, one has the T matrix given in terms of the low-momentum interaction $V_{\rm lowk}$. Since by RG invariance, $\frac{d}{d\Lambda}T_{\rm lowk} = 0$, we have the RG equation

$$\frac{d}{d\Lambda}V_{\rm lowk}(p', p; p^2) = \beta([V_{\rm lowk}], \Lambda). \tag{4.78}$$

Now what is V, the "bare" potential, computed from bsHLS Lagrangian? We give an answer here and apply it in Chapters 8 and 9.

In the vacuum (i.e., $n = 0$), two-nucleon (and three-nucleon if needed) potentials are calculated by irreducible diagrams (as in the SχPT). Putting this potential into (4.77), one can then determine $V_{\rm lowk}$ by fitting phase shifts up to $E_{lab} \sim 350$ MeV. That would determine the vacuum parameters of the Lagrangian that figure in $V_{\rm lowk}$ in the vacuum. Now to apply to dense matter, say, the EoS of compact-star matter, one incorporates the IDDs in the Lagrangian, then incorporate higher order irreducible diagrams in the driving term and sum "reducible graphs" (corresponding to solving the Schrödinger equation for bound states or Lippmann–Schwinger equation for scattering). By means of judicious — and involved though transparent — calculations in which only the forward-going terms contribute with non-forward-going terms suppressed — the so-called "$\omega/Q \to 0$

limit" in Green's function approach [Abrikosov, Gorkov and Dzyaloshinsky (1975)],[13] one arrives at the Fermi-liquid fixed-point potential $V_{\rm FL} = V_{\rm lowk} + \delta V$

$$\frac{d}{d\Lambda} V_{\rm FL}(n, \Lambda) = 0 \tag{4.79}$$

which says that the β function for the effective potential $V_{\rm FL}$ is zero, modulo corrections of $O(\overline{\Lambda}/k_F)$. Given such $V_{\rm FL}$, one can write the Landau parameters (i.e., f, f', g, g' etc.) as linear combinations of $V_{\rm FL}$ in appropriate quantum number channels [Holt, Brown, Holt and Kuo (2007)].

What is found remarkable with the $V_{\rm lowk}$ approach in the matter-free vacuum (that is, $n = 0$) is that various different bare potentials, both high-precision phenomenological and high-order effective field theoretic, yield for $\Lambda = 2.1\ {\rm fm}^{-1}$, converge to a universal $V_{\rm lowk}$. This was seen in Fig. 4.1 for a particular channel. It turns out to be so in all channels. This would suggest that varying the cutoff around $2.1\ {\rm fm}^{-1}$ — corresponding to E_{lab} to which accurate scattering data are available — would not affect much the phase shifts and hence $\frac{d}{d\Lambda} V_{\rm lowk} \approx 0$.

What comes as a total surprise is the tensor channel [Rho (2016)].

The bare tensor force in EFT is essentially given by the sum of a pion exchange and a ρ exchange in bsHLS Lagrangian. Calculations are available with the phenomenological BonnS potential which is similar to the EFT potential. In Fig. 4.5 are shown the bare potential V_{bare} and $V_{\rm lowk}$ for the $^3{\rm S}_1 - {}^3{\rm D}_1$ channel. One sees that there is no Λ dependence around $2 \lesssim \Lambda/{\rm fm} \lesssim 3$. Furthermore the result for the bare potential is identical to $V_{\rm lowk}$. This suggests that the bare tensor force is at the fixed point[14]

$$\frac{d}{d\Lambda} V_{\rm lowk}^{\rm tensor} \approx 0. \tag{4.80}$$

It turns out that the tensor force comes out to be a fixed-point quantity not only in free space but also in medium, which means that the β function vanishes at *any* density. The tensor force involved here is not a local operator, so that it cannot be identified with Landau fixed-point parameters that are local. The possible fixed-point structure of the tensor force has an important impact on the structure of exotic nuclei (nuclei far from stability) and also on the symmetry energy of compact-star matter. These matters are treated in great detail in Chapters 8 and 9.

[13] In [Holt, Brown, Holt and Kuo (2007)], the approach relying on induced interactions is used.

[14] We quote here an unpublished result obtained by Tom Kuo for which we are grateful.

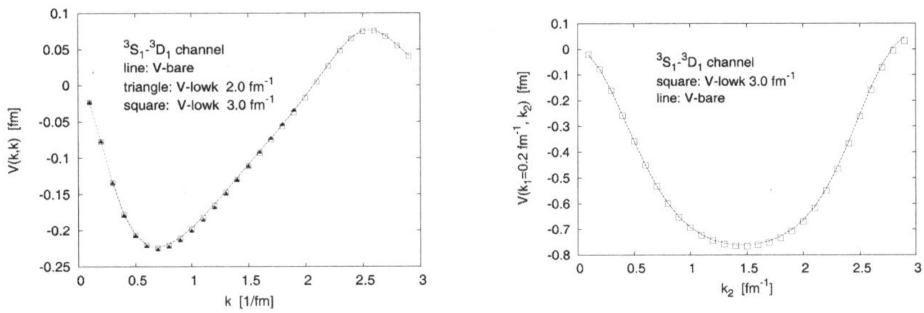

Fig. 4.5 "Bare" BonnS and V_{lowk} tensor forces in matter-free space.

4.5 Energy density functional

To density regime in the vicinity of nuclear matter, n_0, the widely employed method in nuclear studies is RMF approach and improvement thereof. This class of treatments belongs to the class of EDF approach, in close analogy to the density functional theory (DFT). A pertinent question in the point of view taken in this volume is: Is the field theoretic approach related to the DFT? Can the density dependence carried by the "bare" parameters of the effective Lagrangian with condensates of field variables reflecting the "in-medium vacuum" (more properly the ground state) property of QCD, i.e., IDD (including $\mathrm{DD}_{induced}$), be implemented in a DFT for dense matter? The arguments developed in Chapter 8 suggest that the IDD provides an "ab initio" Kohn-Sham potential — linked in a way to QCD — in energy functional approach, i.e., RMF theory popular in nuclear physics.

Here we give a brief description of the possible implication of IDDs (or rather IDD*s in practical calculations) in what might be taken as a Kohn-Sham potential in density functional approach to many-body problem with the *bs*HLS Lagrangian of Chapter 8.

4.5.1 *Generating functional and Hohenberg–Kohn theorem*

The basic idea of EDF is anchored on Hohenberg-Kohn (HK) theorem in atomic and molecular physics transcribed to hadronic and nuclear physics.[15]

The statement of the HK theorem is this: For a Fermi system, with a non-degenerate ground state, it is possible to find a functional or functionals

[15] For pedagogical review, see [Argaman and Makov (1999)]. For application to nuclear physics, see [Baldo, M., P. Schuck and X. Vinas (2008)].

that depend *only explicitly* on density and which is minimized at the ground state energy with the ground state density $n_0(\boldsymbol{r})$. In other words, the ground-state energy can be expressed as a functional of density $n(\boldsymbol{r})$ *only*. Such a functional reaches its variational minimum when evaluated with the *exact ground state density* $n_0(\boldsymbol{r})$.[16] What's given in terms of the effective potential follows it closely.

4.5.1.1 *Effective field theory approach to energy density functional*

In EFT approaches to dense matter, it is natural to phrase the EDF in terms of generating functional employed in field theory. Let us start with the generating functional (in Minkowski space) focusing on baryon density dependence

$$\mathcal{Z}[J] = e^{iW[J]} = \int [\mathcal{D}\psi^\dagger][\mathcal{D}\psi] e^{i\int d^4x(\mathcal{L}+J\psi^\dagger\psi)}. \tag{4.81}$$

Here J is a canonical variable to $\psi^\dagger\psi$, a function of $x(t,\boldsymbol{r})$, replacing the chemical potential μ in thermodynamics.[17] One can of course introduce external potentials of other quantum numbers, but we will first focus on density since the main action that we will treat is density dependence. We will come to other quantum numbers below. The density in the presence of J is

$$n(x) = n(\boldsymbol{r},t) = \langle\psi^\dagger(x)\psi(x)\rangle_J = \frac{\delta W[J]}{\delta J(x)}. \tag{4.82}$$

Taking the Legendre transform, one obtains the *quantum effective action*

$$\Gamma[n] = W[J] - \int d^4x J(x)n(x) \tag{4.83}$$

with

$$J(x) = \frac{\delta\Gamma[n]}{\delta n(x)}. \tag{4.84}$$

Note that $\Gamma[n]$ has no explicit dependence on J. The existence of such a functional is anchored on the fact that $W[J]$ is concave [Valiev and Fernando (1997)] and that guarantees that (4.82) can be inverted to give $J = J[n]$.

[16] How to determine the *exact ground state density?*. This problem is illustrated with air and water in [Argaman and Makov (1999)]. The proof of the HK theorem is given for the Coulomb system but we are not aware of a proof for nuclear dynamics.

[17] To consider excitations one would have to keep time dependence in it. We are focusing however on the ground state property so we will leave out the time variable.

We are interested in ground-state properties, in particular the EDF $E[n(\boldsymbol{r})]$,

$$E[n(\boldsymbol{r})] = -\Gamma[n(\boldsymbol{r})] \equiv -\Gamma[n] \times \left[\int_{\infty}^{\infty} dt\right]^{-1}. \tag{4.85}$$

From (4.83) we have

$$\frac{\delta W[J]}{\delta J(\boldsymbol{x})} = n(\boldsymbol{x}) = \int d^3\boldsymbol{y} \left(\frac{\delta\Gamma[n]}{\delta n(\boldsymbol{y})} + J(\boldsymbol{y})\right)\left(\frac{\delta n(\boldsymbol{y})}{\delta J(\boldsymbol{x})}\right) + n(\boldsymbol{x}) \tag{4.86}$$

or

$$\int d^3\boldsymbol{y} \left(\frac{\delta\Gamma[n]}{\delta n(\boldsymbol{y})} + J(\boldsymbol{y})\right)\left(\frac{\delta n(\boldsymbol{y})}{\delta J(\boldsymbol{x})}\right) = 0. \tag{4.87}$$

Now since $\frac{\delta n(\boldsymbol{x})}{\delta J(\boldsymbol{x})} \neq 0$ as they can be inverted, we get

$$\frac{\delta\Gamma[n]}{\delta n(\boldsymbol{x})} = -J(\boldsymbol{x}). \tag{4.88}$$

This says that in the absence of the source $J = 0$, the action Γ is extremized. In fact one can go further and say that the energy density (4.85) — which is also called the effective potential — is minimized. That such a functional exists such that the energy functional is minimized when evaluated at the exact expectation value of the density n_0 is HK theorem.[18] The corollary is that there is a *universal* term E_0 which is independent of the external potential that is arbitrary $v(\boldsymbol{r})$.[19]

$$E[n] = E[n]_{v=0} - \int d^3\boldsymbol{x} v(\boldsymbol{x}) n(\boldsymbol{x}). \tag{4.89}$$

The mapping between $v(\boldsymbol{x})$ and $n(\boldsymbol{x})$ is fixed after the stationarity equation

$$\frac{\delta E[n]_{v=0}}{\delta n(\boldsymbol{x})} = v(\boldsymbol{x}). \tag{4.90}$$

The $E[n]_{v=0}$ is the HK energy functional, E_{HK}.

---

[18]The point is that density can be used as a fundamental variable in describing quantum many-body problem instead of the wave function. It leads, e.g., to collective modes in RPA[Runge and Gross (1984)]. This reasoning implies that one can do a similar manipulation with currents (both vector, axial vector etc.) rather than charges. More on this later.

[19]The implication that concerns us is that if one takes the external potential to be gravitational, then the EoS of the compact-star matter is exactly given by the HK energy density functional. The gist of the matter is how to calculate the HK functional closest to the EFT à la FT.

It is easy to arrive at this result. To see it, start with

$$-E[n]T = W[J] - \int d^3\boldsymbol{x} J(\boldsymbol{x}) \frac{\delta W[J]}{\delta J(\boldsymbol{x})}. \tag{4.91}$$

By the nature of Legendre transform, one can rewrite

$$\begin{aligned}
-E[n]T &= W[v+J] - \int d^3\boldsymbol{x} J(\boldsymbol{x}) \frac{\delta W[v+J]}{\delta J(\boldsymbol{x})} \\
&= W[v+J] - \int d^3\boldsymbol{x} (J(\boldsymbol{x}) + v(\boldsymbol{x})) \frac{\delta W[v+J]}{\delta J(\boldsymbol{x})} \\
&\quad + \int d^3\boldsymbol{x} J(\boldsymbol{x}) \frac{\delta W[v+J]}{\delta J(\boldsymbol{x})}.
\end{aligned} \tag{4.92}$$

If we invert $n(\boldsymbol{x})T = \frac{\delta W[v+J]}{\delta J(\boldsymbol{x})}$, then $v+J$ is given as a functional of n which is independent of v. So the first two terms of (4.92) make up just an energy functional *independent* of v, that is, F_{HK} from which Eq. (4.89) follows.

The extension of the HK theorem to relativistic systems requires covariant formulation with four-current density. For nuclear EFT, this is discussed in [Speicher, Dreizler and Engel (1992)]. One has to extend the density n to the current density $J_\mu(x)$ where the time component of the current density is the nucleon density $j_0(x) = n(x)$ and to the scalar density $j_s(x) = n_s(x)$:

$$\begin{aligned}
j_\mu(x) &= \bar{\psi}\gamma_\mu\psi, \\
j_s(x) &= \bar{\psi}\psi.
\end{aligned}$$

The HK theorem then goes through by replacing $J\psi^\dagger\psi$ in the generating functional (4.81) by

$$V_\mu\bar{\psi}\gamma^\mu\psi + V_s\bar{\psi}\psi \tag{4.93}$$

and working with the sources V_μ and V_s.

4.5.1.2 *Kohn–Sham equation*

The practical calculation in DFT approach is done in the approximation of Kohn and Sham [Kohn and Sham (1965)]. For this we start with the Hamiltonian

$$H_s = \int d^3x \left[\bar{\psi}(-i\gamma^i\partial_i + M) + j^\mu(x)v_\mu(x) - j_s(x)v_s(x)\right]. \tag{4.94}$$

Here v_μ and v_s are the external potentials that are to be set equal to zero at the end of the day for the ground state. Let the ground state of the system be given by $|\check{0}[j^\mu, j_s]\rangle$. The kinetic energy functional is

$$T_s[j_\mu, j_s] \equiv \int d^3x \langle \check{0}[j_\mu, j_s]|\bar{\psi}(-i\gamma^i\partial_i + M)\psi|\check{0}[j_\mu, j_s]\rangle. \tag{4.95}$$

By the HK theorem we know that this functional — independent of the external potentials — exists. To go to the Kohn–Sham potential, write the EDF

$$E_{v_\mu, v_s}[j_\mu, j_s] = T_s[j_\mu, j_s] + R[j_\mu, j_s] + \int d^3x \left\{ j^\mu(x) v_\mu(x) - j_s(x) v_s(x) \right\}. \tag{4.96}$$

Here R is defined as what's not included in T_s. This may sound like a tautology. It just means the whole works, mean field term, correlations etc., are supposed to be in it. Now impose the stationary conditions

$$\frac{\delta}{\delta j^\mu} E_{v_\mu, v_s}[j_\mu, j_s] = \frac{\delta}{\delta j_s} E_{v_\mu, v_s}[j_\mu, j_s] = 0 \tag{4.97}$$

which gives the relativistic Kohn–Sham equation

$$\Big(- i\boldsymbol{\alpha} \cdot \boldsymbol{\nabla} + \beta[M - V_s^{KS}(x) + \gamma^\mu V_\mu^{KS}(x)] \Big) \phi_i(x) = \epsilon_i \phi_i(x), \tag{4.98}$$

where

$$V_\mu^{KS} = v_\mu(x) + \frac{\delta}{\delta j^\mu(x)} R[j_\mu, j_s], \tag{4.99}$$

$$V_s^{KS} = v_s(x) + \frac{\delta}{\delta j_s(x)} R[j_\mu, j_s]. \tag{4.100}$$

Nuclei and nuclear matter are self-bound systems, so we have to set $v_\mu = v_s = 0$. The densities are given in terms of the Kohn–Sham equation for the orbitals

$$j_s(x) = \frac{1}{2} \left\{ \sum_{i, e_i < e_F} \phi_i^*(x) \gamma^0 \phi_i(x) - \sum_{i, e_i > e_F} \phi_i^*(x) \gamma^0 \phi_i(x) \right\},$$

$$j^\mu(x) = \frac{1}{2} \left\{ \sum_{i, e_i < e_F} \phi_i^*(x) \gamma^0 \gamma^\mu \phi_i(x) - \sum_{i, e_i > e_F} \phi_i^*(x) \gamma^0 \gamma^\mu \phi_i(x) \right\}. \tag{4.101}$$

Thus the burden falls into how to determine $R[j_\mu, j_s]$. Once gotten, it is an iterative process.

4.5.1.3 *Skyrme energy density functional*

For a simplified illustration, we look at the connection between the Skyrme energy density functional and the Kohn–Sham equation [Drut, Furnstahl and Platter (2010)]. The energy density functional is given by[20]

$$E = \int d\boldsymbol{r}\mathcal{E}[n(\boldsymbol{r}), \tau(\boldsymbol{r})] \tag{4.102}$$

where

$$n(\boldsymbol{r}) = \sum_j |\phi_j(\boldsymbol{r})|^2, \tag{4.103}$$

$$\tau(\boldsymbol{r}) = \sum_j |\nabla\phi_j(\boldsymbol{r})|^2. \tag{4.104}$$

The Kohn–Sham equation is obtained by varying Skyrme EDF with respect to ϕ_i that gives the density $n(\boldsymbol{r})$, Eq. (4.103),

$$\left(-\nabla\frac{1}{2M^*(\boldsymbol{r})}\nabla + U(\boldsymbol{r})\right)\phi_j(\boldsymbol{r}) = 0 \tag{4.105}$$

with

$$n(\boldsymbol{r}) = \sum_j |\phi_j(\boldsymbol{r})|^2 \tag{4.106}$$

where U and M^* are in general nonlinear functionals of $n(\boldsymbol{r})$ in consistency with the symmetries involved, containing numerous parameters to be fit phenomenologically depending on the sophistication of the fitting.[21] These make the set of equations to be solved self-consistently by iteration. *Although the wave function enters here, it is the density that plays the principal role via Eq. (4.103).*

[20]The Skyrme energy density is typically of the form

$$\begin{aligned}\mathcal{E}[n, \tau, ...] = &(1/2M)\tau + (3/8)t_0 n^2 + (1/16)t_3 n^{2+\alpha} + (1/16)(3t_1 + 5t_2)n\tau \\ &+ (1/64)(9t_1 - 5t_2)(\nabla n)^2 + \cdots\end{aligned}$$

[21]Just to give an idea how many parameters can be involved, one has typically ~ 10 parameters in the Skyrme energy functional for accessing the symmetry energy $E_{\rm sym}$ that figures importantly in the EoS described in Chapter 9. See e.g., [Nazarewicz, Reinhard, Satula and Vretenar (2014)].

Now the Skyrme EDF is equivalent to a mean field of a given effective Lagrangian. It could be a mean field of a *bs*HLS-type Lagrangian with IDDs inherited from the matching. (The solution obtained in this way is verified to give an exact ground-state energy in atomic and molecular systems with the Coulomb interaction. Whether one gets something similar with *bs*HLS is to be clarified.) What is significant is that a mean-field quantity of EFT Lagrangian plays a role in the Kohn–Sham approach and the IDDs — that enter in the MF — set the stage for the RG program.

4.5.2 *Variety of energy density functionals for applications*

There are a large number of EDFs in application to normal nuclear matter (including finite nuclei) as well as to dense matter including compact-star matter. There are the well-known Skyrme-type density functionals and RMF density functionals etc. Let us consider the latter since our approach can be considered to be in that class. It is not in the spirit of this volume to go into details of what they are and how they fare with nature. However it is meaningful to have a general idea of what they are in connection with what we develop in this volume.

Recall that the underlying premise of RMF models is that written in terms of *all* relevant boson fields together with the baryonic fields, constrained to the symmetries consistent with QCD, faithful, in some sense, to the philosophy of Weinberg's FT, the effective Lagrangian in the mean-field approximation captures Landau Fermi-liquid fixed-point theory discussed above (i.e., Section 4.4). This is how Walecka's linear RMF model was conceived [Matsui (1981)]. A specific illustration was given above in Section 4.4.1.2 with a Lagrangian greatly simplified from the more sophisticated and realistic one developed in the chapters that follow.

There are in the literature hundreds of such models with an equally large number of free parameters, with varying achievements as reviewed extensively in the literature, e.g., [Meng (2016)]. They can be broadly classified in three categories of Lagrangians to be treated in mean-field approximations. This scheme does not render a systematic corrections, the success or failure depending on the astuteness of the parameter choice. The first is a Lagrangian ($\mathcal{L}_I$) with multi-dimension meson field operators with the parameters of the Lagrangian density-independent, the second is a linear meson-field Lagrangian ($\mathcal{L}_{II}$) but with density-dependent parameters and the third a multi-dimension fermion field Lagrangian ($\mathcal{L}_{III}$) with density-independent parameters but containing no meson fields. In simplified notations, they are of the form[22]

$$
\begin{aligned}
\mathcal{L}_I &= \bar{\psi}(i\gamma^\mu\partial_\mu - m_N + g_\sigma\sigma - g_\omega\gamma^\mu\omega_\mu - g_\rho\gamma^\mu\vec{\rho}_\mu\cdot\vec{\tau}/2)\psi \\
&\quad + \frac{1}{2}(\partial^\mu\sigma\partial_\mu\sigma - m_\sigma^2\sigma^2) - \frac{1}{4}\omega_{\mu\nu}^2 + \frac{1}{2}m_\omega^2\omega^2 - \frac{1}{4}\vec{\rho}_{\mu\nu}^2 + \frac{1}{2}m_\rho^2\vec{\rho}^2 \\
&\quad - a\sigma^3 - c\sigma^4 + bg_\omega^2\omega^4 + \sigma(c\omega^2 + d\vec{\rho}^2) + \sigma^2(e\omega^2 + f\vec{\rho}^2) \\
\mathcal{L}_{II} &= \bar{\psi}(i\gamma^\mu\partial_\mu - m_N^* + g_\sigma^*\sigma - g_\omega^*\gamma^\mu\omega_\mu - g_\rho^*\gamma^\mu\vec{\rho}_\mu\cdot\vec{\tau}/2)\psi \\
&\quad + \frac{1}{2}(\partial^\mu\sigma\partial_\mu\sigma - {m_\sigma^*}^2\sigma^2) - \frac{1}{4}\omega_{\mu\nu}^2 + \frac{1}{2}{m_\omega^*}^2\omega^2 - \frac{1}{4}\vec{\rho}_{\mu\nu}^2 + \frac{1}{2}{m_\rho^*}^2\vec{\rho}^2 \\
\mathcal{L}_{III} &= \bar{\psi}(i\gamma^\mu\partial_\mu - m_N)\psi - \alpha_S(\bar{\psi}\psi)^2 - \beta_S(\bar{\psi}\psi)^3 - \gamma_S(\bar{\psi}\psi)^4 \\
&\quad - \alpha_v(\bar{\psi}\gamma_\mu\psi)^2 - \gamma_v(\bar{\psi}\gamma_\mu\psi)^4 - \gamma_{\tau v}(\bar{\psi}\gamma_\mu\vec{\tau}\psi)^2 - \gamma_{\tau v}(\bar{\psi}\gamma_\mu\vec{\tau}\psi)^4
\end{aligned}
$$

where a, b, c, α, β and γ are density-independent constants and the $*$-ed parameters are density-dependent. Here ψ stands for the Dirac field for the nucleons, σ is an effective scalar field which is chiral singlet (later to be associated with a dilaton of scale symmetry), $\omega_{\mu\nu}$ and $\rho_{\mu\nu}$ the appropriate field tensors.

Though generally considered in the literature as different formulations, these Lagrangians, treated in the mean field, are actually related to each other. One can go from one to the others in the framework of mean field approximation. $\mathcal{L}_{II}$ can be gotten by rewriting mean-filed quantities of $\mathcal{L}_I$ and $\mathcal{L}_{III}$ is related to $\mathcal{L}_{II}$ by using equations of motion for the meson fields. The point is that all three Lagrangians in the mean field purport to capture Landau Fermi-liquid theory, albeit approximately, in the sense described in Section 4.4.

As is well-understood, the linear Walecka's mean-field model [Serot and Walecka (1986)] — $\mathcal{L}_I$ with the multi-dimension terms omitted — is significantly improved upon by the nonlinear terms. The Lagrangian $\mathcal{L}_{III}$ in the mean field with the fermionic terms of dimension higher than quartic ignored is equivalent to Walelck's linear model [Gelmini and Ritzi (1995)]. The sextet terms play the role of higher-dimension meson-field terms in $\mathcal{L}_I$. This means that it is $\mathcal{L}_{II}$ with density-dependent "bare" parameters that captures the other classes of mean-field Lagrangians.

It should be possible to arrive at a Lagrangian of $\mathcal{L}_{II}$-type from chiral perturbative theory with the density dependence systematically controlled [Rho (1991); Holt, Rho and Weise (2016)].* The approach presented in this volume with scale-invariant HLS theory *bs*HLS has the potential to

[22]There are too numerous references on this matter to cite in a fair way. One can find all the relevant references in [Dutra *et al.* (2013)].

fix the density-dependent "bare" parameters of the Lagrangian in terms of IDDs from QCD supplemented with $\mathrm{DD_{induced}}$ with which one can do systematic renormalization-group calculations. It should be stressed that the "effective" IDDs will consist of the Lorentz-invariant one inherited from QCD plus the induced non-Lorentz-invariant one — such as from integrating out the three-body forces present in chiral perturbation theory.

The simplified model (4.23) analyzed above presages this development that follows in the volume.

4.5.3 *Energy density functional from bsHLS theory*

Let us now see how the EDF $E[n]_{v=0} \equiv E[n]_{HK}$ can be calculated from the *bs*HLS theory with the bare parameters fixed from QCD developed in Chapters 5, 6 and 8. A quantitative application to massive compact stars will be made in Chapter 9.

The steps involved in implementing the strategy of density functional with the *bs*HLS Lagrangian, equivalent to the Kohn–Sham scheme, are as follows.

(1) With the *bs*HLS Lagrangian obtained in Chapter 8 with the IDDs and $\mathrm{DD_{induced}}$ incorporated, one derives the equations of motion for each of the degrees of freedom involved in the mean field limit. They will be the mean fields of the boson fields $\Phi = (V_\mu, \chi)$ and fermion bilinears $\bar{\psi}\Gamma\psi$ etc.
(2) The nucleons satisfy a Dirac equation in the presence of mean field potentials having Lorentz scalar and vector character generated from the classical meson fields. The meson fields satisfy Klein–Gordon equations with the relevant baryon and other densities acting as sources.
(3) The solution of the Dirac equation for the appropriate orbitals gives single-particle energies and wave functions which are then used to construct new one-body densities. This procedure is iterated until self-consistency is achieved.

This mean field procedure should then be equivalent to Fermi-liquid fixed-point theory (à la Matsui). In Chapter 8, the Fermi-liquid theory will be formulated in terms of RG decimation using V_{lowk}.

Now the key point of this section: Given that the IDD — supplemented with possible induced corrections ($\mathrm{DD_{induced}}$) — which is essentially the BR scaling plays a principal role via density dependence *inherited from QCD* in the bare parameters of the EFT Lagrangian and Landau Fermi-liquid fixed-point approach provides the EDF with in-medium vacuum structure of QCD, *the HK theorem states that the EDF so obtained should correspond to QCD by the FT.* The V_{lowk} approach developed in Chapter 8 translated into this EDF form should allow us to discuss in terms of a Kohn–Sham potential.

4.5.3.1 *BR scaling in energy density functional*

It should be feasible to show that the Dirac equation in the mean field potential given by the VeV — or more precisely ground-state expectation value — of the dilaton and vector fields with the parameters of the simplified *bs*HLS Lagrangian carrying BR scaling corresponds to a Kohn–Sham-type equation. In nonrelativistic form, the mean field containing BR scaling can be identified with the potential U in (4.105). In fact the V_{lowk} with the double-decimation RG approach with much fewer parameters, presented below (in Chapter 8), goes far beyond what's done in the Skyrme–Hatree–Fock approach and RMF approach [Nazarewicz, Reinhard, Satula and Vretenar (2014)].

4.5.3.2 *Skyrme potential and V_{lowk}RG*

It has been shown that by fine-tuning the parameters of a given Skyrme potential[23] one should be able to capture the BR-scaling-implemented V_{lowk} RG calculation; the two approaches should give more or less the same results [Kuo, Holt and Osnes (2016)]. The essential reasoning is that as long as there are no (Ginzburg–Landau–Wilsonian) phase transitions, which is a good assumption up to the vicinity of nuclear matter density, a Skyrme potential can *always* be concocted to agree with the V_{lowk} result.

Following [Kuo, Holt and Osnes (2016)], write the Skyrme potential as

$$V_{skyrme} = V_{sk}(\boldsymbol{r}_1 - \boldsymbol{r}_2) + t_3\delta(\boldsymbol{r}_1 - \boldsymbol{r}_2)\delta(\boldsymbol{r}_2 - \boldsymbol{r}_3), \qquad (4.107)$$

where V_{sk} is a zero-range two-body interaction. The second term is a zero-range three-body interaction, reduced to a density-dependent two-body force by integrating out one participating nucleon over the Fermi

[23]One word for nuclear theorists familiar with "Skyrme potentials" but not with "skyrmions". They are not the same although, as will become clear in this volume, the former could be derived from the latter. The latter is Skyrme's discovery for baryons derived from meson fields via topology and unifies mesons and baryons in nuclear interactions. This approach, discussed in Chapter 7, plays the central role in leading to the EoS relevant to compact star physics nearly independent of arbitrary parameters.

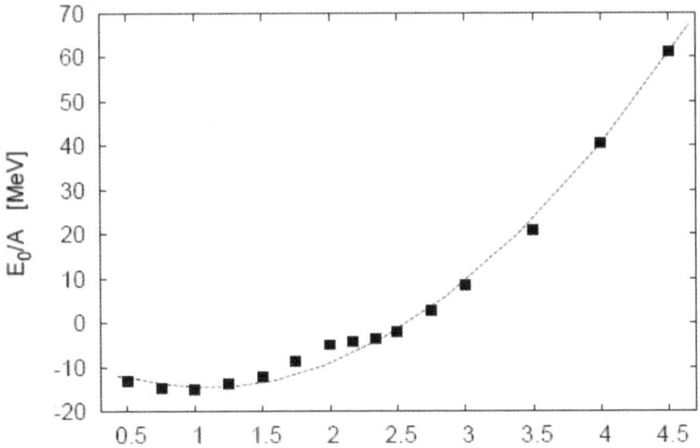

Fig. 4.6 Comparison of ring-diagram EoS for symmetric nuclear matter calculated with BR-scaled- V_{NN} (dark square) and $V_{NN} + D_{sk}$ (dotted line) where V_{NN} is the potential without BR scaling and the density dependence is brought in by D term. The latter corresponds to the Skyrme EDF.

sea, denoted D_{sk}

$$\begin{aligned} V_{skyrme} &= V_{sk}(\boldsymbol{r}_1 - \boldsymbol{r}_2) + D_{sk}(\boldsymbol{r}_1 - \boldsymbol{r}_2), \\ D_{sk} &= \frac{1}{6}(1 + x_3 P_\sigma) t_3 \delta(\boldsymbol{r}_1 - \boldsymbol{r}_2) n(\vec{r}_{av}), \end{aligned} \tag{4.108}$$

where $\vec{r}_{av} \equiv (\boldsymbol{r}_1 + \boldsymbol{r}_2)/2$ and both t_3 and x_3 are strength parameters. In calculations with Skyrme interactions, the inclusion of D_{sk} is essential for nuclear matter saturation.

In Fig. 4.6, both approaches are compared with the V_{sk} and V_{lowk} used in ring-diagram calculation (i.e., the "second decimation"). They are essentially equivalent, modulo some fine-tuning in the Skyrme potential, in the whole range of densities involved in compact stars. This calculation illustrates how the effect of three-body potentials is captured in the BR scaling through the $\mathrm{DD_{induced}}$ effect.

It is interesting — and perhaps very significant — that the Skyrme potential calculation agrees with the V_{lowk}RG that has a topology change, as will be shown in Chapter 7, at $n_{1/2} \approx 2n_0$ with the soft-to-hard changeover in the EoS. This means that the topology change in the (topological) skyrmion description may be subsumed into the parameters of the nonlinear Skyrme potential, implying that the topology change does not belong to the standard Ginzburg–Landau–Wilson paradigm of phase transitions. This issue will re-emerge later in the volume.

Chapter 5

Hidden Symmetries in Dense Matter

5.1 Symmetries hidden in QCD

There are certain symmetries that are not visible, or stated more precisely, hidden in QCD in the matter-free space. Among them, two will play a crucial role in the approach to dense baryonic matter developed in this volume. They are scale invariance[1] and local flavor symmetry. They will be referred to in this volume as "hidden scale symmetry" and "hidden local symmetry" (HLS for short). The former is visible in the QCD Lagrangian at the classical level in the chiral limit, i.e., "QCDLite," but is not present at the quantum level due to the trace anomaly. Therefore there is no apparent scale symmetry in the strong interactions. Similarly while gauge-invariant in color, QCD has no apparent local symmetry in flavors. There are no massless vector bosons in the vacuum, hence the low-lying vector mesons are not gauge bosons per se. In this chapter, we will invoke and exploit the notion that both can, however, "emerge" and become visible under certain conditions in nuclear dynamics. This notion is totally novel in nuclear physics and makes nuclear matter at high density extremely interesting to explore for new physics associated with these symmetries. The argument will be developed in this volume that both hidden symmetries could be made, albeit partially, "emergent" or "unhidden" at high density, particularly in compact-star matter. Even if they were not explicitly made visible, the notion of hidden scale symmetry and HLS would facilitate, within a fairly well-defined approximation scheme, to access high-density regime in baryonic matter. In this chapter, we will discuss, in as simple a

[1]Scale symmetry is a part of more general conformal symmetry. In this volume, we will be specializing only in the former, *i.e.*, "dilatation" symmetry.

way as possible, the conceptual aspect for specific and detailed applications to be made in Chapter 9.

5.2 Hidden scale symmetry

Scale symmetry is an extremely subtle and, presently, highly controversial issue in all areas of physics. It touches on cosmology, physics beyond the Standard Model, strongly correlated condense matter and other areas. In this volume, it will be seen to have an equally important impact on nuclear structure and specially on the core of massive compact stars. This aspect is not shared by most of the currently actively pursued works in the field. This matter was alluded to briefly in Chapter 2 and will be addressed more in depth in Chapter 6 to which we relegate for precise definitions. Here we will simply quote some of the points elaborated in Chapter 6. What we are concerned here is scale symmetry, a part of a larger symmetry, i.e., conformal invariance, in gauge theory, in particular QCD. Currently there are at least two issues involved with scale symmetry in gauge theories. One is for beyond the Standard Model with the structure of Higgs boson for which a large number of flavors, $N_f \sim 8$, and a conformal window are involved. The other is QCD for low $N_f \sim 3$. In the latter at issue is the scalar listed in the Particle Data Booklet, $f_0(500)$, commonly referred to as "σ meson."

A scalar meson of such a low mass, extremely elusive in QCD, remains highly controversial [Pelaez (2015)]. For nuclear physics, a scalar degree of freedom of mass $\sim 500 - 700$ MeV is badly, if not indispensably, needed. There are a large number of ideas for such a scalar in the literature, but none has emerged as an established candidate. The option taken in this volume is that it is a dilaton, a pseudo-NG boson of spontaneously broken scale symmetry. As explained in detail in Chapter 6, whether or not an infrared (IR) fixed point associated with scale invariance is present in QCD for N_f as low as 3 is far from settled. Our strategy, in the given unsettled situation, will be an intuitively appealing option that is based on the possibility that independently of whether such an IR fixed point is present in QCD in the matter-free vacuum, the symmetry can emerge in dense matter under extreme conditions offering predictable phenomena.

First we note the subtlety in the manifestation of scale symmetry in gauge theories. The spontaneous breaking of scale symmetry is quite unlike the spontaneous breaking of chiral symmetry. There are broadly two ways of spontaneous breaking of scale symmetry. The first is to start with a scale invariant theory, containing a dilaton with a shift-invariant potential, coupled to matter fields. The shift-symmetry is broken by Yukawa coupling to the matter fields. The vacuum expectation value (VeV) of the dilaton field gives masses to the matter field but the energy-momentum tensor remains traceless. That leads to a scale invariant theory with massless dilaton but massive matter fields.

The alternative is to start with massive matter fields and include a dilaton with a shift-invariant potential, but with Yukawa couplings to matter that break the shift symmetry. The energy-momentum tensor is not traceless, so the scale invariance will be broken. However one can find a combination of the scale current and the dilaton shift current that is conserved. This means that there is a hidden symmetry.

In both, there is a caveat we are ignoring, namely, that there is a quantum anomaly that breaks the scale symmetry *explicitly* and makes the situation much subtler. This is because as will be seen in Chapter 6, scale symmetry, quite unlike global symmetry, can be spontaneously broken *only if* it is explicitly broken.[2] Let us ignore here that caveat — which does not invalidate what's argued below — and return to it in the next chapter.

The second way described above is closer to our strategy. For this we argue the well-known linear sigma model can be transformed to nonlinear sigma model with no scalar mesons *and* to scale-invariant non-linear sigma model — with a scalar — by just dialing one parameter. This matter was briefly alluded to in Chapter 2 in transforming the linear sigma model to a non-linear realization of chiral symmetry. Here we follow the reasoning given in [Fukano *et al.* (2015); Yamawaki (2015, 2018)]. A more formal reasoning applicable generally to conformal symmetry can be found in [Di Vecchia, Marotta and Mojaza (2017)]. This discussion illustrates simply that the scale symmetry can be hidden in one sector but can be exposed by dialing a parameter in another sector, and vice versa. The parameter to dial in baryonic matter, it is suggested, is the baryon density. Therefore the baryonic matter with varying densities is a probe for various different symmetry regimes. How to systematically bring scale symmetry into nuclear dynamics will be relegated to Chapter 6.

---

[2]There is a loop-hole to this conundrum and that is to have scale-invariance at the quantum level [Shaposhnikov and Zenhausern (2009)]. This, however, does not apply to QCD.

5.2.1 *Linear sigma model*

5.2.1.1 *Flavor $SU(2)$*

In Chapter 2, we showed, starting from linear sigma model in the chiral limit, how one arrives, by field redefinitions, at the form

$$\mathcal{L}_{\mathrm{L}\sigma\mathrm{M}} = \frac{1}{4}\partial_\mu\sigma\partial^\mu\sigma + \frac{1}{2}\sigma^2\mathrm{Tr}(\partial_\mu U(x)\partial^\mu U^\dagger(x)) - V_0\left(\sigma^2\right) \tag{5.1}$$

where $V_0(\sigma)$ is a potential that governs chiral symmetry. If the symmetry is spontaneously broken with $\sigma \to f_\pi$, then we wind up with the non-linear sigma model

$$\mathcal{L}_{\mathrm{L}\sigma\mathrm{M}}|_{\sigma\to f_\pi} = \mathcal{L}_{\mathrm{NL}\sigma} = \frac{f_\pi^2}{4}\mathrm{Tr}\big(\partial_\mu U(x)\partial^\mu U^\dagger(x)\big). \tag{5.2}$$

This manipulation, however, hides a scale symmetry.

In order to expose the hidden scale symmetry, we return to the $SU(2)_L\times SU(2)_R$ linear sigma model with two parameters μ and λ,

$$\mathcal{L}_{L\sigma M} = \frac{1}{2}\left[(\partial_\mu\hat{\sigma})^2 + (\partial_\mu\hat{\pi}_a)^2\right] - \frac{\mu^2}{2}(\hat{\sigma}^2+\hat{\pi}_a^2) - \frac{\lambda}{4}(\hat{\sigma}^2+\hat{\pi}_a^2)^2\,. \tag{5.3}$$

With a 2×2 matrix defined as

$$M = \frac{1}{\sqrt{2}}\left(\hat{\sigma}\cdot 1_{2\times 2} + 2i\hat{\pi}\right) \quad \left(\hat{\pi}\equiv\hat{\pi}_a\frac{\tau_a}{2}\right) \tag{5.4}$$

that transforms under $G = SU(2)_L\times SU(2)_R$ as

$$M \to g_L\, M\, g_R^\dagger\,, \quad g_{R,L}\in SU(2)_{R,L} \tag{5.5}$$

we can rewrite the Lagrangian as

$$\mathcal{L}_{L\sigma M} = \frac{1}{2}\mathrm{Tr}\left(\partial_\mu M\partial^\mu M^\dagger\right) - \frac{\mu^2}{2}\mathrm{Tr}\left(MM^\dagger\right) - \frac{\lambda}{4}\left(\mathrm{Tr}\left(MM^\dagger\right)\right)^2\,. \tag{5.6}$$

Under the scale transformation of an operator $\mathcal{O}(x)$ with the scale dimension $d_{\mathcal{O}}$, $\delta\mathcal{O}(x) = (d_{\mathcal{O}} + x^\nu\partial_\nu)\mathcal{O}(x)$, the action $S = \int d^4x\mathcal{L}(x)$ for a given Lagrangian $\mathcal{L}(x)$ becomes

$$\delta S = \int d^4x(d_{\mathcal{L}}+x^\nu\partial_\nu)\mathcal{L} = \int d^4x[(d_{\mathcal{L}}-4)\mathcal{L}+\partial_\nu(x^\nu\mathcal{L})] = \int d^4x(d_{\mathcal{L}}-4)\mathcal{L}\,. \tag{5.7}$$

Thus the action will be scale-invariant only if $d_{\mathcal{O}} = 4$ in the Lagrangian $\mathcal{L} = \sum_i\mathcal{O}_i$. This means that the term $\propto\mu^2$ in (5.6), i.e., the mass term, with scale dimension 2 is not scale-invariant while the other terms are. Now any complex matrix, such as M, can be decomposed into a hermitian matrix H — which is diagonalizable — and a unitary matrix U as $M = HU$. In the chiral symmetry spontaneously broken phase, i.e., Nambu–Goldstone (NG) phase, write

$$M = H \cdot U\,, \quad H = \frac{1}{\sqrt{2}}\begin{pmatrix} \sigma & 0 \\ 0 & \sigma \end{pmatrix}, \quad U = \exp\left(\frac{2i\pi}{f}\right) \tag{5.8}$$

which under the chiral transformation, transform as

$$U \to g_L\, U\, g_R^\dagger\,, \quad H \to H\,. \tag{5.9}$$

In the chiral symmetry spontaneously broken phase, i.e., NG phase, write with $\langle\sigma\rangle \equiv f$. It will be seen that the constant f can be related to the pion decay constant f_π under certain condition. This identification will be important in what follows. Note that H is a chiral singlet and $UU^\dagger = 1$, meaning $\langle U\rangle = \langle\exp\left(\frac{2i\pi}{f}\right)\rangle = 1 \neq 0$, so the spontaneous symmetry breaking is properly taken into account. The Lagrangian then takes the form

$$\begin{aligned} \mathcal{L}_{L\sigma} &= \frac{1}{2}\left(\partial_\mu\sigma\right)^2 + \frac{1}{4}\sigma^2 \cdot \mathrm{Tr}\left(\partial_\mu U \partial^\mu U^\dagger\right) - V(M) \\ V(M) &= \frac{\mu^2}{2}\sigma^2 + \frac{\lambda}{4}\sigma^4\,. \end{aligned} \tag{5.10}$$

Minimizing the potential (5.10) at $\sigma = f$, we rewrite it as

$$V(M) = \frac{\lambda}{4}\left[\left(\sigma^2 - f^2\right)^2 - f^4\right]. \tag{5.11}$$

We therefore have at our disposal one coupling constant λ that can be dialed to get at different limits.

Let us consider two extreme limits: strong coupling limit and weak coupling limit.

(1) In the strong coupling limit, $\lambda \to \infty$, $\langle\sigma\rangle \to f = f_\pi$, so one simply gets the familiar non-linear sigma model

$$\mathcal{L}_{L\sigma M} \stackrel{\lambda\to\infty}{\longrightarrow} \mathcal{L}_{NL\sigma} = \frac{f_\pi^2}{4} \cdot \mathrm{Tr}\left(\partial_\mu U \partial^\mu U^\dagger\right). \tag{5.12}$$

Note that the breaking of the scale invariance gets shoved into the kinetic term for the pion, $f_\pi^2 \cdot \mathrm{Tr}\left(\partial_\mu U \partial^\mu U^\dagger\right)$, which being of scale dimension 2 is no longer scale invariant. The constant f is identified with the pion decay constant, $f = f_\pi$.

(2) Now we turn to the weak coupling limit $\lambda \to 0$. Define the scale-dimension-1 and mass-dimension-1 field χ, the conformal compensator

field,[3]

$$\chi = f_\phi e^{\phi/f_\phi} \,. \tag{5.13}$$

Under scale transformation, χ transforms linearly while ϕ transforms nonlinearly

$$\delta\chi = (1 + x^\mu \partial_\mu)\chi \,, \qquad \delta\phi = f_\phi + x^\mu \partial_\mu \phi \,. \tag{5.14}$$

Here f_ϕ is the decay constant for the scalar ϕ. Expressed in terms of the field χ, the Lagrangian (5.10) can be written as

$$\mathcal{L}_{L\sigma M} = \mathcal{L}_{\rm sinv} - V(\chi) \tag{5.15}$$

with

$$\mathcal{L}_{\rm sinv} = \frac{1}{2}(\partial_\mu \chi)^2 + \frac{f_\pi^2}{4}\left(\frac{\chi}{f_\phi}\right)^2 \cdot {\rm Tr}\left(\partial_\mu U \partial^\mu U^\dagger\right) \,, \tag{5.16}$$

$$V(\chi) = \frac{\lambda}{4} f_\phi^4 \left[\left(\left(\frac{\chi}{f_\phi}\right)^2 - 1\right)^2 - 1\right] \,, \tag{5.17}$$

with $\frac{\partial}{\partial\chi} V(\chi)|_{\langle\chi\rangle = f_\phi} = 0$.. The first term of (5.15) is scale-invariant with scale breaking lodged entirely in the potential (5.17). It is important to note that scale invariance is obtained in the limit $\lambda \to 0$ from a linear sigma model.

Under the scale transformation, the potential transforms

$$\delta V(\chi) = +\lambda f_\phi^4 \left(\frac{\chi}{f_\phi}\right)^2 + \text{total derivative} \,, \tag{5.18}$$

which yields

$$\partial^\mu D_\mu = \theta_\mu^\mu = -\delta V(\chi) = -\lambda f_\phi^4 \left(\frac{\chi}{f_\phi}\right)^2 \,, \tag{5.19}$$

where D_μ is the dilatation current and θ_μ^μ is the trace of energy-momentum tensor. Then this leads to the partially conserved dilatation current (PCDC) analogous to PCAC

$$m_\phi^2 f_\phi^2 = -\langle 0 | f_\phi \partial^\mu D_\mu | \phi \rangle = -d_\theta \langle \theta_\mu^\mu \rangle = 2\lambda f_\phi^4 \langle \left(\frac{\chi}{f_\phi}\right)^2 \rangle = 2\lambda f_\phi^4 \,, \tag{5.20}$$

[3]A warning on the notation. Here ϕ stands for the nonlinearly transforming dilaton field which is denoted as σ in Chapter 6 and elsewhere. In the discussion given here, $\hat\sigma$ appears as the scalar of linear sigma model, namely, the fourth component of the chiral four vector, which gets "transmuted" into the chiral singlet "σ" by the limiting process. At the end of the day, the ϕ in the conformal compensator can be identified with the dilaton, chiral singlet "σ". In what follows, this gymnastic of notation will be eschewed.

where θ_μ^μ has a scale dimension $d_\theta = 2$ as one can see from Eq. (5.19)). The potential (5.17) is the first term in a more general potential that is anchored on the trace anomaly reflecting on the dimensional transmutation, an intrinsic property of QCD,

$$V(\phi)\Big|_{\rm anomaly} = \frac{m_\phi^2 f_\phi^2}{4}\left(\frac{\chi}{f_\phi}\right)^4\left(\ln\frac{\chi}{f_\phi} - \frac{1}{4}\right) \tag{5.21}$$

which yields $\langle\delta V\rangle = -\langle\theta_\mu^\mu\rangle = m_\phi^2 f_\phi^2 \frac{\langle\chi^4\rangle}{f_\phi^4}/4 = m_\phi^2 f_\phi^2/4$ and has a minimum at $\langle\chi\rangle = f_\phi$.

A key point we will develop in what follows, which is arguably a drastic proposition, is that density in baryonic matter does the dialing $\lambda = \infty$ to $\lambda = 0$ in going from low density to high density, approaching what's called "dilaton limit fixed point (DLFP)" which figures prominently in our discussions.

5.2.1.2 *Flavor $SU(3)$*

The argument given above that scale symmetry is hidden in linear sigma model is specifically for two flavors. It is simple and in some sense unique. The situation is different for three flavors. It is a lot more complicated. This is because in general for the flavor $SU(3)$, there can be nonet of scalars, unlike the sole singlet scalar for the flavor $SU(2)$. Depending on how the scalars manifest themselves, there can be many options of how the linear sigma model can be transformed, driven by dialing parameters of the model. In fact this is one of the issues addressed in hadron physics, say, concerning the spectra of scalars at high temperature and/or high density [Lenaghan, Rischke and Schaffner-Bielich (2000)]. In our scheme, we are interested in an emerging scale symmetry with a dilaton and hence we are picking, among others, the particular scenario that follows closely the $SU(2)$ scenario with only the singlet scalar intervening and with the octet scalars integrated out. This is essentially what we are extracting from the Crewther–Tunstall (CT) model.

5.2.2 *Nambu–Jona-Lasinio model*

One can uncover the same hidden scale symmetry from the celebrated Nambu–Jona-Lasinio (NJL for short) model [Yamawaki (2015)]. This exemplifies the generic nature of both chiral symmetry and scale symmetry hidden in QCD.

As is well known, the NJL model is equivalent to linear sigma model. Start with the NJL Lagrangian with the fermion (quark) in the fundamental representation in the $SU(N_f)$ flavor and $SU(N_c)$ color,

$$\mathcal{L}_{NJL} = \bar{\psi} i\gamma^\mu \partial_\mu \psi + \frac{G}{2}[(\bar{\psi}\psi)^2 + (\bar{\psi} i\gamma_5 \tau^a \psi)^2]. \tag{5.22}$$

A quantum-loop analysis is feasible in the large N_c limit. In the limit $N_c \to \infty$ and $N_c G \neq 0$ and constant, the quarks are found to condense $\langle\bar{\psi}\psi\rangle \neq 0$ for $G > G_{cr}$, with $G_{cr} = \frac{4\pi^2}{N_c \Lambda^2}$ the critical strength. Then (5.22) takes the form

$$\mathcal{L}_{NJL} - \bar{\psi} i\gamma^\mu \partial_\mu \psi = G\langle\bar{\psi}\psi\rangle\bar{\psi}\psi + \cdots = -m_F \bar{\psi}\psi + \cdots \tag{5.23}$$

with the gap equation for the dynamical quark mass m_F

$$m_F = -G\langle\bar{\psi}\psi\rangle = 4GN_c \int \frac{d^4p}{i(2\pi)^4} \frac{m_F}{m_F^2 - p^2} \tag{5.24}$$

which has a solution for $m_F \neq 0$ provided

$$\frac{1}{G} - \frac{1}{G_{cr}} = -\frac{1}{4\pi^2} N_c m_F^2 \ln\left(\frac{\Lambda^2}{m_F^2}\right). \tag{5.25}$$

Next we rewrite the NJL Lagrangian in terms of the boson fields σ and π by the Hubbard–Stratonovich transformation

$$\mathcal{L}_{NJL} = \bar{\psi}(i\gamma^\mu \partial_\mu + \sigma + i\gamma_5 \tau^a \pi^a)\psi - \frac{1}{2G}(\sigma^2 + \pi_a^2). \tag{5.26}$$

Upon integrating out the fermion field, to one-loop order, this Lagrangian yields for the σ and π sector the sigma model Lagrangian (5.3) with [Yamawaki (2015)]

$$\begin{aligned} \mu^2 &= \left(\frac{1}{G} - \frac{1}{G_{cr}}\right) Z^{-1}, \\ \lambda &= \frac{1}{Z} \end{aligned} \tag{5.27}$$

with

$$Z = \frac{N_c}{8\pi^2} \ln \frac{\Lambda^2}{m_F^2}. \tag{5.28}$$

Thus one recovers the same limits found with the linear model with $\lambda \to \infty$ to arrive at the nonlinear sigma model and $\lambda \to 0$ for conformal symmetric model. In this approach, both the scalar and pseudoscalars are bound states of constituent quarks.

5.2.3 Dilaton-limit fixed point: Emerging scale symmetry

We briefly discuss here how one can dial the coupling constant λ in Eq. (5.3) in dense medium. For this we take bsHLS constructed precisely in Chapter 8. It has been shown in terms of Wilsonian renormalization group (RG) equation [Paeng, Lee, Rho and Sasaki (2011, 2013)] that there is a possible IR fixed point in bsHLS[4]

$$(m_N,\, g_{V\rho} - 1,\, g_A - g_{V\rho}) \to (0,\, 0,\, 0) \tag{5.29}$$

for the nucleon mass m_N, the ρNN coupling $g_{\rho NN} = (g_{V\rho} - 1)\, g_\rho$ — with g_ρ the hidden gauge coupling — and g_A for the axial-vector current. This is referred to as dilaton-limit fixed point (DLFP). The parameters of the Lagrangian satisfy a complicated set of coupled renormalization-group (RG) equations and can flow in various different directions as in the case of HLS analyzed in [Harada and Yamawaki (2003)].

We consider how this fixed point can be reached in dense medium. As described in detail in [Lee, Paeng and Rho (2015); Paeng, Kuo, Lee and Rho (2015)], when embedded in nuclear medium with the vacuum modified by density, the condensates that figure in the "bare" parameters of bsHLS, i.e., dilaton condensate, quark condensate etc., pick up density dependence, i.e., intrinsic density dependence (IDD) inherited from QCD. Denoting the IDD by $*$, the "bare masses" in the Lagrangian will be controlled by three parameters $\langle\chi\rangle^*$, g_ρ^* and g_ω^*, where g_V's are the HLS gauge couplings for $V = \rho,\, \omega$. At work here is a hitherto unnoticed relation, between the scalings of $\langle\chi\rangle^*$ and g^*.

The resulting Lagrangian has a Walecka-type form with the scaling masses at the leading order. A crucial point to note here is that there are singular terms when $\langle\chi\rangle^*$ goes to zero

$$\mathcal{L}' = \frac{f_\sigma^2 - f_\pi^2}{\langle\chi\rangle^{*2}} \mathcal{A}(\bar\sigma,\, \bar\pi) + \frac{g_A - g_{V\rho}}{\langle\chi\rangle^{*2}} \mathcal{B}(\bar\sigma,\, \bar\pi,\, N) \tag{5.30}$$

arising from the non-linear terms of N, $\bar\sigma \equiv \langle\chi\rangle^*\sigma/f_\sigma$ and $\bar\pi \equiv \langle\chi\rangle^*\pi/f_\sigma$. Taking the limit $\langle\chi\rangle^* \to 0$ corresponds to the DLFP, where $f_\pi = f_\sigma$[5] and $g_A = g_{V\rho}$ are required to avoid the singularity. In this limit, one obtains a Gell-Mann–Lévy (GML)-type linear sigma model with the degerate $O(4)$ multiplet for the scalar and pseudo-scalar fields (massless in the chiral limit) [Beane and van Kolck (1994); Sasaki, Lee, Paeng and Rho (2011);

[4]Up to near the DLFP, $m \to m_0 \neq 0$, but vanishing at $n_{\rm dl}$ where $\langle\chi\rangle = 0$. This point will be understood although not specified.

[5]That $f_\sigma \leftrightarrow f_\pi$ as the DLFP is approached will figure most importantly in Chapters 7, 8 and 9.

Paeng, Lee, Rho and Sasaki (2013)]. There the ρ mesons are decoupled from the nucleons. As $f_\sigma^* \propto \langle\chi\rangle^* \to 0$ and $m_0 \propto \langle\chi\rangle^* \to 0$, one arrives at restored scale-chiral invariance (in the chiral limit). Now when $m_N^*(\mu = \Lambda)$ (where μ is the renormalization scale parameter) approaches zero with $\langle\chi\rangle^* \to 0$ as density increases, $(g_{V\rho}(\mu) - 1,\, g_A(\mu) - g_{V\rho}(\mu))$ flow to $(0, 0)$.

This is a symmetry which is not in QCD in the sense that one cannot zero-in on the infrared (IR) fixed point with the anomalous dimension β' going to zero as discussed in Chapter 6. We will see that this observation has an important impact on compact-star properties in Chapter 9.

5.3 Hidden local symmetry

We briefly summarize what was given in Chapter 2 mainly to recall the definitions of HLS. Together with the hidden scale symmetry, HLS will play the most important and pivotal role in what we uncover in nuclear physics, most significantly in the nuclear tensor forces and the nuclear symmetry energy E_{sym}.

Although QCD does not have explicit gauge symmetries other than color, so there is no à priori reason to consider flavor local fields, it turns, however, that there is nothing to prevent bringing in flavor gauge symmetry in nonlinear sigma model, and more significantly there is predictive power in having vector mesons among the degrees of freedom intervening at high baryonic density. As pointed out in Chapter 2, by resorting to the gauge symmetry that arises in exploiting the redundancy in the chiral field U and making the gauge symmetry dynamical, i.e., HLS, first put forward by Yamawaki and his collaborators [Bando, Kugo and Yamawaki (1988); Harada and Yamawaki (2003)], one gains a totally new insight into what goes on at high density as we will see in Chapter 9.

5.3.1 *Redundancy in the chiral field*

We note that the Lagrangian (5.15) is a nonlinear realization of the spontaneous broken internal symmetry $G = SU(2)_L \times SU(2)_R$ down to $H = SU(2)_{L+R} = SU(2)_V$, based on the manifold G/H as well as that of the spontaneously broken scale symmetry. The model actually has, in addition to the scale symmetry, a hidden symmetry $G_{\rm global} \times H_{\rm local}$ due to the redundancy in the decomposition

$$U(x) = \xi_L^\dagger(x) \cdot \xi_R(x)\,, \tag{5.31}$$

where $\xi_{R,L}(x)$ transform under $G_{\rm global} \times H_{\rm local}$ as

$$\xi_{R,L}(x) \to h(x) \cdot \xi_{R,L}(x) \cdot {g'}^{\dagger}_{R,L}\,, \quad U(x) \to \hat{g}_L U(x) {g'}^{\dagger}_R \tag{5.32}$$

with

$$h(x) \in H_{\rm local}\,, \quad g'_{R,L} \in G_{\rm global}\,. \tag{5.33}$$

This redundancy can be elevated to gauge symmetry by introducing the HLS gauge boson fields $V_\mu(x)$. How the kinetic energy term for the HLS fields arises making the bosons dynamical fields is a matter that has been around for a long time [Yamawaki (2018)]. In lower dimension theories (such as CP^{N-1} model), such kinetic energy terms can be obtained from high-order loop terms. In 4D, its presence is more or less an assumption. As mentioned, this assumption makes sense if the vector mass can be considered to be "light" [Harada and Yamawaki (2003); Komargodski (2011)].

With the covariant derivatives acting on $\xi_{R,L}$ as

$$D_\mu \xi_{R,L}(x) = (\partial_\mu - iV_\mu(x))\xi_{R,L}(x)\,, \tag{5.34}$$

and the two Maurer–Cartan 1-forms that transform homogeneously under $H_{\rm local}$

$$\hat{\alpha}_{\mu||,\perp}(x) \equiv \frac{1}{2}\left(\hat{\alpha}_{\mu R}(x) \pm \hat{\alpha}_{\mu L}(x)\right) \tag{5.35}$$

with

$$\hat{\alpha}_{\mu R,L}(x) \equiv \frac{1}{i} D_\mu \xi_{R,L}(x) \cdot \xi^{\dagger}_{R,L}(x) = \frac{1}{i}\partial_\mu \xi_{R,L}(x) \cdot \xi^{\dagger}_{R,L}(x) - V_\mu(x) \tag{5.36}$$

the HLS Lagrangian reads

$$\mathcal{L}_{HLS} = \mathcal{L}_A + a\mathcal{L}_V + \mathcal{L}_{\rm Kinetic}\left(V_\mu\right)\,, \tag{5.37}$$

where

$$\mathcal{L}_A = f_\pi^2 \cdot {\rm tr}[\hat{\alpha}^2_{\mu\perp}(x)]\,, \qquad \mathcal{L}_V = f_\pi^2 \cdot {\rm tr}[\hat{\alpha}^2_{\mu||}(x)] = f_\pi^2 \cdot {\rm tr}[\left(V_\mu(x) - \alpha_{\mu||}(x)\right)^2]\,. \tag{5.38}$$

This procedure can be applied to the infinite number of redundancies present in the chiral field U to have an infinite tower of hidden local fields which can be "deconstructed" to a five-dimensional (5D) Yang–Mills action capturing QCD [Son and Stephanov (2004)]. The holographic QCD coming from string theory encoding gravity-gauge duality (to be described in Chapter 7 for skyrmions) gives a similar 5D YM Lagrangian. The metric warping due to the (gravity) background is different in the two actions: the latter is fixed by the reduction from holographic construction while in the former, it is left undermined. We will discuss the latter approach in the skyrmion approach.

5.3.2 *Hidden-local-symmetrizing scale symmetry*

For continuity in discussions, we make a brief remark here on implementing scale symmetry in HLS. This may be considered as an hors d'oeuvre to the main course given in Chapter 6.

We note that the scale-invariant version of the non-linear sigma model Lagrangian Eq. (5.16) can be made hidden gauge symmetric (sHLS) by using the conformal compensator field χ

$$\mathcal{L}_{\rm sHLS} = \left(\frac{\chi}{f_\sigma}\right)^2 \cdot \left(\frac{1}{2}\left(\partial_\mu \sigma\right)^2 + \mathcal{L}_A + a\mathcal{L}_V\right) + \mathcal{L}_{\rm Kinetic}\left(V_\mu\right) . \qquad (5.39)$$

If we fix the gauge of HLS as $\xi_L^\dagger = \xi_R = \xi = e^{i\pi/f_\pi}$ such that $U = \xi^2$, then $H_{\rm local}$ and $G_{\rm global}$ get simultaneously broken spontaneously via Higgs mechanism, leaving the diagonal subgroup which is nothing but the subgroup of the original G of G/H. According to the Higgs mechanism, the HLS gauge fields $V_\mu(x)$ acquire the masses $\frac{1}{2}a(g\,f_\pi)^2\,(V_\mu^a(x))^2$ through the invariant $\mathcal{L}_V$ after rescaling the kinetic term of V_μ by the HLS gauge coupling g as $V_\mu(x) \to g\,V_\mu(x)$. Note that the vector boson mass terms are scale-invariant[6] thanks to the nonlinear realization of the scale symmetry. When the kinetic term is ignored in the low energy region $p^2 \ll m_V^2 = a(g\,f_\pi)^2$, the $\mathcal{L}_V$ term yields just 0 when the equation of motion $V_\mu = \alpha_{\mu||}$ is used. Noting that $\mathcal{L}_A = f_\pi^2 \cdot {\rm tr}[\hat{\alpha}_{\mu\perp}^2(x)] = f^2 \cdot {\rm tr}[\alpha_{\mu\perp}^2(x)] = \frac{f_\pi^2}{4} \cdot {\rm tr}[\partial_\mu U \partial^\mu U^\dagger] = \mathcal{L}_{{\rm NL}\sigma}$ in Eq. (5.12) by a straightforward algebraic calculation, Eq. (5.39) becomes identical to the scale-invariant version of the Higgs Lagrangian. Hence in the low energy $p^2 \ll m_V^2 = a(g\,f_\pi)^2$ where the massive vector boson V_μ gets decoupled, the sHLS Lagrangian Eq. (5.39) is in fact reduced back to precisely the original scale-invariant nonlinear sigma model Eq. (5.16).

5.3.3 *Power of hidden local symmetry*

As stated above, the notion of HLS for the vector mesons makes sense if the vector mesons can be taken as light as NG bosons. There are two known ways that this can happen. One is when there is supersymmetry as in supersymmetric QCD [Komargodski (2011)]. The other, the possibility that is exploited in Chapters 8 and 9, is that the vector meson mass is driven to zero by density as in compact-star matter.

This comes about because there is something magical with HLS. Why it is so is being explained [Yamawaki (2018)]. The argument should apply

[6]This is correct, however, only to the leading order in scale-chiral perturbative expansion as described in Chapter 6.

well to $V \in U(2)$ at low density, but as we will see in contact with Nature, there is a clear evidence that $U(2)$ symmetry breaks down at high density. We will therefore, in what follows in this chapter, focus on the ρ meson instead of ρ and ω together in $U(2)$ symmetry. This issue will be taken up in the later chapters since the role of ω at compact-star densities is extremely intricate.

Consider the HLS Lagrangian (5.37). A remarkable feature of HLS is that the resulting Lagrangian — which is written to $\mathcal{O}(p^2)$ in the chiral expansion — makes good prediction for various properties for the ρ meson. Going to the unitary gauge, the ρ mass term comes out as[7]

$$m_\rho^2 = af_\pi^2 g^2. \tag{5.40}$$

And the $\rho\pi\pi$ and ρ-photon coupling are

$$g_{\rho\pi\pi} = \frac{1}{2}ga, \quad g_{\rho\gamma} = gaf_\pi^2 = 2g_{\rho\pi\pi}f_\pi^2. \tag{5.41}$$

Note that the $\rho\gamma$ coupling is independent of a . Now since the vector current is conserved, the Noether construction gives the vector current

$$J_{V\mu}^a = 2af_\pi^2\rho_\mu^a + 2f_\pi^2(a-2)\epsilon^{abc}\pi^b\partial_\mu\pi^c + \cdots . \tag{5.42}$$

The vector meson dominance (VMD) of EM form factors of hadrons at the current level is given by the current-field identity that would correspond to

$$a = 2. \tag{5.43}$$

This leads to the celebrated Kawarabayashi–Susuki–Riazuddin–Fayyazuddin (KSRF) relation [Kawarabayashi and Suzuki (1966); Riazuddin and Fayyazuddin (1966)][8]

$$m_\rho^2 = 2f_\pi^2 g_{\rho\pi\pi}^2 . \tag{5.44}$$

With the experimental value $f_\pi = 92.42 \pm 0.26$ MeV and $g_{\rho\pi\pi} = 6.00 \pm 0.01$, one gets $m_\rho = 784$ MeV to be compared with the experimental value $m_\rho = 775.26 \pm 0.25$ MeV.

The success of (5.44) is well-known. What is not so well-known is that the relation follows *naturally* if the ρ meson is taken to be a gauge boson. Indeed it is a low-energy theorem result valid to all loop orders of the vacuum HLS Lagrangian [Harada, Kugo and Yamawaki (1993)] with the

[7]Unless otherwise noted, we will simply denote the hidden gauge coupling by g in place g_V.

[8]For a simple derivation from current algebras, see Johnson [Johnson (1970)].

corrections of $\mathcal{O}(m_\rho^2/\Lambda_\chi^2)$ and higher, where Λ_χ is the chiral symmetry scale ~ 1 GeV. This means that the relation (5.44) becomes exact as $m_\rho/\Lambda_\chi \to 0$. In the vacuum, the ρ mass is ~ 6 times the pion mass, so local gauge symmetry for it sounds unreasonable. Nonetheless if one were to take — by fiat — the HLS Lagrangian with the ρ mass on the same scale as the pion mass in making the chiral expansion, low-order chiral perturbation calculations give reasonable results comparable to those of non-linear sigma model.[9] What distinguishes the HLS approach from other approaches that possess no local symmetry is that RG analysis of HLS has a flow to the "vector manifestation (VM) fixed point" at which the ρ mass goes to zero — described more in detail in the next subsection — and this flow is picked when the (bilinear) quark condensate $\langle\bar{q}q\rangle$ goes to zero. This implies that in the limit where the light-quark masses are taken to be zero, the vector meson mass goes to zero at the chiral restoration in the same multiplet as the NG pion.

In the strict sense of Weinberg's folk theorem, there are higher-order chiral/derivative terms in HLS and of course sHLS. In Chapter 7, terms of $O(p^4)$ are included in formulating the skyrmion structure of dense matter. Given the increased number of parameters, the numerical results tend to be blurred, but they do support the qualitatively robust feature of the magical $O(p^2)$ HLS.

5.3.4 *Vector manifestation: Emerging local symmetry*

In matter-free space, the vector meson mass $m_\rho = 770$ MeV is not "light" in the sense that the pion mass is light. Since the notion of HLS makes sense only if the vector meson is light, one may ask in what way it is relevant in nuclear physics. The answer is in the way the VM fixed point can be approached in dense medium.

Starting with the HLS Lagrangian (5.37) with the "bare" parameters endowed with the intrinsic density dependence (IDD) determined at the matching scale Λ_M (discussed more precisely below), one performs Wilsonian RG decimation down to the scale at which nuclear dynamics is treated. This yields the RG equations for the parameters of the Lagrangian, namely, the hidden gauge coupling g, the pion decay constant $f_\pi = f$ and the parameter a.

To one-loop order with HLS, as the scale parameter is varied, one gets

[9]This is demonstrated by Harada and Yamawaki [Harada and Yamawaki (2003)].

the RG equations [Harada and Yamawaki (2003)][10]

$$\mu\frac{d}{d\mu}g^2 = -\frac{N_f}{2(4\pi)^2}\frac{87-a^2}{6}g^4, \tag{5.45}$$

$$\mu\frac{d}{d\mu}a = -\frac{N_f}{2(4\pi)^2}(a-1)\left[3a(a+1)g^2-(3a-1)\frac{\mu^2}{f_\pi^2}\right], \tag{5.46}$$

$$\mu\frac{d}{d\mu}f_\pi^2 = -\frac{N_f}{2(4\pi)^2}\left[3a^2g^2f_\pi^2+2(2-a)\mu^2\right]. \tag{5.47}$$

One sees immediately that there are fixed-point solutions $g=0$ and $a=1$. On the other hand there is no obvious fixed point for f_π. In fact $(g,a)=(0,1)$ is the VM fixed point.

In general, Eqs. (5.45)-(5.47) have four fixed points — three in the physical space and one in the unphysical space (with $a<0$ and $\mu^2/f_\pi^2<0$) — and one fixed line. The flows to those fixed quantities are complicated, so in what way they are relevant to physical quantities are difficult to see. One thing that transpires from the analysis, the most important point of the model based on HLS, is that when the baryonic system approaches the restoration of chiral symmetry, that is, approaching the Wigner–Weyl phase, in the idealistic chiral limit, the flow goes to the fixed point $(g,a)=(0,1)$, the VM fixed point. Physically what this would imply is that the vector meson mass m_ρ approaches zero, the point around which the HLS theory we are discussing has its most validity.

When confronting Nature with this model, an important point to take into account is that in reality, the light quark masses are not zero, although small. Furthermore the symmetry at the VM fixed point is doubled (or enhanced), which is not in QCD in the vacuum. *Therefore if the hadronic*

[10]In this reference, the RG flow equations were derived for the $\mathcal{O}(p^4)$ terms in HLS Lagrangian. We do not need them for this section.

Just to illustrate how the RG equation is gotten — for those who are not familiar with matters of this type, consider the $\rho\pi\pi$ coupling which gives the renormalization of the coupling g. Calculating one-loop graphs with (5.37), the renormalization condition is

$$g_{bare}^{-2} - \frac{N_f}{2(4\pi)^2}\frac{87-a^2}{12}\ln\Lambda^2 = \text{constant}.$$

where g_{bare} is the "bare" hidden gauge coupling and Λ is the cutoff for the momentum integral. Wilsonian RG equation is obtained by taking derivative of both sides of this equation with $\Lambda\frac{d}{d\Lambda}$. It gives

$$\Lambda\frac{d}{d\Lambda}g_{bare}^{-2} - \frac{N_f}{2(4\pi)^2}\frac{87-a^2}{12} = 0.$$

Noting that $\mu=-\Lambda$, one gets (5.45). This is the field-theory derivation of Wilsonian RG equation. See [Peskin and Schroeder (1990)].

system does approach the VM fixed point, it cannot go on top and furthermore the flow must be highly distorted from what it would be in the chiral limit. This means that how the system goes to the vicinity of the VM fixed point will be extremely difficult, if not impossible, to map out. However what density can do is to dial the quark condensate toward zero as chiral symmetry is putatively restored. In this approach, given that the KSRF formula with $a \to 1$ becomes more accurate, the mass $m_\rho^* \sim g^* f_\pi^* \to 0$. At that limit, one can think of the vector meson as a local gauge field.

- **The role of the a_1 meson:**

Now what about the a_1 meson which is a chiral partner to the ρ? What happens to it as density approaches the VM fixed point? The a_1 meson does not play a direct role in the chapters that follow. Its role is only implicit in nuclear matter and compact-star matter. But it could figure at chiral restoration, as in "mended symmetries" discussed below.

To study this issue, we follow [Harada and Sasaki (2006); Hidaka, Morimatsu and Ohtani (2006)]. To introduce the a_1 degree of freedom, we generalize the chiral field (5.31) increasing the redundancy with an extra three matrix-valued field ξ_M which transforms the same as $\xi_{L,R}$ under $SU(N_f)_{L+R}$[11]

$$U = \xi_L^\dagger \xi_M \xi_R \,, \tag{5.48}$$

and write the Lagrangian possessing the symmetry $G_{\rm global} \times G_{\rm local}$ where $G_{\rm global} = [SU(N_f)_L \times SU(N_f)_R]_{\rm global}$ is the chiral symmetry and $G_{\rm local} = [SU(N_f)_L \times SU(N_f)_R]_{\rm local}$ is the generalized HLS with the local gauge fields $L_\mu \in [SU(N_f)]_{\rm local}$ and $R_\mu \in [SU(N_f)]_{\rm local}$. The symmetry $G_{\rm global} \times G_{\rm local}$ is spontaneously broken down to flavor diagonal $SU(N_f)_V$.

Analyses along the same line — one-loop renormalization group — as in HLS [Harada and Yamawaki (2003)], with the assumption that the well-established Weinberg sum rules based on precise relations between the vector and axial-vector correlators [Weinberg (1967)] are satisfied, leads to three fixed points: VM, Ginzburg–Landau (GL) and hybrid (HB).

$$\text{VM} : m_\rho^2/m_{a_1}^2 \to 0\,, \tag{5.49}$$

$$\text{GL} : m_\rho^2/m_{a_1}^2 \to 1\,, \tag{5.50}$$

$$\text{HB} : m_\rho^2/m_{a_1}^2 \to 1/3 \tag{5.51}$$

[11] We will specialize below to $N_f = 2$.

while

$$m_\rho \to 0\,, \quad m_{a_1} \to 0\,. \tag{5.52}$$

Here, (5.49) represents the generalized VM scenario where the ρ mass goes to zero faster than the a_1 mass, (5.50) the standard sigma model scenario which is called "Ginzburg–Landau" where the ρ joins the a_1 as a chiral partner, and (5.51) a hybrid of VM and LG. These three fixed points are characterized by the multiplet structure

$$\begin{aligned}
|s\rangle &= |(N_f, N_f^*) \oplus (N_f^*, N_f)\rangle\,,\\
|\pi\rangle &= |(N_f, N_f^*) \oplus (N_f^*, N_f)\rangle \sin\psi\\
&\quad + |(1, N_f^2 - 1) \oplus (N_f^2 - 1, 1)\rangle \cos\psi\,,\\
|\rho\rangle &= |(1, N_f^2 - 1) \oplus (N_f^2 - 1, 1)\rangle\,,\\
|a_1\rangle &= |(N_f, N_f^*) \oplus (N_f^*, N_f)\rangle \cos\psi\\
&\quad - |(1, N_f^2 - 1) \oplus (N_f^2 - 1, 1)\rangle \sin\psi\,,
\end{aligned} \tag{5.53}$$

where $N_f = 2$ for the case we are dealing with and s is the scalar that is the fourth component of the chiral four-vector in linear sigma model, denoted σ in the literature and below.[12] Here ψ is the mixing angle that is given in terms of the parameters that intervene in the generalized HLS Lagrangian. At the fixed points, we find the multiplet structure

$$\begin{aligned}
&\text{VM} \;:\; \sin\psi \;\to\; 0\,,\\
&\text{GL} \;:\; \cos\psi \;\to\; 0\,,\\
&\text{HB} \;:\; \sin\psi \;\to\; \sqrt{\frac{1}{3}}\,, \quad \cos\psi \;\to\; \sqrt{\frac{2}{3}}\,.
\end{aligned} \tag{5.54}$$

5.4 Mended symmetries

As we will encounter in a specific way, there are two limits that may be exposed as density goes up. One is the VM fixed point and the other is the dilaton-limit fixed point. A question is how they could be related if any.

HLS — in the presence or absence of baryons with the assumed matching of the correlators to QCD at the matching scale — predicts that as the quark condensate is dialed to zero, it is the hidden gauge coupling $g \sim \langle\bar{q}q\rangle \to 0$ making the ρ mass go to zero as $m_\rho \sim g \to 0$, i.e., the VM fixed point. The multiplet structure here is that the pion and the longitudinal ρ come together while the scalar there — the fourth component of the chiral

[12]Not to be confused with the dilaton σ that appears in various places in the volume.

four vector with the triplet of pions — is to join the a_1 meson. On the other hand, the movement toward the DLFP discussed above with the ρNN coupling, g_{VNN}, going to zero independently of the hidden gauge coupling g, involves the dilaton scalar σ joining the triplet of pions. Matching with QCD at a suitable matching scale should give an ultraviolet fixed point from which the IR flow will be on a different trajectory from the VM ultraviolet fixed point of HLS.

So how does one go over from the trajectory on which the flow to the DLFP lies to the VM fixed point? Since the multiplet structures are different involving different symmetries, one the Gell-Mann–Lévy sigma model and the other HLS, it looks that there has to be a phase transition going from the DLFP to the VM fixed point. This question is suggested to be tied to the issue of "mended symmetries" [Weinberg (1990)].

Which of the three fixed points are relevant in dense matter is unknown. An appealing possibility is one of the first two, namely VM or GL. They are IR fixed points, but the third is not, and are arrived at only by fine-tuning. In all cases, both ρ and a_1 become massless at the restoration point, since the generalized HLS gauge coupling goes to zero.

The analysis does not indicate which fixed point is arrived at in which condition. For instance, it cannot answer the question as to whether and how the transition from the dilaton-limit fixed point in *bs*HLS to one of the fixed points (5.49) or (5.50) is triggered, i.e., by density or temperature or both? There may be a variety of different flows possible depending on probes and conditions. However, a highly plausible possibility of what might happen at the chiral restoration point seems to be offered by Weinberg's mended symmetries [Weinberg (1990)]. It has been shown using a collinear current algebra for Goldstone bosons of a broken symmetry group that under certain assumptions in the context of QCD such as large N_c and the high-energy behavior of exclusive inelastic scattering amplitude, the quartet mesons, ρ, a_1, π and σ, become massless at a second-order phase transition at which the chiral symmetry becomes unbroken. This implies the emergence of massless hidden gauge bosons ρ and a_1.

What this scenario of mended symmetries is pointing to is that the approach to the chiral phase transition in dense matter we are concerned with is most likely highly complex and intricate at the same time. The dilaton-limit fixed point (DLFP) and the VM/GL fixed point discussed above could be just a glimpse of the tip of an iceberg of what might lie in the hadronic phase diagram in the process that goes in the direction of dense matter relevant to compact stars. It also implies a variety of scenarios in the response of vector mesons to photon as has been explored in relativistic heavy-ion collisions. At present there is neither experimental nor theoretical solid evidence for such a symmetry pattern, an open issue in nuclear physics. We will see in Chapter 9 how the combination of vector manifestation fixed point and dilaton limit fixed point impacts on the properties of massive compact stars.

Chapter 6

Scale-Chiral Symmetry

In Chapter 2, a scalar σ as dilaton, a Nambu–Goldstone (NG) boson of spontaneously broken scale symmetry, was introduced as an essential degree of freedom for constructing an effective field theory (EFT) Lagrangian for nuclear physics along the line of the folk theorem (FT). In Chapter 5, we argued that scale symmetry is in fact hidden in nonlinear sigma model and can arise under certain conditions as an emergent symmetry. In this chapter, we enter into more details and develop how scale symmetry can be incorporated in a systematic perturbative scheme along with chiral symmetry. Simplified and approximated, what's discussed in this chapter will figure in confronting Nature in Chapter 9.

6.1 $f_0(500)$ and scalar conundrum in nuclear physics

We devote this chapter to the role of the lightest scalar meson $f_0(500)$ or σ in nuclear physics. We will identify this lightest scalar meson as a dilaton and focus on the applicability to nuclei and nuclear matter of the notion that the dilaton is associated with a possible infrared fixed point (IRFP) in the strongly coupled gauge theory, QCD, at low energy. Although the presence of such a dilaton in QCD with the number of flavors $N_f \sim 3$ — relevant to what we are concerned with — is not yet settled, it will play in the framework introduced in the chapters above and in this chapter a role as important as that of vector mesons in the dense matter of compact stars.

A scalar meson with a mass around 600 MeV has figured importantly in nuclear physics since a very long time. The only scalar mesonic excitation with such a low mass currently known in particle physics is $f_0(500)$ with a broad width [Patrignani (2016)]. Despite the observed large width, when taken as a local bosonic field, it has fairly successfully accounted for the attractive scalar channel in nucleon-nucleon potentials such as the well-known Bonn boson-exchange potential and, perhaps more significantly, in the highly popular relativistic mean field (RMF) theories for nuclei, nuclear matter and dense (compact-star) matter as described in Chapters 4, 8 and 9. There has been a long-lasting controversy as to whether the scalar excitation with such a large width can be considered as a point particle depicted in terms of a local field in phenomenological or effective-theory approaches. If it is a hadronic particle, then the question is, what is its QCD structure? Can it be described in quark model, as, say, quarkonium of $(q\bar{q})^m$ complex with $m \geq 1$, a gluonium (glueball) or a mixture thereof [Pelaez (2015)]? In certain models such as linear sigma model, there is a natural scalar, which is the fourth component of the chiral four-vector with the triplet pions. However it must have a mass $\gtrsim 1$ GeV to be compatible with low-energy pionic processes, e.g., soft-pion theorems, so is not relevant far below the chiral scale $4\pi f_\pi \sim 1$ GeV. Furthermore what figures in Walecka model-type [Serot and Walecka (1986)] RMF theories — which are successful in heavy nuclei and nuclear matter — must be a predominantly, if not entirely, chiral singlet scalar, not the scalar of linear sigma model. If it were the sigma model scalar with a mass < 1 GeV as needed in nuclear phenomenology, it would tend to trigger the destabilization of nuclear matter due to strong attractive many-body forces at the mean-field level. As a way out, it has been suggested that in densities near that of nuclear matter, the scalar is a chiral singlet scalar, but at high density as the system approaches chiral restoration, it should transmute to a $\bar{q}q$ configuration that is expected to dominate at the large N_c limit. This feature has been discussed in terms of the "dilaton-limit fixed point" (DLFP)[Beane and van Kolck (1994); Sasaki, Lee, Paeng and Rho (2011)].

6.1.1 *Scalar σ as a Nambu–Goldstone boson*

The principal thesis we develop here is that the scalar needed to resolve the above "conundrums" is a (pseudo-)NG boson, say, dilaton, arising from the spontaneous breaking of scale symmetry which is explicitly broken by QCD trace anomaly, focusing on its role in the dynamics of nuclei and nuclear matter. We will adopt the notion that the dilaton *can be associated in the precise sense specified* with a possible IR fixed point, following the suggestions by Crewther and Tunstall [Crewther and Tunstall (2015); Catà, Crewther and Tunstall (2018)] (CT for short).

It must be reiterated that whether an IR fixed point exists in the (matter-free) vacuum of QCD for three flavors ($N_f = 3$) is presently a totally open question. For large N_f, as in the techni-dilaton approach to Higgs physics, there is an indication, supported by lattice calculations, that such an IR fixed point could exist. The nature of IR fixed point involved is basically different here from what's presumed to be the QCD case. Nonetheless a dilaton chiral perturbation approach similar to what is advocated by CT (and also GS) and adopted here has been formulated [Matsuzaki and Yamawaki (2014)].[1] It is also fair to say that there is, up to date, neither a clear-cut lattice indication nor strong theoretical argument anchored on QCD proper that falsifies it either. On the contrary, there is a stochastic numerical perturbation calculation that "votes for the existence" of an IR fixed point for two massless quark flavors [Horsley, Perlt, Takow, Schierholtz and Schiller (2014)] although it needs to be confirmed along the line advocated in [Crewther and Tunstall (2015)]. Also, a calculation of the full QCD beta function (including both the perturbative and nonperturbative effects) in the light-front holography with two flavors indicates the existence of an IR fixed point [Brodsky, de Teramond and Deur (2010)]. What is more appropriate for our purpose, it is highly plausible that such an IR fixed point could be generated in medium as an *emergent symmetry* even if it were absent in the matter-free vacuum. The reasoning leading to this observation is based on that an IR fixed point of a sort, dubbed "dilaton-limit fixed point" [Beane and van Kolck (1994); Sasaki, Lee, Paeng and Rho (2011)], can emerge through strong nuclear correlations. The existence of such a symmetry emergence has the potential of resolving several long-standing conceptual problems in nuclear physics and offering a possibility to probe strongly compressed neutron-star matter that is otherwise hard to access.

That there is a scalar NG on par with a pseudo-scalar NG boson (pion) in low-energy hadronic interactions is an old idea dating way back to early 1970's [Isham, Salam and Strathee (1970); Ellis (1970)]. What is new in the current development — and potentially powerful for nuclear physics — is that it can be formulated in QCD. In terms of the QCD trace anomaly

$$\theta_\mu^\mu = \frac{\beta(\alpha_s)}{4\alpha_s} G^a_{\mu\nu} G^{a\mu\nu} + (1+\gamma_m(\alpha_s)) \sum_{q=u,d,s} m_q \bar{q} q \tag{6.1}$$

[1]It should be mentioned that Yamawaki [Yamawaki (2017)] has given a strong argument against the notion of $f_0(500)$ as a dilaton associated with a possible IR fixed point in the matter-free vacuum but has not ruled out the notion of dilatonic scalar in nuclear medium.

there can be an exact scale invariance if, in the chiral limit $m_q \to 0$, there is an IR fixed point $\beta(\alpha_{IR}) = 0$. The suggestion by CT is that such an IR fixed point with a *non-vanishing chiral condensate*, nonperturbative in character, is highly plausible and far below from the chiral scale $4\pi f_\pi \sim 1$ GeV, the QCD β function flows along the trajectory leading to the IR fixed point. The scale symmetry associated with the vanishing of θ_μ^μ is then assumed to be spontaneously broken, giving rise to a NG boson. Note that the chiral symmetry is spontaneously broken at the IR fixed point as long as the quark condensate $\langle \bar{q}q \rangle$ is non-vanishing. That the two spontaneous broken symmetries are intimately locked to each other is the key point of our development. The scalar $f_0(500)$ is identified with that scalar NG boson associated with the spontaneous scale symmetry breaking with the mass generated by "explicit" symmetry breaking encoded in both the departure of α_s from α_{IR} (with a non-zero gluon condensate) *and* the current quark mass. Thus the dilaton σ joins the pseudo-scalars, pions and kaons, to form the pseudo-NG multiplet. What ensues is then a potentially more powerful EFT that combines both chiral symmetry and scale symmetry with the possibility of doing systematic expansions both in the chiral counting and in the scale counting, i.e., "scale-chiral" counting.

Among the advantages in this approach in particle physics is a simple explanation of the $\Delta I = 1/2$ rule for kaon decays that is accomplished by elevating next-to-leading order (loop) terms in three-flavor chiral perturbation theory χPT_3 into the leading tree order in terms of the σ field in χPT_σ. What we are interested in is what this scheme with a NG scalar put together with the NG pseudo-scalars does in nuclear phenomena.

For applying to nuclear matter, the key degrees of freedom are the nucleons — proton and neutron — and the pions, the degrees of freedom figuring in the usual 2-flavor chiral perturbation theory χPT_2. Implementing the scalar σ in the CT scheme, on the other hand, requires three flavors including the strangeness. This is because the mass of the scalar is comparable rather to the kaon mass ~ 500 MeV than the pion mass. In what follows, however, we will be focusing on non-strange phenomena that take place in nuclear systems, so we will be projecting out the two-flavor sector from $SU(3)$ for most of the consideration, apart from the structure of the σ. To apply to strange hadrons, hyperons and kaons need to, and can be, straightforwardly incorporated.

Before going to the explicit breaking of the scale and chiral symmetries, let us consider the case of exact scale-chiral symmetry.[2] Both the

[2]The meaning of exact scale symmetry is subtle and will be specified below.

charged axial current J^{5a}_μ and the dilatation current D_μ related to the energy-momentum tensor $\theta_{\mu\nu}$ by

$$D_\mu = x^\nu \theta_{\mu\nu} \tag{6.2}$$

will be conserved

$$\partial^\mu J^{5a}_\mu = 0, \quad \partial^\mu D_\mu = \theta^\mu_\mu = 0. \tag{6.3}$$

The conserved axial current leads to the celebrated Goldberger–Treiman (GT_π) relation for the axial current

$$m_N = f_\pi g_{\pi NN}/g_A \tag{6.4}$$

where f_π is the pion decay constant and $g_{\pi NN}$ is the pion-nucleon coupling constant. Similarly, the conserved dilatation current gives the analog that we will call GT_σ relation

$$m_N = f_\sigma g_{\sigma NN} \tag{6.5}$$

where f_σ is the σ decay constant and $g_{\sigma NN}$ is the σ-nucleon coupling constant [Pagels (1966); Caruthers (1970)]. We note that one can obtain the same relations (6.4) and (6.5) by using partially conserved currents, PCAC and PCDC.

One can also write down a scalar analog to Gell-Mann–Oakes–Renner (GMOR) relation (for the pion) $f_\pi^2 m_\pi^2 = -\frac{1}{2}(m_u+m_d)\langle(\bar{u}u+\bar{d}d)\rangle$ [Crewther and Tunstall (2015)]:

$$f_\sigma^2 m_\sigma^2 = -(4+\beta')\left\langle \frac{\beta(\alpha_s)}{4\alpha_s} G^2 \right\rangle \\ -(3-\gamma_m)(1+\gamma_m) \sum_{q=u,d,s} \langle m_q \bar{q} q \rangle \tag{6.6}$$

where $\beta' = \partial\beta(\alpha_s)/\partial\alpha_s$ is the anomalous dimension of $G^2_{\mu\nu}$ and γ_m is the anomalous dimension of $\bar{q}q$. Note that apart from the gluon condensate, this involves the kaon mass in addition to the pion mass that figures in the GMOR relation for the pion.

The GT relations and GMOR relations for π and σ, being low-energy theorems, are to constitute the leading terms in scale-chiral perturbation theory $\chi\mathrm{PT}_\sigma$. What makes sense in two-flavor chiral perturbation theory is the small up and down quark masses resulting in small pion mass. However the σ mass is comparable to the kaon mass, so in addressing kaon decays, it could make sense to consider them, as is done in the CT approach, on the same footing as the pseudo-scalar NG bosons. In addressing nuclear phenomena, on the other hand, one may wonder whether the σ

mass, $\sim 1/2$ GeV, is small enough to be considered on the same footing as the pion mass. There are, up to date, no systematic χPT_σ calculations in non-strange sectors, so one cannot say anything, but there are cases where the mass difference of the size in question does not obstruct chiral-perturbative approach. For instance in HLS theory, the ρ meson of mass 770 MeV can be treated on the same footing as the pion mass in formulating chiral perturbation theory [Harada and Yamawaki (2003)]. This aspect is more relevant in nuclear medium since there is a possible symmetry limit, at high temperature or high density, called "vector manifestation (VM) fixed point" [Harada and Yamawaki (2003)] at which the ρ and π become degenerate. As suggested in [Sasaki, Lee, Paeng and Rho (2011)], the dilaton can also join the pion at the DLFP. This matter will be discussed in somewhat more detail below. Here we simply mention that this points to a possible existence of an IR fixed point of the type suggested by CT in dense medium regardless of whether such a fixed point is present in the QCD vacuum. These observations suggest, as stated in Chapter 5, that π, ρ, a_1 and σ could be put on the same footing as in Weinberg's mended symmetries [Weinberg (1990)], where there is a sense in which they all become massless with ρ and a_1 becoming local gauge bosons at the scale-chiral symmetry restoration point.

6.2 Scale-chiral Lagrangian

In the preceding chapter (Chapter 5), it was shown that scale symmetry is actually hidden in standard Higgs-type model belonging to the class of linear sigma model (and Nambu–Jona-Lasinio model) and by "dialing" one coupling constant controlling the vacuum structure, with or without baryonic matter, one can probe systems ranging from nonlinear sigma model without scale symmetry to scale-chiral invariant model. In this section we write down the scale symmetric chiral Lagrangian of Crewther and Tunstall [Crewther and Tunstall (2015)] and develop scale-chiral perturbative Lagrangian denoted as χPT_σ. It will then be generalized with HLS implemented and baryon fields added. We adopt this model since we find it more adaptable to the dense-matter system we are interested in. We will give a brief discussion on the relation between this model and the GS model [Golterman and Shamir (2016)] at the end of this chapter. The objective of this chapter is to set up a scale-chiral perturbation scheme by incorporating scale symmetry into the standard chiral expansion that is now very well developed with baryons included. We shall first write down the Lagrangian for the meson sector of the octet π and the dilaton, then incorporate hidden local fields and finally octet baryons.

We construct the Lagrangian in terms of the conformal compensator field χ which has a mass-dimension 1 and, under scale transformation $x \to \lambda^{-1}x$, transforms linearly as

$$\chi(x) \to \lambda\chi(\lambda^{-1}x). \tag{6.7}$$

Under chiral transformation $SU(3)_L \times SU(3)_R$, χ is invariant. Then, in terms of the NG boson σ, the dilaton compensator χ is written as

$$\chi(x) = f_\sigma e^{\sigma/f_\sigma}, \tag{6.8}$$

where f_σ is the σ decay constant and, under scale transformation, the σ field transforms *nonlinearly* as

$$\sigma(x) \to \sigma(\lambda^{-1}x) + f_\sigma \ln \lambda. \tag{6.9}$$

For the NG bosons associated with the spontaneous breaking of chiral symmetry, we use the scale- and mass-dimension zero $U(x)$ with

$$U(x) = e^{2i\pi_a T_a/f_\pi} \tag{6.10}$$

which transforms under chiral transformation as

$$U(x) \to g_L U(x) g_R^\dagger \tag{6.11}$$

where $g_{L,R} \in SU(3)_{L,R}$ and under scale transformation,

$$U(x) \to U(\lambda^{-1}x). \tag{6.12}$$

6.2.1 χPT_σ at the leading order

Using the conformal compensator field given above, we write the effective Lagrangian valid to the leading scale-chiral order $O(p^2)$ as [Crewther and Tunstall (2015)]

$$\mathcal{L}^{\rm LO}_{\chi{\rm PT}_\sigma} = \mathcal{L}^{d=4}_{\chi{\rm PT}_\sigma;{\rm inv}} + \mathcal{L}^{d>4}_{\chi{\rm PT}_\sigma;{\rm anom}} + \mathcal{L}^{d<4}_{\chi{\rm PT}_\sigma;{\rm mass}}, \tag{6.13}$$

with

$$\mathcal{L}^{d=4}_{\text{inv,LO}} = c_1 \frac{f_\pi^2}{4}\left(\frac{\chi}{f_\sigma}\right)^2 \operatorname{Tr}\left(\partial_\mu U \partial^\mu U^\dagger\right) + \frac{1}{2} c_2 \partial_\mu \chi \partial^\mu \chi + c_3 \left(\frac{\chi}{f_\sigma}\right)^4, \tag{6.14a}$$

$$\mathcal{L}^{d>4}_{\text{anom,LO}} = (1-c_1)\frac{f_\pi^2}{4}\left(\frac{\chi}{f_\sigma}\right)^{2+\beta'} \operatorname{Tr}\left(\partial_\mu U \partial^\mu U^\dagger\right) + \frac{1}{2}(1-c_2)\left(\frac{\chi}{f_\sigma}\right)^{\beta'} \partial_\mu \chi \partial^\mu \chi + c_4 \left(\frac{\chi}{f_\sigma}\right)^{4+\beta'}, \tag{6.14b}$$

$$\mathcal{L}^{d<4}_{\text{mass,LO}} = \frac{f_\pi^2}{4}\left(\frac{\chi}{f_\sigma}\right)^{3-\gamma_m} \operatorname{Tr}\left(\mathcal{M}^\dagger U + U^\dagger \mathcal{M}\right), \tag{6.14c}$$

where $\mathcal{M}$ is the current quark matrix with $\mathcal{M} = \text{diag}(m_\pi^2, m_\pi^2, 2m_K^2 - m_\pi^2)$, c_i's are unknown constants of scale-chiral order $O(p^0)$ for $i = 1,2$ and $O(p^2)$ for $i = 3,4$ but mass dimension-zero for $i = 1,2$ and mass dimension-four for $i = 3,4$. It is important to note that the coefficient c_3 figuring in the leading order scale-invariant term is implicitly $O(p^2)$ in scale-chiral counting. This is because it plays a role for the dilaton mass similarly to $\mathcal{M}$ for the pseudo-scalar NG mesons. However differently from $\mathcal{M}$, setting to zero of which corresponds to turning off explicit chiral symmetry breaking, i.e., going to the chiral limit, c_3 does not turn off the explicit scale symmetry breaking: It is $\beta' \to 0$ that does the job. This difference should be kept in mind in keeping track of scale-chiral order and explicit symmetry breaking for the dilaton. We will return to this matter at the end of this subsection.

In (6.13), both the anomalous dimension γ_m of the quark mass and $\beta' = \partial\beta(\alpha_s)/\partial\alpha_s$ are evaluated at the IR fixed point α_{IR}. Equation (6.14a) is invariant under scale transformation, (6.14c) is the current quark mass term including the anomalous dimension of quark mass operator and (6.14b) accounts for the effect of the anomalous dimension of G^2 which depends on the slope of the beta function.

Let us briefly review the structure of the leading order (LO) Lagrangian (6.13). Consider for simplicity the (matter-free) vacuum in the chiral limit. In this case, since the vacuum should be stable in the $\sigma = 0$ direction, or, equivalently, $\chi = f_\sigma$, one has

$$4c_3 + (4+\beta')c_4 = 0. \tag{6.15}$$

The nontrivial solution of this equation with the lower-bound dilaton potential is

$$c_3 = (4 + \beta')c; \quad c_4 = -4c, \tag{6.16}$$

with a positive $c \sim O(p^2)$. By using this solution, the dilaton potential becomes

$$V(\chi) = -(4+\beta')c\left(\frac{\chi}{f_\sigma}\right)^4 + 4c\left(\frac{\chi}{f_\sigma}\right)^{4+\beta'}, \tag{6.17}$$

which explicitly shows that, with $\beta' \neq 0$, the dilaton potential is in the NG mode, i.e., the minima of the potential appears at $\langle\chi\rangle = f_\sigma$. However, if $\beta' = 0$, $V(\chi) = 0$ so the scale symmetry cannot be broken. This simple observation illustrates that the spontaneous breaking and explicit breaking of scale symmetry are correlated and the spontaneous breaking is triggered by explicit breaking, which agrees with that unlike chiral symmetry, spontaneous breaking of scale symmetry cannot take place in the absence of explicit symmetry breaking [Freund and Nambu (1968); Fubini (1976); Serra (2013)]. We could refer to this as "Freund–Nambu theorem." *This theorem implies that it is not possible to "sit on" the IR fixed point with both scale and chiral symmetries spontaneously broken.* This is analogous to that one cannot "sit" on the VM fixed point in HLS theory [Harada and Yamawaki (2003)]. In what follows, *we shall therefore always consider the state of matter near but not on the IR fixed point.*

In the case that $\beta'(\alpha_{\rm IR})$ is a small quantity, i.e., $\beta'(\alpha_{\rm IR}) \ll 1$,[3] the potential (6.17) can be expanded to the leading order in β'

$$V(\chi) = \frac{m_\sigma^2 f_\sigma^2}{4}\left(\frac{\chi}{f_\sigma}\right)^4\left[\ln\left(\frac{\chi}{f_\sigma}\right) - \frac{1}{4}\right], \tag{6.18}$$

which is the familiar Coleman–Weinberg type potential used in the literature [Goldberger, Grinstein and Skiba (2008)]. In the derivation of (6.18), we have used that, in the chiral limit, $m_\sigma^2 f_\sigma^2 = 4\beta'(4+\beta')c \simeq 16\beta' c$ for a small β', which, with $\beta' c \sim O(p^2)$, is the dilaton analog to the Gell-Mann–Oakes–Renner relation for the pion with $m_\pi^2 \sim O(p^2)$ [Crewther and Tunstall (2015)].

[3]There is a preliminary indication in dense matter that the system becomes unstable for $\beta' < 2$ so this limit may not be applicable in our system. We will come back to this point below.

Moreover, when $\beta'(\alpha_{\rm IR})$ is a small parameter, in the chiral limit, one can rewrite the sum of $\mathcal{L}^{d=4}_{\rm anom,LO}$ and $\mathcal{L}^{d>4}_{\rm anom,LO}$ as

$$\begin{aligned}
\mathcal{L}^{\chi\rm limit}_{\rm LO} &= \mathcal{L}^{d=4}_{\rm anom,LO} + \mathcal{L}^{d>4}_{\rm anom,LO} \\
&= \frac{f_\pi^2}{4}\left(\frac{\chi}{f_\sigma}\right)^2 {\rm Tr}\left(\partial_\mu U \partial^\mu U^\dagger\right) + \frac{1}{2}\partial_\mu\chi\partial^\mu\chi \\
&\quad + \left[(1-c_1)\frac{f_\pi^2}{4}\left(\frac{\chi}{f_\sigma}\right)^2 {\rm Tr}\left(\partial_\mu U \partial^\mu U^\dagger\right)\right. \\
&\quad \left. + (1-c_2)\frac{1}{2}\left(\frac{\chi}{f_\sigma}\right)^2 \partial_\mu\chi\partial^\mu\chi\right]\sum_{n=1}^{\infty}(\beta'\Sigma)^n \\
&\quad + c_3\left(\frac{\chi}{f_\sigma}\right)^4\left[1 + \frac{c_4}{c_3}\sum_{n=1}^{\infty}(\beta'\Sigma)^n\right],
\end{aligned} \tag{6.19}$$

where $\Sigma = \ln(\chi/f_\sigma) = \sigma/f_\sigma$. The Lagrangian $\mathcal{L}^{\chi\rm limit}_{\rm LO}$ explicitly shows the effect of β' which accounts for the explicit breaking of scale symmetry in the chiral limit. There are two cases that would enable us to proceed further to confront nature: (A) $\beta' \approx 0$ except in the dilaton potential and (B) $c_i \to 1$ for $i = 1, 2$ with $\beta' \neq 0$. We then have what we shall refer in what follows to as "leading order scale symmetry (LOSS)" approximation

$$\begin{aligned}
\mathcal{L}^{\chi\rm limit}_{\rm LO} &= \frac{f_\pi^2}{4}\left(\frac{\chi}{f_\sigma}\right)^2 {\rm Tr}\left(\partial_\mu U \partial^\mu U^\dagger\right) + \frac{1}{2}\partial_\mu\chi\partial^\mu\chi \\
&\quad + c_3\left(\frac{\chi}{f_\sigma}\right)^4\left[1 + \frac{c_4}{c_3}\beta'\ln\frac{\chi}{f_\sigma} + \cdots\right].
\end{aligned} \tag{6.20}$$

This Lagrangian is what results from the linear sigma model in the limit $\lambda \to 0$ in Yamawaki's argument for hidden scale symmetry [Yamawaki (2015)]. It is also of the form obtained in the leading order in scale-chiral symmetry in [Golterman and Shamir (2016)]. We will exploit this Lagrangian in the applications discussed in Chapters 8 and 9. For practical applications, the option (B) without requiring $\beta' = 0$ will be adopted.

Next, we consider the infinitesimal scale transformation $x \to \lambda^{-1}x \equiv (1+\theta)^{-1}x$ with θ being an infinitesimal parameter. Under such transformation, the fields in Lagrangian (6.13) transform as

$$\begin{aligned}
\delta_\theta U(x) &= \theta x_\mu \partial^\mu U(x), \\
\delta_\theta \chi(x) &= \theta\left(1 + x_\mu\partial^\mu\right)\chi(x).
\end{aligned} \tag{6.21}$$

From these transformation rules, one can derive the partially conserved dilatation current (PCDC) relation

$$\langle \theta_\mu^\mu \rangle = \langle \partial_\mu D^\mu \rangle = 4\beta' c = \frac{m_\sigma^2 f_\sigma^2}{4}, \tag{6.22}$$

an analog to the familiar PCAC.

Let us pause briefly to make a side remark highly pertinent to the application to nuclear physics discussed below. In a medium of baryonic matter as discussed in [Lee, Paeng and Rho (2015); Paeng, Kuo, Lee and Rho (2015)], the vacuum is modified by density. Hence we expect that the decay constant of σ deviates from its vacuum value, i.e., $f_\sigma^* \neq f_\sigma = \langle\chi\rangle$. In [Lee, Paeng and Rho (2015)], it was argued that $f_\pi^* = f_\pi \langle\chi\rangle^*/f_\sigma$, which is consistent with the suggestion of [Crewther and Tunstall (2015)] that chiral symmetry breaking is locked to scale symmetry $f_\pi \approx f_\sigma$.

At the tree order with the Lagrangian (6.13), apart from the phenomenological values of masses and constants, there are seven parameters $f_\sigma, \beta', \gamma_m$ and $c_i (i = 1, \ldots, 4)$. The σ meson decay constant f_σ is roughly estimated to be ~ 100 MeV in [Crewther and Tunstall (2015)] requiring that the power counting of the sigma loop be the same as that of pion loop. This estimation is supported by the simulation of the dense nuclear matter from the skyrmion approach [Ma and Rho (2017)]. Through the minimal condition of the potential and the mass of σ, two relations among β', γ_m, c_3 and c_4 can be established. By using the $\sigma \to \pi\pi$ decay width, one can obtain a relation among β', γ_m and c_1. Consequently, we are left with three unknown constants, for example, β', γ_m and c_2 relating to multi-σ interaction. It should be feasible to fix these parameters by a comprehensive analysis of hadron and nuclear phenomena that involve the scalar effect in the nucleon-nucleon interactions.

6.2.2 χPT_σ at the-next-to-leading order

To the leading order, both β' and γ_m are evaluated at the IR fixed point $\alpha_{\rm IR}$ which are the leading order terms in the expansions of these functions around $\Delta\alpha_s \equiv \alpha_{\rm IR} - \alpha_s$. Therefore, the next-to-leading (NLO) Lagrangian includes, in addition to the higher scale-chiral order corrections due to the current quark mass and derivatives on the NG bosons, the leading terms in the expansion of β' and γ_m in $\Delta\alpha_s$. Take for instance the anomalous dimension of the mass operator. After expanding γ_m with respect to $\alpha_{\rm IR}$, one should make the following substitution:

$$\chi^{\gamma_m(\alpha_{\rm IR})} \to \chi^{\gamma_m(\alpha_{\rm IR})}\left[1+\sum_{n=1}^{\infty}C_n\left(\Delta\alpha_s\Sigma\right)^n\right], \tag{6.23}$$

where C_n are constants evaluated at $\alpha_{\rm IR}$. And the same applies to $\chi^{\beta'(\alpha_{\rm IR})}$.

In the CT approach, from the σ-to-vacuum matrix element of the divergence of the dilatation current, one can conclude the power counting

$$\Delta\alpha_s \sim O(p^2). \tag{6.24}$$

Therefore, in the effective theory, in terms of the degrees of freedom considered in the present section, the effect of $\Delta\alpha_s$ is represented by the following scale-chiral dimension-two chiral invariant operators:

$$\mathrm{Tr}\left(\partial_\nu U\partial^\nu U^\dagger\right);\ \ \partial_\nu\chi\partial^\nu\chi;\ \ \mathrm{Tr}\left(\mathcal{M}^\dagger U + U^\dagger\mathcal{M}\right). \tag{6.25}$$

Since the LO terms are already chiral and Lorentz invariant, the terms listed above responsible for $\Delta\alpha_s$ should be chiral and Lorentz invariant. In addition, because of the factor Σ in the expansion (6.23), each contribution from (6.25) is accompanied by a σ field.

It should be noted that even though in the expansion (6.23) the term proportional to Σ includes only one unknown parameter combination $C_1\Delta\alpha_s$ in the microscopic theory, in the effective theory expressed in terms of the macroscopic degrees of freedom, the contribution of $C_1\Delta\alpha_s$ is expressed in terms of a combination of the quantities in (6.25) and there is, in front of each term, an unknown low-energy constant that cannot be fixed by using only the symmetry argument. We should emphasize that the contribution from all the quantities in Eq. (6.25) get smaller as the coupling constant α_s runs nonperturbatively closer to the IR fixed point and vanishes at the IR fixed point. This is because, for on-shell processes, the effect of the derivative on the NG bosons translates into the momentum of the external NG bosons, therefore the mass of the NG bosons, which goes toward zero as the IR fixed point is approached.

Based on what is stated above, one can straightforwardly write down the NLO Lagrangian $\mathcal{L}_{\chi\mathrm{PT}_\sigma}^{\rm NLO}$. It includes two sectors: one, $\mathcal{L}_{\chi\mathrm{PT}_\sigma}^{\mathrm{LO}\times\Delta\alpha_s}$, from the $\Delta\alpha_s$ expansion of the leading order Lagrangian $\mathcal{L}_{\chi\mathrm{PT}_\sigma}^{\mathrm{LO};d>4}$ and $\mathcal{L}_{\chi\mathrm{PT}_\sigma}^{\mathrm{LO};d<4}$ and the other one, $\mathcal{L}_{\chi\mathrm{PT}_\sigma}^{O(p^4)}$, from the $O(p^4)$ of counting from the standard effective theory construction:

$$\mathcal{L}_{\chi\mathrm{PT}_\sigma}^{\rm NLO} = \mathcal{L}_{\chi\mathrm{PT}_\sigma}^{\mathrm{LO}\times\Delta\alpha_s} + \mathcal{L}_{\chi\mathrm{PT}_\sigma}^{O(p^4)}. \tag{6.26}$$

For $\mathcal{L}_{\chi\mathrm{PT}_\sigma}^{\mathrm{LO}\times\Delta\alpha_s}$, one can write it as

$$\mathcal{L}_{\chi\mathrm{PT}_\sigma}^{\mathrm{LO};\Delta\alpha_s} = \left[(1-c_1)\frac{f_\pi^2}{4}\left(\frac{\chi}{f_\sigma}\right)^2 \mathrm{Tr}(\partial_\mu U\partial^\mu U^\dagger)+\frac{1}{2}(1-c_2)\partial_\mu\chi\partial^\mu\chi+c_4\left(\frac{\chi}{f_\sigma}\right)^4\right]$$
$$\times \left(\frac{\chi}{f_\sigma}\right)^{\beta'} \ln\left(\frac{\chi}{f_\sigma}\right)\Big[D_1 \mathrm{Tr}\left(\partial_\nu U\partial^\nu U^\dagger\right) + D_2\partial_\nu\left(\frac{\chi}{f_\sigma}\right)\partial^\nu\left(\frac{\chi}{f_\sigma}\right)$$
$$+ D_3\left(\frac{\chi}{f_\sigma}\right)^{1-\gamma_m} \mathrm{Tr}\left(\mathcal{M}^\dagger U + U^\dagger\mathcal{M}\right)\Big]$$
$$+ \mathrm{Tr}\left(\mathcal{M}^\dagger U + U^\dagger\mathcal{M}\right)$$
$$\times \chi^{3-\gamma_m} \ln\left(\frac{\chi}{f_\sigma}\right)\Big[D_4 \mathrm{Tr}\left(\partial_\nu U\partial^\nu U^\dagger\right) + D_5\partial_\nu\left(\frac{\chi}{f_\sigma}\right)\partial^\nu\left(\frac{\chi}{f_\sigma}\right)$$
$$+ D_6\left(\frac{\chi}{f_\sigma}\right)^{1-\gamma_m} \mathrm{Tr}\left(\mathcal{M}^\dagger U + U^\dagger\mathcal{M}\right)\Big]. \tag{6.27}$$

Here D_is are unknown constants in addition to the c_is appearing in the LO terms. There will figure many of such unknown terms in what follows at NLO,

Note that the Lagrangian (6.27), due to the factor $\ln(\chi/f_\sigma)$, does not transform homogeneously under scale transformation. This Lagrangian serves as renormalization counter terms in the loop expansion of $\chi\mathrm{PT}_\sigma$ [Crewther and Tunstall (2015)]. This term vanishes when the explicit breaking of scale symmetry is absent. Unlike chiral symmetry, it cannot be "mechanically" turned off by, say, dialing β' to zero in the Lagrangian (6.27). When the explicit symmetry breaking is absent, it amounts to setting $\Sigma = \ln(\chi/f_\sigma)$ to zero, a feature that we pointed out before, that scale symmetry cannot be spontaneously broken in the absence of explicit symmetry breaking. This reflects the unfamiliar separation of roles for the chiral order counting in the coefficients $c_{3,4}$ and for the explicit symmetry breaking in β' that carries no order counting. It is easier to see the effect of β' if we go back to the Lagrangian (6.20) which explicitly shows that, when the scale symmetry is restored with $\beta' = 0$, what's left are scale-invariant, so $\mathcal{L}_{\chi\mathrm{PT}_\sigma}^{\mathrm{LO}\times\Delta\alpha_s}$ is no longer present.

As for $\mathcal{L}_{\chi\mathrm{PT}_\sigma}^{O(p^4)}$, similarly to the LO, it can be divided into the scale-invariant term $\mathcal{L}_{\chi\mathrm{PT}_\sigma;\mathrm{inv}}^{O(p^4);d=4}$, the trace anomaly term $\mathcal{L}_{\chi\mathrm{PT}_\sigma;\mathrm{anom}}^{O(p^4);d>4}$ and the explicit chiral symmetry breaking term $\mathcal{L}^{O(p^4);d<4}_{\chi\mathrm{PT}_\sigma;\mathrm{mass}}$:

$$\mathcal{L}^{O(p^4)}_{\chi\mathrm{PT}_\sigma} = \mathcal{L}^{O(p^4);d=4}_{\chi\mathrm{PT}_\sigma;\mathrm{inv}} + \mathcal{L}^{O(p^4);d>4}_{\chi\mathrm{PT}_\sigma;\mathrm{anom}} + \mathcal{L}^{O(p^4);d<4}_{\chi\mathrm{PT}_\sigma;\mathrm{mass}}, \tag{6.28}$$

where

$$\begin{aligned}
\mathcal{L}^{d=4(p^4)}_{\chi\mathrm{PT}_\sigma;\mathrm{Inv}} = {} & s_1 L_1 \left[\mathrm{Tr}\left(\partial_\mu U^\dagger \partial^\mu U\right)\right]^2 + s_2 L_2 \mathrm{Tr}\left(\partial_\mu U^\dagger \partial_\nu U\right) \mathrm{Tr}\left(\partial^\mu U^\dagger \partial^\nu U\right) \\
& + s_3 L_3 \mathrm{Tr}\left(\partial_\mu U^\dagger \partial^\mu U \partial_\nu U^\dagger \partial^\nu U\right) \\
& + x_1 J_1 \left(\partial_\nu \Sigma \partial^\nu \Sigma\right) \mathrm{Tr}\left(\partial_\mu U \partial^\mu U^\dagger\right) \\
& + x_2 J_2 \left(\partial_\mu \Sigma \partial^\nu \Sigma\right) \mathrm{Tr}\left(\partial_\nu U \partial^\mu U^\dagger\right) \\
& + x_3 J_3 \partial_\mu \Sigma \partial^\mu \Sigma \partial_\nu \Sigma \partial^\nu \Sigma,
\end{aligned} \tag{6.29}$$

$$\begin{aligned}
\mathcal{L}^{d>4;(p^4)}_{\chi\mathrm{PT}_\sigma;\mathrm{Inv}} = {} & \Big\{ (1-s_1) L_1 \left[\mathrm{Tr}\left(\partial_\mu U^\dagger \partial^\mu U\right)\right]^2 \\
& + (1-s_2) L_2 \mathrm{Tr}\left(\partial_\mu U^\dagger \partial_\nu U\right) \mathrm{Tr}\left(\partial^\mu U^\dagger \partial^\nu U\right) \\
& + (1-s_3)) L_3 \mathrm{Tr}\left(\partial_\mu U^\dagger \partial^\mu U \partial_\nu U^\dagger \partial^\nu U\right) \\
& + (1-x_1) J_1 \left(\partial_\nu \Sigma \partial^\nu \Sigma\right) \mathrm{Tr}\left(\partial_\mu U \partial^\mu U^\dagger\right) \\
& + (1-x_2) J_2 \left(\partial_\mu \Sigma \partial^\nu \Sigma\right) \mathrm{Tr}\left(\partial_\nu U \partial^\mu U^\dagger\right) \\
& + (1-x_3) J_3 \partial_\mu \Sigma \partial^\mu \Sigma \partial_\nu \Sigma \partial^\nu \Sigma \Big\} \left(\frac{\chi}{f_\sigma}\right)^{\beta'},
\end{aligned} \tag{6.30}$$

$$\begin{aligned}
\mathcal{L}^{d<4;(p^4)}_{\chi\mathrm{PT}_\sigma;\mathrm{Inv}} = {} & L_4 \chi^{1-\gamma_m} \mathrm{Tr}\left(\partial_\mu U^\dagger \partial^\mu U\right) \mathrm{Tr}\left(\mathcal{M}^\dagger U + U^\dagger \mathcal{M}\right) \\
& + L_5 \left(\frac{\chi}{f_\sigma}\right)^{1-\gamma_m} \mathrm{Tr}\left[\partial_\mu U^\dagger \partial^\mu U \left(\mathcal{M}^\dagger U + U^\dagger \mathcal{M}\right)\right] \\
& + L_6 \left(\frac{\chi}{f_\sigma}\right)^{2(1-\gamma_m)} \left[\mathrm{Tr}\left(\mathcal{M}^\dagger U + U^\dagger \mathcal{M}\right)\right]^2 \\
& + L_7 \left(\frac{\chi}{f_\sigma}\right)^{2(1-\gamma_m)} \left[\mathrm{Tr}\left(\mathcal{M}^\dagger U - U^\dagger \mathcal{M}\right)\right]^2 \\
& + L_8 \left(\frac{\chi}{f_\sigma}\right)^{2(1-\gamma_m)} \mathrm{Tr}\left(\mathcal{M}^\dagger U \mathcal{M}^\dagger U + U^\dagger \mathcal{M} U^\dagger \mathcal{M}\right) \\
& + H_2 \left(\frac{\chi}{f_\sigma}\right)^{2(1-\gamma_m)} \mathrm{Tr}\left(\mathcal{M}^\dagger \mathcal{M}\right) \\
& + J_4 \left(\frac{\chi}{f_\sigma}\right)^{1-\gamma_m} \left(\partial_\mu \Sigma \partial^\mu \Sigma\right) \mathrm{Tr}\left(\mathcal{M}^\dagger U + U^\dagger \mathcal{M}\right).
\end{aligned} \tag{6.31}$$

Before closing this subsection, we should comment on the low energy constants of χPT_σ at NLO. In Lagrangian (6.26), the low energy constants $L_i (i = 1, \ldots, 8)$ and H_2 take the same values as those in χPT_3 since, when the dilaton field is switched-off, $\chi = \langle\chi\rangle = f_\sigma$. In addition to $L_i (i = 1, \ldots, 8)$ and H_2, we have more than fifty additional parameters, such as x_i, s_i etc. Obviously there are too many unknown constants to allow a reliable higher-order treatments. As noted, in the LOSS approximation, the Lagrangian can be considerably reduced to a manageable form. It is in this approximation that χPT_σ will be applied to the EoS of compact-star matter in Chapters 8 and 9.

6.3 Scale-invariant hidden local symmetry

We next extend the above discussion of χPT_σ to chiral effective theory including the lowest-lying vector mesons via the HLS strategy. We shall refer to it as sHLS. Here we use the conformal compensator field in the HLS Lagrangian to make it gauge equivalent to χPT_σ. We are employing $[SU(3)]_{\text{HLS}}$ for HLS. The extension to other HLS, such as $[U(3)]_{\text{HLS}}$ or $[SU(3) \times U(1)]_{\text{HLS}}$, is straightforward.

The HLS Lagrangian is constructed in terms of the quantities defined in Chapters 2 and 6

$$\hat{\alpha}_{\parallel\mu}, \quad \hat{\alpha}_{\parallel\mu}, \quad \frac{1}{g} V_{\mu\nu}, \tag{6.32}$$

where g is the gauge coupling constant of HLS.[4] These quantities are $O(p)$ of the chiral counting. To generate the boson masses, we take the unitary gauge

$$\xi_L^\dagger = \xi_R \equiv \xi = e^{i\pi/f_\pi}. \tag{6.33}$$

Then the 3-flavor massive vector meson fields are

$$V_\mu = \frac{g}{\sqrt{2}} \begin{pmatrix} \frac{1}{\sqrt{2}}(\rho_\mu^0 + \omega_\mu) & \rho_\mu^+ & K_\mu^{*,+} \\ \sqrt{2}\rho_\mu^- & -\frac{1}{\sqrt{2}}(\rho_\mu^0 - \omega_\mu) & K_\mu^{*,0} \\ K_\mu^{*,-} & \bar{K}_\mu^{*,0} & \phi_\mu \end{pmatrix}. \tag{6.34}$$

Given the HLS Lagrangian, one can implement scale symmetry by using the conformal compensator field used before. As in the χPT_σ case, the

[4] Whenever the QCD coupling constant figures in the discussion, we denote the hidden gauge coupling as g_V. The QCD gauge coupling will always be denoted as g_s. Unless there is confusion, we will ignore the subscript V in the hidden gauge coupling.

leading-order sHLS Lagrangian consists of three components:

$$\mathcal{L}^{\rm LO}_{{\rm HLS}_\sigma} = \mathcal{L}^{d=4}_{{\rm HLS}_\sigma;{\rm inv}} + \mathcal{L}^{d>4}_{{\rm HLS}_\sigma;{\rm anom}} + \mathcal{L}^{d<4}_{{\rm HLS}_\sigma;{\rm mass}}, \tag{6.35}$$

with

$$\begin{aligned}\mathcal{L}^{d=4}_{{\rm HLS}_\sigma;{\rm inv}} &= f_\pi^2 h_1 \left(\frac{\chi}{f_\sigma}\right)^2 {\rm Tr}[\hat{\alpha}_{\perp\mu}\hat{\alpha}_\perp^\mu] + a f_\pi^2 h_2 \left(\frac{\chi}{f_\sigma}\right)^2 {\rm Tr}[\hat{\alpha}_{\parallel\mu}\hat{\alpha}_\parallel^\mu] \\ &\quad - \frac{1}{2g^2} h_3 {\rm Tr}[V_{\mu\nu}V^{\mu\nu}] + \frac{1}{2} h_4 \partial_\mu\chi\partial^\mu\chi + h_5 \left(\frac{\chi}{f_\sigma}\right)^4, \end{aligned} \tag{6.36a}$$

$$\begin{aligned}\mathcal{L}^{d>4}_{{\rm HLS}_\sigma;{\rm anom}} &= f_\pi^2(1-h_1) \left(\frac{\chi}{f_\sigma}\right)^{2+\beta'} {\rm Tr}[\hat{\alpha}_{\perp\mu}\hat{\alpha}_\perp^\mu] \\ &\quad + (1-h_2) a f_\pi^2 \left(\frac{\chi}{f_\sigma}\right)^{2+\beta'} {\rm Tr}[\hat{\alpha}_{\parallel\mu}\hat{\alpha}_\parallel^\mu] \\ &\quad - \frac{1}{2g^2}(1-h_3) \left(\frac{\chi}{f_\sigma}\right)^{\beta'} {\rm Tr}[V_{\mu\nu}V^{\mu\nu}] \\ &\quad + \frac{1}{2}(1-h_4) \left(\frac{\chi}{f_\sigma}\right)^{\beta'} \partial_\mu\chi\partial^\mu\chi + c_6 \left(\frac{\chi}{f_\sigma}\right)^{4+\beta'}, \end{aligned} \tag{6.36b}$$

$$\mathcal{L}^{d<4}_{{\rm HLS}_\sigma;{\rm mass}} = \frac{f_\pi^2}{4} \left(\frac{\chi}{f_\sigma}\right)^{3-\gamma_m} {\rm Tr}\left(\hat{\mathcal{M}} + \hat{\mathcal{M}}^\dagger\right), \tag{6.36c}$$

where $\hat{\mathcal{M}} \equiv \xi_L \mathcal{M} \xi_R^\dagger$ transforming under HLS as $\hat{\mathcal{M}} \to h(x)\hat{\mathcal{M}}h^\dagger(x)$.

To construct the NLO sHLS one should include all the possible operators expressed in terms of the macroscopic degrees of freedom in the effective theory representing $\Delta\alpha_s$ in the expansion of γ_m and β'. For $[{\rm SU}(3)]_{\rm HLS}$, the following scale-chiral dimension-two HLS-invariant quantities are needed:

$$\begin{aligned} &{\rm Tr}(\hat{a}_{\perp\mu}\hat{a}_\perp^\mu); \qquad {\rm Tr}(\hat{a}_{\parallel\mu}\hat{a}_\parallel^\mu); \qquad \frac{1}{2g^2}{\rm Tr}(V_{\mu\nu}V^{\mu\nu}); \\ &\partial_\mu\chi\partial^\mu\chi; \qquad {\rm Tr}\left(\hat{\mathcal{M}} + \hat{\mathcal{M}}^\dagger\right). \end{aligned} \tag{6.37}$$

It is straightforward, though tedious, to construct the NLO sHLS following the approach used in $\chi{\rm PT}_\sigma$. We will not write the explicit form here.

For completeness, we indicate how many parameters are involved in the LO sHLS (6.35). Apart from the parameter a which can be fixed to $a = 2$ from the VMD, the KSRF relation and the pion decay constant

f_π, there are nine parameters $f_\sigma, \beta', \gamma_\mu$ and $h_i (i = 1, \ldots, 6)$. In the CT scheme, the σ decay constant f_σ is taken $f_\sigma \simeq 100$ MeV. Using the three relations: the minimization of the potential, the σ mass and the $\sigma \to \pi\pi$ decay, calculated in the $\chi\mathrm{PT}_\sigma$ approach, we can reduce the number of free parameters to five, $\beta', \gamma_m, h_2, h_3$ and h_4. The parameters h_2 and h_3 can be estimated or constrained with the processes that involve the vector mesons, such as K^*.

6.4 Scale-invariant hidden local symmetry with baryon octet

We next turn to the dilaton compensated baryonic chiral perturbation theory including the vector mesons via HLS, *bs*HLS. The procedure is straightforward. It suffices to first write down baryonic chiral perturbation theory with HLS, denoted *b*HLS, with the minimal number of derivatives and, then, couple the dilaton to it by means of the conformal compensator, denoted *bs*HLS. We then apply heavy-baryon formalism, as before, to *bs*HLS to construct the dilatonic baryon chiral perturbation with HLS, *hbs*HLS. This would give a well-defined power counting rule.

As mentioned, in the CT scheme, the strangeness flavor needs to be included in the meson sector because of the dilaton. Thus the symmetry pattern concerned is $[SU(3)_L \times SU(3)_R]_{\mathrm{chiral}} \times [SU(3)]_{\mathrm{HLS}}$. The octet baryon field with quantum numbers $\frac{1}{2}^+$ was written down explicitly in Chapter 2, i.e., Eq. (2.104). As noted there, the baryon field B transforms under HLS as

$$B(x) \to h(x) B(x) h^\dagger(x). \tag{6.38}$$

Under scale transformation, we take it transforming as

$$B(x) \to \lambda^{3/2} B(\lambda^{-1} x). \tag{6.39}$$

We are now ready to write down the effective baryonic scale-HLS theory. Similarly to the mesonic case, the baryonic Lagrangian, at the leading order, includes the scale-invariant part and the trace anomaly part. It should be noted that, at the leading order, i.e., with the first derivative terms, there are no $d < 4$ terms, because the light quark mass is counted as $O(p^2)$ in $\chi\mathrm{PT}_\sigma$. The Lagrangian consists of two terms,

$$\mathcal{L}_{bs\mathrm{HLS}} = \mathcal{L}^{d=4}_{bs\mathrm{HLS;inv}} + \mathcal{L}^{d>4}_{bs\mathrm{HLS;anom}}, \tag{6.40}$$

with

$$\begin{aligned}\mathcal{L}_{bs\mathrm{HLS;inv}}^{\mathrm{LO};d=4} &= g_1 \mathrm{Tr}\left(\bar{B}i\gamma_\mu D^\mu B\right) - \tilde{\mathring{m}}_B \frac{\chi}{f_\sigma} \mathrm{Tr}\left(\bar{B}B\right) \\ &\quad - \tilde{g}_{A_1} \mathrm{Tr}\left(\bar{B}\gamma_\mu\gamma_5 \left\{\hat{\alpha}_\perp^\mu, B\right\}\right) - \tilde{g}_{A_2} \mathrm{Tr}\left(\bar{B}\gamma_\mu\gamma_5 \left[\hat{\alpha}_\perp^\mu, B\right]\right) \\ &\quad + \tilde{g}_{V_1} \mathrm{Tr}\left(\bar{B}\gamma_\mu \left\{\hat{\alpha}_\parallel^\mu, B\right\}\right) + \tilde{g}_{V_2} \mathrm{Tr}\left(\bar{B}\gamma_\mu \left[\hat{\alpha}_\parallel^\mu, B\right]\right), \end{aligned} \tag{6.41a}$$

$$\begin{aligned}\mathcal{L}_{bs\mathrm{HLS;anom}}^{\mathrm{LO};d>4} &= \Big[(1-g_1)\mathrm{Tr}\left(\bar{B}i\gamma_\mu D^\mu B\right) - (\mathring{m}_B - \tilde{\mathring{m}}_B)\frac{\chi}{f_\sigma}\mathrm{Tr}\left(\bar{B}B\right) \\ &\quad - (g_{A_1} - \tilde{g}_{A_1})\mathrm{Tr}\left(\bar{B}\gamma_\mu\gamma_5 \left\{\hat{\alpha}_\perp^\mu, B\right\}\right) \\ &\quad - (g_{A_2} - \tilde{g}_{A_2})\mathrm{Tr}\left(\bar{B}\gamma_\mu\gamma_5 \left[\hat{\alpha}_\perp^\mu, B\right]\right) \\ &\quad + (g_{V_1} - \tilde{g}_{V_1})\mathrm{Tr}\left(\bar{B}\gamma_\mu \left\{\hat{\alpha}_\parallel^\mu, B\right\}\right) \\ &\quad + (g_{V_2} - \tilde{g}_{V_2})\mathrm{Tr}\left(\bar{B}\gamma_\mu \left[\hat{\alpha}_\parallel^\mu, B\right]\right)\Big]\left(\frac{\chi}{f_\sigma}\right)^{\beta'}, \end{aligned} \tag{6.41b}$$

where $\mathring{m}_B$ and $\tilde{\mathring{m}}_B$ account for the baryon mass in the chiral limit, all the coupling constants are dimensionless and the covariant derivative is defined as

$$D^\mu B = \partial^\mu B - i\left[V^\mu, B\right]. \tag{6.42}$$

Note that, due to the conservation of the vector currents, there is no term $\mathrm{Tr}\left(\bar{B}\gamma_\mu\partial^\mu\chi B\right)$ in Eq. (6.41a), hence no term $\mathrm{Tr}\left(\bar{B}\gamma_\mu\partial^\mu\chi B\right)(\chi/f_\sigma)^\beta$ in Eq. (6.41b).

In $\mathcal{L}_{bs\mathrm{HLS}}$, both $\mathring{m}_B$ and $\tilde{\mathring{m}}_B$ are taken as scale invariant quantities, so we couple the dilaton to the baryon mass term in terms of the conformal compensator to make the term $\tilde{\mathring{m}}_B(\chi/f_\sigma)\mathrm{Tr}\left(\bar{B}B\right)$ in Eq. (6.41a) scale-invariant. It should be noted that there is an ambiguity in the meaning of the baryon mass in the chiral limit when a dilaton is present à la CT. This is because in the CT theory, the quark condensate and the dilaton condensate are locked to each other. This means that in the chiral limit, $\mathring{m}_B$ remains as is but the fluctuating dilaton field couples to the baryon as $g_{\sigma BB}\bar{\psi}\psi$.

As is well known, the chiral counting in chiral Lagrangian including baryon is subtle because of the large baryon mass in the chiral limit. One can avoid the difficulty by resorting to heavy-baryon formalism [Jenkins and Manohar (1990); Bernard, Kaiser, Kambor and Meissner (1992)]. To illustrate how the heavy-baryon formalism works, consider the terms contributing to the baryon mass in (6.40):

$$\mathcal{L}_{bs\mathrm{HLS}}^{\mathrm{mass}} = \left[\tilde{\mathring{m}}_B\frac{\chi}{f_\sigma} + (\mathring{m}_B - \tilde{\mathring{m}}_B)\frac{\chi}{f_\sigma}\left(\frac{\chi}{f_\sigma}\right)^{\beta'}\right]\mathrm{Tr}\left(\bar{B}B\right). \tag{6.43}$$

From this term, one can easily see that, after spontaneous breaking of scale symmetry, i.e., $\langle\chi\rangle = f_\sigma$, the baryon mass becomes $\mathring{m}_B$. For making heavy baryon mass $\mathring{m}_B$ expansion, we first consider a small β', such that

$$\mathcal{L}_{bs\text{HLS}}^{\text{mass}} = \left[\mathring{m}_B + (\mathring{m}_B - \tilde{\mathring{m}}_B)\sum_{n=1}^{\infty}\frac{1}{n!}\left(\beta'\Sigma\right)^n\right]\frac{\chi}{f_\sigma}\text{Tr}\left(\bar{B}B\right). \quad (6.44)$$

We further expand the dilaton field χ around its vacuum expectation value. Then $\mathcal{L}_{bs\text{HLS}}^{\text{mass}}$ is expressed as

$$\begin{aligned}\mathcal{L}_{bs\text{HLS}}^{\text{mass}} &= \mathring{m}_B\text{Tr}\left(\bar{B}B\right) + \mathring{m}_B\sum_{m=1}^{\infty}\frac{1}{m!}\Sigma^m\text{Tr}\left(\bar{B}B\right)\\ &\quad + (\mathring{m}_B - \tilde{\mathring{m}}_B)\sum_{n=1}^{\infty}\frac{1}{n!}\left(\beta'\Sigma\right)^n\\ &\quad \times\left(1 + \sum_{m=1}^{\infty}\frac{1}{m!}\Sigma^m\right)\text{Tr}\left(\bar{B}B\right). \end{aligned} \quad (6.45)$$

In this Lagrangian, the first term is the baryon mass term in the chiral limit while the other terms are the $\sigma^n\bar{B}B\,(n = 1, 2, \ldots)$ interaction terms. For example, the σ-B-B coupling is written as

$$g_{\sigma BB} = \mathring{m}_B\frac{1}{f_\sigma} + (\mathring{m}_B - \tilde{\mathring{m}}_B)\beta'\frac{1}{f_\sigma}, \quad (6.46)$$

which, after removing the explicit scale symmetry breaking by setting $\beta' = 0$, is the Goldberger–Trieman type relation [Caruthers (1970)] in the dilaton sector in the chiral limit.

To finalize the heavy-baryon expansion, we should set up the scale-chiral counting of the interaction terms. Since the dilaton couples to baryons non-derivatively, one cannot do the usual power counting as with the derivative in pion-nucleon coupling. In the present work, in the absence of first-principle guidance, we establish the power counting using a numerical estimation. If we take the nucleon mass in the chiral limit $\mathring{m}_B \simeq 900$ MeV [Li *et al.* (2009)], by taking $f_\sigma \simeq f_\pi$, we obtain $g_{\sigma BB} \simeq 10$ which is close to $g_{\pi NN} = 13$ [Schroder *et al.* (2001)]. This suggests that the other terms in (6.45) could be considered as of scale-chiral order $O(p)$. That is, in terms of the compensator χ

$$\begin{aligned} S &\equiv \mathring{m}_B\left(\frac{\chi}{f_\sigma} - 1\right) + (\mathring{m}_B - \tilde{\mathring{m}}_B)\left[\left(\frac{\chi}{f_\sigma}\right)^{\beta'} - 1\right]\frac{\chi}{f_\sigma}\\ &\equiv \mathring{m}_B\left(\frac{\chi}{f_\sigma} - 1\right) + (\mathring{m}_B - \tilde{\mathring{m}}_B)\frac{\chi}{f_\sigma}S'\\ &\sim O(p), \end{aligned} \quad (6.47)$$

where $S' = \sum_{n=1}^{\infty} \frac{1}{n!} (\beta'\Sigma)^n$. The NG boson σ appears in the similar fashion as the pion in $U(x)$. Once the scale-chiral counting discussed here is accepted, the heavy-baryon formalism of bsHLS can be derived straightforwardly.

It is more convenient for our treatment to rewrite Lagrangian (6.40) as

$$\begin{aligned}
\mathcal{L}_{bs\text{HLS}}^{\text{LO}} = {} & \text{Tr}\left(\bar{B} i\gamma_\mu D^\mu B\right) - \mathring{m}_B \text{Tr}\left(\bar{B}B\right) \\
& - \text{Tr}\left(\bar{B}B\right) S + (1-g_1)\text{Tr}\left(\bar{B}i\gamma_\mu D^\mu B\right) S' \\
& + \text{Tr}\left(\bar{B}\gamma_\mu\gamma_5 A_1^\mu B\right) + \text{Tr}\left(\bar{B}\gamma_\mu\gamma_5 B A_2^\mu\right) \\
& + \text{Tr}\left(\bar{B}\gamma_\mu V_1^\mu B\right) + \text{Tr}\left(\bar{B}\gamma_\mu B V_2^\mu\right),
\end{aligned} \tag{6.48}$$

where

$$\begin{aligned}
S &= \mathring{m}_B \left(\frac{\chi}{f_\sigma} - 1\right) + (\mathring{m}_B - \tilde{\mathring{m}}_B)\frac{\chi}{f_\sigma} S' \\
A_1^\mu &= \left\{ - g_A^{(+)} - \left[g_A^{(+)} - \tilde{g}_A^{(+)}\right] S' \right\} \hat{\alpha}_\perp^\mu, \\
A_2^\mu &= \left\{ - g_A^{(-)} - \left[g_A^{(-)} - \tilde{g}_A^{(-)}\right] S' \right\} \hat{\alpha}_\perp^\mu, \\
V_1^\mu &= \left\{ g_V^{(+)} + \left[g_V^{(+)} - \tilde{g}_V^{(+)}\right] S' \right\} \hat{\alpha}_\parallel^\mu, \\
V_2^\mu &= \left\{ g_V^{(-)} + \left[g_V^{(-)} - \tilde{g}_V^{(-)}\right] S' \right\} \hat{\alpha}_\parallel^\mu
\end{aligned} \tag{6.49}$$

with $g_{A,V}^{(\pm)} = g_{A_1,V_1} \pm g_{A_2,V_2}$. All the quantities in Eq. (6.49) are of scale-chiral order $O(p)$. As a result, we can formulate the heavy baryon expansion along the standard approach available in the literature. From Eq. (6.48), one obtains the equation of motion of $B(x)$ as

$$\begin{aligned}
0 = {} & i\gamma_\mu D^\mu B - \mathring{m}_B B - SB + i(1-g_1)S'\gamma_\mu D^\mu B \\
& + \gamma_\mu\gamma_5 A_1^\mu B + \gamma_\mu\gamma_5 B A_2^\mu + \gamma_\mu V_1^\mu B + \gamma_\mu B V_2^\mu .
\end{aligned} \tag{6.50}$$

We next define

$$B(x) = e^{-i\mathring{m}_B v\cdot x}\left[\mathcal{B}(x) + b(x)\right], \tag{6.51}$$

where the four-vector v_μ is the velocity of baryon and $\mathcal{B}(x)$ and $b(x)$ are, respectively, the heavy and light components of $B(x)$,

$$\begin{aligned}
\mathcal{B}(x) &= e^{i\mathring{m}_B v\cdot x}\frac{1+\not{v}}{2}B(x), \\
b(x) &= e^{i\mathring{m}_B v\cdot x}\frac{1-\not{v}}{2}B(x).
\end{aligned} \tag{6.52}$$

Substituting (6.51) to (6.50) one can express the small component $b(x)$ in terms of the large component $\mathcal{B}(x)$ as

$$
\begin{aligned}
b(x) = &-\{[2\mathring{m}_B + iv\cdot D + S + i(1-g_1)S'v\cdot D - (1-g_1)\mathring{m}_B S' \\
&- \gamma^{\perp}_{\mu}\gamma_5 A^{\mu}_1 + v\cdot V_1\delta^{bi} + [\gamma^{\perp}_{\mu}\gamma_5 A^{\mu}_{2,bi} + v\cdot V_{2,bi}]\}^{-1} \\
&\times [i\gamma^{\perp}_{\mu}D^{\mu}\mathcal{B} - S\mathcal{B} - (1-g_1)\mathring{m}_B S'\mathcal{B} \\
&+ i(1-g_1)S'\gamma^{\perp}_{\mu}D^{\mu}\mathcal{B} - v\cdot A_1\gamma_5\mathcal{B} + \mathcal{B}v\cdot A_2\gamma_5 + \gamma^{\perp}_{\mu}(V^{\mu}_1\mathcal{B} + \mathcal{B}V^{\mu}_2)],
\end{aligned}
\tag{6.53}
$$

which shows that the small component is suppressed by $1/(2\mathring{m}_B)$. Then the effective Lagrangian in terms of the large component of $\mathcal{B}(x)$ is expressed as

$$
\begin{aligned}
\mathcal{L} = &\,\overline{\mathcal{B}}\,[iv\cdot D\mathcal{B} - S\mathcal{B} + i(1-g_1)S'v\cdot D\mathcal{B} - (1-g_1)\mathring{m}_B S'\mathcal{B} \\
&+ \gamma^{\perp}_{\mu}\gamma_5(A^{\mu}_1\mathcal{B} + \mathcal{B}A^{\mu}_2) + v\cdot V_1\mathcal{B} + \mathcal{B}v\cdot V_2] \\
&- \overline{\mathcal{B}}\,[i\gamma^{\perp}_{\mu}D^{\mu} - 2\mathring{m}_B \\
&- S + i(1-g_1)S'\gamma^{\perp}_{\mu}D^{\mu} - (1-g_1)\mathring{m}_B S' + v\cdot A_1\gamma_5 + \gamma^{\perp}_{\mu}V^{\mu}_1] \\
&\times [2\mathring{m}_B - (1-g_1)\mathring{m}_B S' \\
&+ iv\cdot D + S + i(1-g_1)S'v\cdot D - \gamma^{\perp}_{\mu}\gamma_5 A^{\mu}_1 + v\cdot V_1]^{-1} \\
&\times [i\gamma^{\perp}_{\mu}D^{\mu}\mathcal{B} - S\mathcal{B} - (1-g_1)\mathring{m}_B S'\mathcal{B} + i(1-g_1)S'\gamma^{\perp}_{\mu}D^{\mu}\mathcal{B} \\
&- v\cdot A_1\gamma_5\mathcal{B} + \mathcal{B}v\cdot A_2\gamma_5 + \gamma^{\perp}_{\mu}(V^{\mu}_1\mathcal{B} + \mathcal{B}V^{\mu}_2)].
\end{aligned}
\tag{6.54}
$$

At the leading order of $1/\mathring{m}_B$, up to $O(p)$ in the chiral order, one has

$$
\begin{aligned}
\mathcal{L}_{1/\mathring{m}_B} = &\,\overline{\mathcal{B}}\,[iv\cdot D\mathcal{B} - S\mathcal{B} + i(1-g_1)S'v\cdot D\mathcal{B} - (1-g_1)\mathring{m}_B S'\mathcal{B} \\
&+ \gamma^{\perp}_{\mu}\gamma_5(A^{\mu}_1\mathcal{B} + \mathcal{B}A^{\mu}_2) + v\cdot V_1\mathcal{B} + \mathcal{B}v\cdot V_2] \\
&+ \overline{\mathcal{B}}\,[i\gamma^{\perp}_{\mu}D^{\mu}\mathcal{B} - S\mathcal{B} - (1-g_1)\mathring{m}_B S'\mathcal{B} + i(1-g_1)S'\gamma^{\perp}_{\mu}D^{\mu}\mathcal{B} \\
&- v\cdot A_1\gamma_5\mathcal{B} + \mathcal{B}v\cdot A_2\gamma_5 + \gamma^{\perp}_{\mu}(V^{\mu}_1\mathcal{B} + \mathcal{B}V^{\mu}_2)]\,.
\end{aligned}
\tag{6.55}
$$

The extension to higher orders can be written down in a straight forward way. We will not write down the explicit Lagrangian here.

6.5 Dilatonic chiral effective theory for non-abelian gauge theory in the Veneziano limit

In this section the alternative approach to scale symmetry with an IR fixed point discussed by Golterman and Shamir [Golterman and Shamir (2016)] is described for completeness. As discussed in [Li, Wen, Ma and Rho (2018)], this approach, although scale expansion is different, is related in scale-chiral perturbation scheme to the CT approach. In fact to the leading order (LO) in the scale-chiral counting, the two are the same in structure.

Golterman and Shamir (GS) posit that there is an IR fixed point at a critical number of flavors N_f^c and consider approaching N_f^c from below. This is similar to approaching the "sill" to the conformal window for large N_f as is done for dilatonic Higgs. Here one simply assumes that there is such N_f^c, not necessarily as large as for dilatonic Higgs. GS develop a scale-chiral symmetric EFT involving a dilaton and pions taking the limit of both N_f and N_c going to ∞ with N_f/N_c held fixed $n_f = N_f/N_c$, which is what is known as the Veneziano limit [Veneziano (1976)]. We make the assumption in nuclear physics that taking $N_f = 3$ to be "large" makes as much sense as taking $N_c = 3$ to be large. Whether this is a valid assumption is an open question. It appears that this assumption holds true in the analyses thus far made in nuclear processes [Li, Wen, Ma and Rho (2018)]. In what follows, we examine the GS scheme, which appears to be quite different physics-wise from the CT scheme, set up the systematic power counting up to NLO incorporating baryons and HLS fields as was done in [Li, Ma and Rho (2017)] for the CT approach. We will then show that in the leading order in scale-chiral symmetry, we get exactly the same LOSS Lagrangian as in the CT scheme. This suggests that the property of LOSS in medium is generic, independent of the nature of how scale symmetry arises. We suggest this as an indication that the notion of emergent scale symmetry and the role of dilaton for the scalar in dense baryonic matter are promising for the physics of dense matter.

GS assume that in the Veneziano limit, the beta function for α_s scales as

$$\beta(N_c\alpha_s) = O(\Delta) + O(1/N_c) \tag{6.56}$$

where $\Delta = |n_f^c - n_f|$. With the chiral counting taken the same as in χPT, the power counting with scale symmetry is then given by $\Delta \sim O(\partial^2)$ — in place of $|\alpha_s - \alpha_{IR}|$ in the CT — in expanding around the IR fixed point. Thus the scale-chiral counting is

$$O(p^2) \sim O(\partial^2) \sim O(m_\pi^2) \sim O(\Delta). \tag{6.57}$$

We take the scale-invariant Lagrangian of GS [Golterman and Shamir (2016)][5]

$$\mathcal{L}_{\chi\mathrm{PT}_\sigma}^{\mathrm{LO;GS}} = \mathcal{L}_{\chi\mathrm{PT}_\sigma;\pi}^{\mathrm{LO;GS}} + \mathcal{L}_{\chi\mathrm{PT}_\sigma;\sigma}^{\mathrm{LO;GS}} + \mathcal{L}_{\chi\mathrm{PT}_\sigma;m}^{\mathrm{LO;GS}} + \mathcal{L}_{\chi\mathrm{PT}_\sigma;d}^{\mathrm{LO;GS}}, \tag{6.58}$$

[5]Note that we are using the notations that are different from those of GS. They are closer to the notations used in LMR [Li, Ma and Rho (2017)] for the CT scheme.

with

$$\mathcal{L}^{\mathrm{LO;GS}}_{\chi\mathrm{PT}_\sigma;\pi} = \frac{f_\pi^2}{4} V_\pi(\sigma-\eta)\left(\frac{\chi}{f_\sigma}\right)^2 \mathrm{Tr}\left(\partial_\mu U \partial^\mu U^\dagger\right), \tag{6.59a}$$

$$\mathcal{L}^{\mathrm{LO;GS}}_{\chi\mathrm{PT}_\sigma;\sigma} = \frac{1}{2} V_\sigma(\sigma-\eta)\partial_\mu \chi \partial^\mu \chi, \tag{6.59b}$$

$$\mathcal{L}^{\mathrm{LO;GS}}_{\chi\mathrm{PT}_\sigma;m} = \frac{f_\pi^2}{4} V_M(\sigma-\eta)\left(\frac{\chi}{f_\sigma}\right)^{3-\gamma_m} \mathrm{Tr}\left(\mathcal{M}^\dagger U + h.c.\right), \tag{6.59c}$$

$$\mathcal{L}^{\mathrm{LO;GS}}_{\chi\mathrm{PT}_\sigma;d} = \left(\frac{\chi}{f_\sigma}\right)^4 V_d(\sigma-\eta), \tag{6.59d}$$

where U is the chiral (NG pion) field $U(x) = \exp\left(2i\pi_a T_a/f_\pi\right)$ and $\chi(x) = f_\sigma e^{\sigma/f_\sigma}$ is the conformal compensator defined above. The "invariant potentials" $V_{\pi,\sigma,M,d}$ are unknown potentials that are functionals of the dilaton field σ and the spurion field η that transforms like σ. These potentials, chiral invariant by construction, are scale-invariant since the inhomogeneous terms in the transformations of σ and η cancel. It turns out that the quantities dependent on $(\sigma-\eta)$ are suppressed by the power of Δ which is a small quantity in the Veneziano limit [Golterman and Shamir (2016, 2018)]. Now the trace anomaly, i.e., explicit breaking of scale symmetry, is generated, while chiral symmetry remains intact, when the external source is turned off. To make a systematic expansion in Δ, one expands the potential V_I terms of the argument $\sigma-\eta$

$$V_I = \sum_{n=0}^{\infty} V_I^{(n)} = \sum_{n=0}^{\infty} \frac{c_{I,n}}{n!}(\sigma-\eta)^n, \tag{6.60}$$

and then makes explicit the dependence on $(n_f - n_f^*)^n$ by expanding $c_{I,n}$ as

$$c_{I,n} = \sum_{k=0}^{\infty} \tilde{c}_{I,nk}(n_f - n_f^*)^k. \tag{6.61}$$

It was shown in [Golterman and Shamir (2016)] by matching correlation functions of the low-energy effective theory to those of the microscopic theory that $V_I^{(n)}$ scales as

$$V_I^{(n)}(\sigma, \eta = 0) = O(\Delta^n) \tag{6.62}$$

with $\tilde{c}_{I,nk} = 0$ for $k < n$. It should be noted that in contrast to ordinary chiral perturbation theory in which the low-energy constants are $O(p^0)$ in the chiral counting, the coefficients of the scale-invariant potentials $c_{I,n}$ carry power counting. This is the same as in the Crewther and Tunstall approach [Crewther and Tunstall (2015)] with a nonperturbative IR fixed point in QCD.

6.5.1 *Leading-order scale symmetry (LOSS)*

We first look at the leading-order scale-chiral symmetry in this GS scheme.

The terms multiplying $V_{\pi,\sigma,M}$ are of $O(p^2)$ in the chiral counting. Therefore to the LO in scale-chiral counting, we can take $V_\pi = V_\tau = V_M = 1$ since the correction terms are suppressed by the powers of Δ. This is the consequence of the Veneziano limit not present in the large N_c limit. Thus we have

$$\mathcal{L}^{O(p^2)}_{\chi\mathrm{PT}_\sigma;\pi} = \frac{f_\pi^2}{4}\left(\frac{\chi}{f_\sigma}\right)^2 \mathrm{Tr}\left(\partial_\mu U \partial^\mu U^\dagger\right), \tag{6.63a}$$

$$\mathcal{L}^{O(p^2)}_{\chi\mathrm{PT}_\sigma;\sigma} = \frac{1}{2}\partial_\mu \chi \partial^\mu \chi, \tag{6.63b}$$

$$\mathcal{L}^{O(p^2)}_{\chi\mathrm{PT}_\sigma;m} = \frac{f_\pi^2}{4}\left(\frac{\chi}{f_\sigma}\right)^{3-\gamma_m} \mathrm{Tr}\left(\mathcal{M}^\dagger U + U^\dagger \mathcal{M}\right). \tag{6.63c}$$

Now as for the term (6.59d) which gives the dilaton potential, scale symmetry is not broken spontaneously if one takes $V_d = 1$ because the dilaton potential $\mathcal{V} \equiv \left(\frac{\chi}{f_\sigma}\right)^4 V_d(\sigma - \tau) \sim \left(\frac{\chi}{f_\sigma}\right)^4$ cannot pick up a non-zero vacuum expectation value. This is associated with that scale symmetry cannot be spontaneously broken unless it is explicitly broken. Thus one is required to go to $O(\Delta) \sim O(p^2)$ in scale-chiral counting,

$$\mathcal{L}^{O(\Delta)}_{\chi\mathrm{PT}_\sigma;d} = \left(\frac{\chi}{f_\sigma}\right)^4 \left(c_{d,0} + \tilde{c}_{d,11}(n_f - n_f^*) \ln\frac{\chi}{f_\sigma}\right) \tag{6.64}$$

where we have written $\sigma = \ln \chi/f_\sigma$ and, $c_{d,0}$ and $\tilde{c}_{d,11}$ are both unknown constants. In addition, $(n_f - n_f^*)$ which is also an unknown constant, accounts for the deviation from the IR fixed point, or equivalently, the contribution from the conformal symmetry breaking. From the saddle-point condition, the dilaton potential with lower bound is written as

$$4c_{d,0} + \tilde{c}_{d,11}(n_f - n_f^*) = 0, \tag{6.65}$$

then the dilaton mass in the chiral limit is expressed as

$$\begin{aligned} m_\sigma^2 &= \frac{1}{f_\sigma^2}\left[12 c_{d,0} + 7\tilde{c}_{d,11}(n_f - n_f^*)\right] \\ &= -\frac{4}{f_\sigma^2}\tilde{c}_{d,11}(n_f - n_f^*). \end{aligned} \tag{6.66}$$

So we get the result that the coefficients $c_{d,0}$ and $\tilde{c}_{d,11}(n_f - n_f^*)$ are both $O(\Delta) \sim O(p^2)$ in this chiral-scale counting. Moreover using the saddle point condition, the dilaton potential with lower bound can be written as

$$\tilde{\mathcal{L}}^{O(\Delta)}_{\chi\mathrm{PT}_\sigma;d} = -\frac{m_\sigma^2 f_\sigma^2}{4}\left(\frac{\chi}{f_\sigma}\right)^4 \left(\ln\frac{\chi}{f_\sigma} - \frac{1}{4}\right). \tag{6.67}$$

This is the same form of dilaton potential we obtained in the CT scheme. Here the explicit scale-symmetry breaking sits in the σ mass term, analogous to the pion mass term for chiral symmetry explicit breaking. The Lagrangian consisting of (6.63a), (6.63b), (6.63c) and (6.67) corresponds to the LOSS approximation employed in [Li, Ma and Rho (2017)].

It is interesting to see whether one cannot make a connection between $(n_f - n_f^*)$ and the anomalous dimension β' of $\mathrm{tr}\mathrm{G}_{\mu\nu}\mathrm{G}^{\mu\nu}$ that figures in the CT approach. To do this, we rewrite the potential (6.64)

$$\begin{aligned}\mathcal{L}^{(\Delta)}_{\chi\mathrm{PT}_\sigma;d} &= \left(\frac{\chi}{f_\sigma}\right)^4 \left[(c_{d,0} - \kappa\tilde{c}_{d,11}) + \kappa\tilde{c}_{d,11}\left(1 + \frac{n_f - n_f^*}{\kappa}\ln\frac{\chi}{f_\sigma}\right)\right] \\ &\simeq \left(\frac{\chi}{f_\sigma}\right)^4 \left[(c_{d,0} - \kappa\tilde{c}_{d,11}) + \kappa\tilde{c}_{d,11}\left(\frac{\chi}{f_\sigma}\right)^{\frac{n_f - n_f^*}{\kappa}}\right],\end{aligned} \tag{6.68}$$

where κ is a dimensionless quantity of $O(\Delta)$ such that $(n_f - n_f^*)/\kappa \ll 1$ is $O(1)$ in the scale-chiral counting. If one takes $(n_f - n_f^*)/\kappa \propto \beta'$, then the GS and CT are *formally* the same. The simlar relation can also be checked by looking at the hadron coupling, such as $\sigma\pi\pi$ coupling, in both approaches with $\kappa \sim m_\sigma^2/f_\sigma^2$. [Li, Ma and Rho (2017)]. Although the two approaches may be different in basic premise as pointed out in [Catà, Crewther and Tunstall (2018)], their structure looks very similar.

6.5.2 *Going beyond the LOSS*

6.5.2.1 *Next-to-leading order*

We have seen that to the leading order in scale symmetry, the two approaches CT [Crewther and Tunstall (2015)] and GS [Golterman and Shamir (2016)] are related if one assumes fluctuating near the respective IR fixed point. In this section, we extend the analysis to the next-to-leading order (NLO) in chiral scale counting. We will also include both hidden local symmetry and baryons. We will argue that applied to nuclear interactions, the concept of scale symmetry manifested in the process involving dilaton as the scalar relevant in nuclear physics is the same in the two schemes at the NLO.

What we will do is to detail how to go to the NLO in general in the GS scheme and then summarize what happens when HLS and baryons are involved. For the latter the argumentation made is essentially the same as what's developed in the CT scheme [Li, Ma and Rho (2017)].

To the NLO, there are two terms to consider:

$$\mathcal{L}^{\rm NLO}_{\chi{\rm PT}_\sigma} = \mathcal{L}^{{\rm GS};O(p^4)}_{\chi{\rm PT}_\sigma} + \mathcal{L}^{{\rm GS;LO}\times\Delta n_f}_{\chi{\rm PT}_\sigma}, \tag{6.69}$$

where $\mathcal{L}^{O(p^4)}_{\chi{\rm PT}_\sigma}$ comes from counting the number of the derivative operators acting on the NGBs, here pion and σ, and the quark mass operator including its anomalous dimension γ_m and $\mathcal{L}^{{\rm LO}\times\Delta n_f}_{\chi{\rm PT}_\sigma}$ comes from the next to leading order of the V_I in the Lagrangian (6.58).

The Lagrangian $\mathcal{L}^{{\rm GS};O(p^4)}_{\chi{\rm PT}_\sigma}$ can be written down straightforwardly

$$\begin{aligned}
\mathcal{L}^{{\rm GS};O(p^4)}_{\chi{\rm PT}_\sigma} = {} & L_1\left[{\rm Tr}\left(\partial_\mu U^\dagger\partial^\mu U\right)\right]^2 + L_2{\rm Tr}\left(\partial_\mu U^\dagger\partial_\nu U\right){\rm Tr}\left(\partial^\mu U^\dagger\partial^\nu U\right)\\
& + L_3{\rm Tr}\left(\partial_\mu U^\dagger\partial^\mu U\partial_\nu U^\dagger\partial^\nu U\right)\\
& + L_4\left(\frac{\chi}{f_\sigma}\right)^{1-\gamma_m}{\rm Tr}\left(\partial_\mu U^\dagger\partial^\mu U\right){\rm Tr}\left(\mathcal{M}^\dagger U + U^\dagger\mathcal{M}\right)\\
& + L_5\left(\frac{\chi}{f_\sigma}\right)^{1-\gamma_m}{\rm Tr}\left[\partial_\mu U^\dagger\partial^\mu U\left(\mathcal{M}^\dagger U + U^\dagger\mathcal{M}\right)\right]\\
& + L_6\left(\frac{\chi}{f_\sigma}\right)^{2(1-\gamma_m)}\left[{\rm Tr}\left(\mathcal{M}^\dagger U + U^\dagger\mathcal{M}\right)\right]^2\\
& + L_7\left(\frac{\chi}{f_\sigma}\right)^{2(1-\gamma_m)}\left[{\rm Tr}\left(\mathcal{M}^\dagger U - U^\dagger\mathcal{M}\right)\right]^2\\
& + L_8\left(\frac{\chi}{f_\sigma}\right)^{2(1-\gamma_m)}{\rm Tr}\left(\mathcal{M}^\dagger U\mathcal{M}^\dagger U + U^\dagger\mathcal{M}U^\dagger\mathcal{M}\right)\\
& + H_2\left(\frac{\chi}{f_\sigma}\right)^{2(1-\gamma_m)}{\rm Tr}\left(\mathcal{M}^\dagger\mathcal{M}\right)\\
& + J_1\partial_\nu\Sigma\partial^\nu\Sigma{\rm Tr}\left(\partial_\mu U\partial^\mu U^\dagger\right)\\
& + J_2\partial_\mu\Sigma\partial^\nu\Sigma{\rm Tr}\left(\partial_\nu U\partial^\mu U^\dagger\right)\\
& + J_3\partial_\mu\Sigma\partial^\mu\Sigma\partial_\nu\left(\frac{\chi}{f_\sigma}\right)\partial^\nu\left(\frac{\chi}{f_\sigma}\right)\\
& + J_4\left(\frac{\chi}{f_\sigma}\right)^{1-\gamma_m}\partial_\mu\Sigma\partial^\mu\Sigma{\rm Tr}\left(\mathcal{M}^\dagger U + U^\dagger\mathcal{M}\right)
\end{aligned} \tag{6.70}$$

where L_i, H_i and J_i's are, as before, arbitrary constants. By comparing with what's given at the same order in Ref. [Li, Ma and Rho (2017)] in the CT approach, one can see that this Lagrangian is the same as $\mathcal{L}_{\chi\text{PT}_\sigma}^{O(p^4)}$ in Ref. [Li, Ma and Rho (2017)].

To construct $\mathcal{L}_{\chi\text{PT}_\sigma}^{\text{LO}\times\Delta n_f}$, let us look at V_I, $I = (\pi, \sigma, M)$. Using the power counting given above, we can extract the $O(\Delta)$ terms of V_I

$$V_I(\sigma) = \left(\tilde{c}_{I,01} + \tilde{c}_{I,11} \ln \frac{\chi}{f_\sigma}\right)(n_f - n_f^*). \tag{6.71}$$

We thus obtain the following contributions to the NLO Lagrangian

$$\begin{aligned}
\mathcal{L}_{\chi\text{PT}_\sigma;\pi}^{\text{NLO;GS}} &= \frac{f_\pi^2}{4}\left(\tilde{c}_{\pi,01} + \tilde{c}_{\pi,11} \ln \frac{\chi}{f_\sigma}\right)(n_f - n_f^*)\left(\frac{\chi}{f_\sigma}\right)^2 \\
&\quad \times \mathrm{Tr}\left(\partial_\mu U \partial^\mu U^\dagger\right), &(6.72\text{a})\\
\mathcal{L}_{\chi\text{PT}_\sigma;\sigma}^{\text{NLO;GS}} &= \frac{1}{2}\left(\tilde{c}_{\sigma,01} + \tilde{c}_{\sigma,11} \ln \frac{\chi}{f_\sigma}\right)(n_f - n_f^*) \\
&\quad \partial_\mu \chi \partial^\mu \chi, &(6.72\text{b})\\
\mathcal{L}_{\chi\text{PT}_\sigma;m}^{\text{NLO;GS}} &= \frac{f_\pi^2}{4}\left(\tilde{c}_{M,01} + \tilde{c}_{M,11} \ln \frac{\chi}{f_\sigma}\right)(n_f - n_f^*) \\
&\quad \times \left(\frac{\chi}{f_\sigma}\right)^{3-\gamma_m} \mathrm{Tr}\left(\mathcal{M}^\dagger U + h.c.\right). &(6.72\text{c})
\end{aligned}$$

In this Lagrangian, $\Delta n_f = n_f - n_f^*$, $O(p^2)$ in the scale-chiral counting, is the key quantity in the GS theory in close analogy to $\Delta\alpha_s$ in the CT theory and it should be replaced with the following $O(p^2)$ chiral, parity and Lorentz invariant quantities

$$\mathrm{Tr}(\partial_\mu U \partial^\mu U^\dagger), \quad \partial_\mu \chi \partial^\mu \chi, \quad \mathrm{Tr}(\mathcal{M}^\dagger U + h.c.). \tag{6.73}$$

Consequently the $\tilde{c}_{I,01}$ terms can be combined into the $\mathcal{L}_{\chi\text{PT}_\sigma}^{\text{GS};O(p^4)}$ term while the $\tilde{c}_{I,11}$ terms are equivalent to $\mathcal{L}_{\chi\text{PT}_\sigma}^{\text{LO}\times\Delta\alpha_s}$ of Ref. [Li, Ma and Rho (2017)] in terms of β' in the CT approach.

Finally we turn to the dilaton potential term. Up to the order $(n_f - n_f^*)^2$, we have

$$\begin{aligned}
\mathcal{L}_d^{\text{LO;GS}} = {} & \left[\tilde{c}_{d,00} + \left(\tilde{c}_{d,01} + \tilde{c}_{d,11} \ln \frac{\chi}{f_\sigma}\right)(n_f - n_f^*)\right. \\
& + \left(\tilde{c}_{d,02} + \tilde{c}_{d,12} \ln \frac{\chi}{f_\sigma}\right. \\
& \left.\left. + \tilde{c}_{d,22}\left(\ln \frac{\chi}{f_\sigma}\right)^2\right)(n_f - n_f^*)^2\right]\left(\frac{\chi}{f_\sigma}\right)^4.
\end{aligned} \tag{6.74}$$

One can see that, after the rearrangement of different terms, the dilaton potential in GS is of the same form as that in CT with β' expanded to to the second order of $\Delta\alpha_s = \alpha_s - \alpha_{\rm IR}$.

6.5.2.2 *Introducing hidden local symmetry and baryons*

Generalizing the scale-chiral Lagrangian of GS written above to hidden local symmetric bosons and baryons follows the same procedure of Ref. [Li, Ma and Rho (2017)], so we won't give the details here. It suffices to summarize the results. This we will do for the HLS fields. Given the HLS Lagrangian, putting scale symmetry is done straightforwardly using the conformal compensator field.

To the leading order, the scale-symmetric HLS Lagrangian is (using the two-forms defined before)

$$\begin{aligned}\mathcal{L}^{\rm LO;GS}_{\rm HLS_\sigma} = {} & \mathcal{L}^{\rm LO;GS}_{\rm HLS_\sigma;\pi} + \mathcal{L}^{\rm LO;GS}_{\rm HLS_\sigma;\rho} + \mathcal{L}^{\rm LO;GS}_{\rm HLS_\sigma;kin} \\ & + \mathcal{L}^{\rm LO;GS}_{\rm HLS_\sigma;\sigma} + \mathcal{L}^{\rm LO;GS}_{\rm HLS_\sigma;m} + \mathcal{L}^{\rm LO;GS}_{\rm HLS_\sigma;d}.\end{aligned} \tag{6.75}$$

The last three terms are of the same form as Eqs. (6.59b), (6.59c) and (6.59d) with the potentials $V(\sigma-\tau)$ replaced by $V_{\rm HLS}(\sigma-\tau)$. As for the first three terms, the scale invariant potentials $V_{\rm HLS}(\sigma-\tau)$s can be set equal to 1 as discussed above and one gets

$$\mathcal{L}^{O(p^2);\rm GS}_{\rm HLS_\sigma;\pi} = f_\pi^2\left(\frac{\chi}{f_\sigma}\right)^2 {\rm tr}(\hat{\alpha}_{\perp\mu}\hat{\alpha}_\perp^\mu), \tag{6.76a}$$

$$\mathcal{L}^{O(p^2);\rm GS}_{\rm HLS_\sigma;\rho} = a f_\pi^2\left(\frac{\chi}{f_\sigma}\right)^2 {\rm tr}(\hat{\alpha}_{\parallel\mu}\hat{\alpha}_\parallel^\mu), \tag{6.76b}$$

$$\mathcal{L}^{O(p^2);\rm GS}_{\rm HLS_\sigma;kin} = -\frac{1}{2g^2}{\rm tr}(V_{\mu\nu}V^{\mu\nu}). \tag{6.76c}$$

Now to the next-to leading order, there are two terms as in the case of $\chi{\rm PT}_\sigma$,

$$\mathcal{L}^{\rm NLO}_{\rm HLS_\sigma} = \mathcal{L}^{{\rm LO}\times\Delta n_f}_{\rm HLS_\sigma} + \mathcal{L}^{O(p^4)}_{\rm HLS_\sigma}. \tag{6.77}$$

For the first term as for $\chi{\rm PT}_\sigma$, we write the potential to $O(p^2)$

$$V_{\rm HLS_\sigma;I} = (n_f - n_f^*)\left(\tilde{h}_{I,01} + \tilde{h}_{I,11}\ln\frac{\chi}{f_\sigma}\right). \tag{6.78}$$

Here h replaces c as the coefficients for ${\rm HLS}_\sigma$. Then the Lagrangian $\mathcal{L}^{{\rm LO}\times\Delta n_f}_{\rm HLS_\sigma}$ can be written as

$$\begin{aligned}\mathcal{L}^{{\rm LO}\times\Delta n_f}_{\rm HLS_\sigma} = {} & \mathcal{L}^{{\rm LO}\times\Delta n_f}_{\rm HLS_\sigma;\pi} + \mathcal{L}^{{\rm LO}\times\Delta n_f}_{\rm HLS_\sigma;\rho} + \mathcal{L}^{{\rm LO}\times\Delta n_f}_{\rm HLS_\sigma;kin} \\ & + \mathcal{L}^{{\rm LO}\times\Delta n_f}_{\rm HLS_\sigma;\sigma} + \mathcal{L}^{{\rm LO}\times\Delta n_f}_{\rm HLS_\sigma;m} + \mathcal{L}^{{\rm LO}\times\Delta n_f}_{\rm HLS_\sigma;d},\end{aligned} \tag{6.79}$$

where the first three terms are of the same form as (6.59b), (6.59c) and (6.59d) and the next three terms are

$$
\begin{aligned}
\mathcal{L}^{\mathrm{LO}\times\Delta n_f}_{\mathrm{HLS}_\sigma;\pi} &= f_\pi^2\left(\tilde{h}_{\pi,01}+\tilde{h}_{\pi,11}\ln\frac{\chi}{f_\sigma}\right)(n_f-n_f^*)\\
&\quad\times\left(\frac{\chi}{f_\sigma}\right)^2\mathrm{tr}(\hat{\alpha}_{\perp\mu}\hat{\alpha}_\perp^\mu),\\
\mathcal{L}^{\mathrm{LO}\times\Delta n_f}_{\mathrm{HLS}_\sigma;\rho} &= f_\kappa^2\left(\tilde{h}_{\kappa,01}+\tilde{h}_{\kappa,11}\ln\frac{\chi}{f_\sigma}\right)(n_f-n_f^*)\\
&\quad\times\left(\frac{\chi}{f_\sigma}\right)^2\mathrm{tr}(\hat{\alpha}_{\parallel\mu}\hat{\alpha}_\parallel^\mu),\\
\mathcal{L}^{\mathrm{LO}\times\Delta n_f}_{\mathrm{HLS}_\sigma;\mathrm{kin}} &= -\frac{1}{2g^2}\left(\tilde{h}_{v,01}+\tilde{h}_{v,11}\ln\frac{\chi}{f_\sigma}\right)(n_f-n_f^*)\\
&\quad\times\mathrm{tr}(V_{\mu\nu}V^{\mu\nu}).
\end{aligned}
\tag{6.80}
$$

Again the quantity $(n_f-n_f^*)$ can be substituted by the following Lorentz invariant $O(p^2)$ quantities

$$
\partial_\mu\chi\partial^\mu\chi,\ \ \mathrm{tr}(\hat{\alpha}_{\parallel\mu}\hat{\alpha}_\parallel^\mu),\ \ \mathrm{tr}(\hat{\alpha}_{\perp\mu}\hat{\alpha}_\perp^\mu),\ \ \mathrm{tr}\left(\hat{\mathcal{M}}+\hat{\mathcal{M}}^\dagger\right). \tag{6.81}
$$

Next, we turn to the $\mathcal{L}^{O(p^4)}_{\mathrm{HLS}_\sigma}$ sector. This sector includes two parts: The part which has the same form as the standard HLS in the chiral limit (since this part is already scale invariant) and the other part arising with the dilaton field. The former part is obvious. For the latter, it can be written as

$$
\begin{aligned}
\mathcal{L}^{O(p^4)}_{\mathrm{HLS}_\sigma} &= \left[D_{\pi,1}\left(\partial_\mu\frac{\chi}{f_\sigma}\right)^2+D_{\pi,2}\Box\left(\frac{\chi}{f_\sigma}\right)\right]\mathrm{tr}(\hat{\alpha}_{\perp\nu}\hat{\alpha}_\perp^\nu)\\
&\quad+D_{\pi,3}\partial_\mu\left(\frac{\chi}{f_\sigma}\right)\partial_\nu\left(\frac{\chi}{f_\sigma}\right)\mathrm{tr}(\hat{\alpha}_\perp^\mu\hat{\alpha}_\perp^\nu)\\
&\quad+\left[D_{\kappa,1}\left(\partial_\mu\frac{\chi}{f_\sigma}\right)^2+D_{\kappa,2}\Box\left(\frac{\chi}{f_\sigma}\right)\right]\mathrm{tr}(\hat{\alpha}_{\parallel\nu}\hat{\alpha}_\parallel^\nu)\\
&\quad+D_{\kappa,3}\partial_\mu\left(\frac{\chi}{f_\sigma}\right)\partial_\nu\left(\frac{\chi}{f_\sigma}\right)\mathrm{tr}(\hat{\alpha}_\parallel^\mu\hat{\alpha}_\parallel^\nu)\\
&\quad+D_{m,1}\left(\frac{\chi}{f_\sigma}\right)^{(y-2)}\mathrm{tr}\left[\hat{\alpha}_{\perp\mu}\hat{\alpha}_\perp^\mu(\hat{\mathcal{M}}+\hat{\mathcal{M}}^\dagger)\right]\\
&\quad+D_{m,2}\left(\frac{\chi}{f_\sigma}\right)^{(y-2)}\mathrm{tr}\left[\hat{\alpha}_{\parallel\mu}\hat{\alpha}_\parallel^\mu(\hat{\mathcal{M}}+\hat{\mathcal{M}}^\dagger)\right]\\
&\quad+D_{m,3}\left(\frac{\chi}{f_\sigma}\right)^{(y-2)}\mathrm{tr}\left[\left(\hat{\alpha}_\parallel^\mu\hat{\alpha}_{\perp\mu}-\hat{\alpha}_{\perp\mu}\hat{\alpha}_\parallel^\mu\right)\left(\hat{\mathcal{M}}-\hat{\mathcal{M}}^\dagger\right)\right]
\end{aligned}
\tag{6.82}
$$

where $D_{I,n}(I = \pi, \kappa, m$ and $n = 1, 2, 3)$ are the low-energy constants. Note that there are some terms independent of $\hat{\alpha}_{\|\mu}$ and $\hat{\alpha}_{\perp\mu}$. They have the same forms as that in $\chi\mathrm{PT}_\sigma$, so it is straightforward to write down their explicit forms.

Finally, we introduce [Li, Ma and Rho (2017)] the baryon octet field B which transforms under HLS as $B(x) \to h(x)B(x)h^\dagger(x)$ and use the canonical dimension of the baryon field, i.e., under scale transformation, it transforms as $B(x) = \lambda^{3/2}B(\lambda^{-1}x)$.

To the leading order, that is, $O(p)$ or $O(\Delta^{1/2})$, the scale invariant potential $V_{bs\mathrm{HLS}}(\tau - \sigma)$s can also be set to 1 as before. Thus the Lagrangian for baryons takes the form:

$$\begin{aligned}\mathcal{L}_{bs\mathrm{HLS}}^{\mathrm{LO;GS}} &= \mathrm{tr}(\bar{B}i\gamma_\mu D^\mu B) - \mathring{m}_B\frac{\chi}{f_\sigma}\mathrm{tr}(\bar{B}B)\\ &\quad - g_{A_1}\mathrm{tr}\left(\bar{B}\gamma_\mu\gamma_5\{\hat{\alpha}_\perp^\mu, B\}\right) - g_{A_2}\mathrm{tr}\left(\bar{B}\gamma_\mu\gamma_5[\hat{\alpha}_\perp^\mu, B]\right)\\ &\quad - g_{V_1}\mathrm{tr}\left(\bar{B}\gamma_\mu\{\hat{\alpha}_\|^\mu, B\}\right) - g_{V_2}\mathrm{tr}\left(\bar{B}\gamma_\mu[\hat{\alpha}_\|^\mu, B]\right), \qquad (6.83)\end{aligned}$$

which is the LOSS of the baryonic part.

As for the next to leading order Lagrangian, i.e., $O(p^2)$ or $O(\Delta)$, as in the cases of the $\chi\mathrm{PT}_\sigma$ and HLS_σ, there are two quantities contributing to the power counting: The first $\mathcal{L}_{bs\mathrm{HLS}}^{\mathrm{LO}\times\Delta n_f;GS}$ is from the next to leading order expansion of the scale invariant potentials and the second $\mathcal{L}_{bs\mathrm{HLS}}^{O(p^2)}$ is from the expansion of derivative operators and quark mass matrix.

For $\mathcal{L}_{bs\mathrm{HLS}}^{\mathrm{LO}\times\Delta n_f;GS}$, only the first two terms need to be multiplied by the potentials $V_1(\sigma - \tau)$ and $V_{\mathring{m}}(\sigma - \tau)$, respectively. Therefore, we have

$$\begin{aligned}\mathcal{L}_{bs\mathrm{HLS}}^{\mathrm{LO}\times\Delta n_f} &= \left(\tilde{c}_{1,01} + \tilde{c}_{1,11}\ln\frac{\chi}{f_\sigma}\right)(n_f - n_f^*)\mathrm{tr}(\bar{B}i\gamma_\mu D^\mu B)\\ &\quad - \left(\tilde{c}_{\mathring{m},01} + \tilde{c}_{\mathring{m},11}\ln\frac{\chi}{f_\sigma}\right)(n_f - n_f^*)\mathring{m}_B\mathrm{tr}(\bar{B}B). \qquad (6.84)\end{aligned}$$

Since $\tilde{c}_{1,01} = \tilde{c}_{\mathring{m},01}$ and $\tilde{c}_{1,11} = \tilde{c}_{\mathring{m},11}$, this is of the form

$$\mathcal{L}_{bs\mathrm{HLS}}^{\mathrm{LO}\times\Delta n_f} \propto (n_f - n_f^*)\mathrm{tr}\left[\bar{B}\left(i\gamma_\mu D^\mu - \mathring{m}_B\right)B\right] \sim O(p^3). \qquad (6.85)$$

As for $\mathcal{L}_{bs\mathrm{HLS}}^{O(p^2)}$, one can readily write down the most general Lagrangian following the standard method.

6.6 Applications in nuclear systems

We have written down — to the NLO in scale-chiral counting — sHLS and bsHLS Lagrangians valid in the vacuum. As noted, to the LOSS, the two schemes given above [Crewther and Tunstall (2015); Golterman and Shamir (2016)] are of the same form. With the sliding vacuum effect, they can be applied to baryonic matter at normal nuclear matter density n_0 as well as high density beyond n_0. Given the exploding number of parameters [Li, Ma and Rho (2017)], however, it is at present not feasible to do a systematic high-order scale-chiral perturbation calculations similar to SχPT calculations of the type reviewed, e.g., in [Holt, Rho and Weise (2016)]. At present the best one can do is to work with LO and this will be done in Chapters 8 and 9 in a double-decimation RG using V_{lowk}. In this section, we examine how the LOSS fares in the mean field approximations focusing on general properties that will be needed in the detailed calculations described in the later chapters. As mentioned, this procedure — that incorporates both "intrinsic density dependence (IDD) and "induced density dependence ($\text{DD}_{\text{induced}}$) — corresponds to the single-decimation RG. In the succeeding chapters, the treatment will go beyond the mean field.

To be fully self-contained, the discussion requires the precise information on the IDD. It will be defined in Chapter 8. Here it will be simply referred to as in the preceding chapters without giving specification, but indicated by $*$ on the parameters involved.

6.6.1 Nuclear χPT_σ: Landau Fermi liquid from scale-chiral symmetry

Here we give a heuristic derivation of Landau Fermi-liquid fixed-point theory in the LOSS approximation with *bs*HLS written above to NLO.

6.6.1.1 *The advantage of χPT_σ over $S\chi PT$*

The standard χPT (SχPT) as reviewed in, e.g., [Holt, Rho and Weise (2016)] is based on the assumption that more massive meson degrees of freedom than pions — and multi-body nonlocal interactions mediated by the massive modes — can be integrated out with their effects inherited in the parameters of SχPT Lagrangian. This is evidently justifiable for the vector mesons ρ and ω for densities in the vicinity of n_0. It has been unclear, however, where the scalar degree of freedom that has figured in phenomenological nuclear potential models such as Bonn potential and also in a class of density-functional models, in particular, RMF approaches, fit into the SχPT scheme. An implicit assumption was that a scalar of mass around 600 MeV can be treated as a *local scalar field.* On the contrary, in SχPT, such scalar excitations are to be generated at higher order chiral perturbation, invoking dispersion relations, higher-loop unitarity corrections etc. That such SχPT with baryons included seems, at higher chiral order, to successfully describe nuclear matter indicates that multi-loop corrections do satisfactorily generate appropriate scalar attraction necessary for binding in nuclei and for nuclear matter.

As suggested, the power of the χPT_σ approach would in principle be that the (pseudo-)dilaton is to capture the dynamics of the massive scalar meson at the leading (tree) order together with the pion, relegating loop terms to small corrections. Having σ and π (and also K) on the same footing at the same scale-chiral order could make the calculation of nuclear potentials vastly simpler and more efficient. The role of the dilaton would justify the scalar exchange in the popular Bonn one-boson-exchange two-nucleon potential as that of the pion exchange. More significantly it has the potential advantage of enabling one to make systematic error estimates of higher order terms as developed above, a power that is lacking in phenomenological potential models. It will also affect medium-range three-body potentials mediated by σ exchanges in a controllable way. In what follows in this chapter, we will approach this feature exploiting mean-field approximations. Going beyond the mean field will be discussed in Chapters 8 and 9.

6.6.1.2 *RMF treatment of bsHLS*

In this subsection we treat the scale-chiral Lagrangian with the baryon field explicitly incorporated, *bs*HLS. In the next chapter we will introduce baryons as skyrmions from the scale-chiral Lagrangian *s*HLS.

The Lagrangian we take is given by the sum of the leading-order terms (in the chiral limit) from Eqs. (6.36a) and (6.48) plus a dilaton potential

$$\mathcal{L}_1 = \mathcal{L}^M_{\chi PT_\sigma}(\pi, \chi, V_\mu) + \mathcal{L}^B_{\chi PT_\sigma}(\psi, \pi, \chi, V_\mu) + V(\chi)\,. \tag{6.86}$$

The first is the scale-invariant HLS Lagrangian from Eq. (6.36a)

$$\begin{aligned}\mathcal{L}^M_{\chi PT_\sigma}(\pi, \chi, V_\mu) = f_\pi^2 \left(\frac{\chi}{f_\sigma}\right)^2 \mathrm{Tr}[\hat{a}_{\perp\mu}\hat{a}_\perp^\mu] + a f_\pi^2 \left(\frac{\chi}{f_\sigma}\right)^2 \mathrm{Tr}[\hat{a}_{\parallel\mu}\hat{a}_\parallel^\mu] \\ -\frac{1}{2g^2}\mathrm{Tr}[V_{\mu\nu}V^{\mu\nu}] + \frac{1}{2}\partial_\mu\chi\partial^\mu\chi\end{aligned} \tag{6.87}$$

and the second is the baryonic coupling to *s*HLS fields

$$\mathcal{L}^B_{\chi PT_\sigma}(\psi, \pi, \chi, V_\mu) = \mathrm{Tr}(\bar{B}i\gamma_\mu D^\mu B) - \frac{\chi}{f_\sigma} m_N \mathrm{Tr}(\bar{B}B) + \cdots\,. \tag{6.88}$$

We take the dilaton potential given by Eq. (6.18).

It is worthwhile to clearly state what this Lagrangian bsHLS corresponds to in $\chi\mathrm{PT}_\sigma$ with vector mesons and baryons incorporated. This Lagrangian arises as an LO scale-chiral expansion in both CT and GS, so it is in some sense generic in the approach anchored on an IR fixed point. We limit our attention to CT to be specific.

As written, the Lagrangian (6.86) arises when β' is taken to be zero in the expansion, so they are *formally* in the scale-invariance limit except in the potential. Scale symmetry breaking is put entirely in the dilaton potential. As described in Subsection 6.2.1 as option B, this does not imply that β' is zero. This is analogous to chiral symmetry in that explicit chiral symmetry breaking is put in the quark mass term. Since the dilaton mass is comparable to the kaon mass, we will treat it as scale-chiral perturbation as $m_K^2 \sim O(p^2)$. As pointed out above, β' is of $O(1)$ in the scale-chiral counting, but it is accompanied by the dilaton condensate as will be seen below in the dilaton mass.

In the matter-free vacuum, say, "vacuum" $|0_0\rangle$, $\langle 0_0|\chi|0_0\rangle = f_\sigma$. Shifting $\chi \to f_\sigma + \chi'$, we have the usual HLS Lagrangian with the linear coupling of the scalar field to the matter fields, the mass of the scalar coming from the dilaton potential. Now embed the Lagrangian in the medium. The vacuum will change to $|0_n\rangle$, the subscript n denoting density. Then the dilaton condensate in that n-vacuum must be changed to $\langle 0_n|\chi|0_n\rangle \equiv \langle\chi\rangle^* = f_\sigma^*$ as a solution of the minimum of the potential in medium (6.17). In this n-vacuum, shift the dilaton field

$$\chi = f_\sigma^* + \chi' \tag{6.89}$$

and substitute this into (6.86). We get two terms, one without χ' and one with. The first one gives back the original baryon HLS Lagrangian (without conformal compensator field) with two parameters changed as

$$f_\pi \to f_\pi^* = \frac{f_\sigma^*}{f_\sigma} f_\pi \tag{6.90}$$

$$m_\pi \to m_\pi^* = \sqrt{\frac{f_\sigma^*}{f_\sigma}}\, m_\pi\,. \tag{6.91}$$

The hidden gauge coupling g, and a are not affected by the dilaton condensate.[6] As a consequence the vector meson and nucleon masses will be

[6] a is a parameter in HLS, $a = f_{\bar\sigma}/f_\pi$, where $\bar\sigma$ is the would-be NG boson that's eaten up by the ρ to become massive via Higgs mechanism.

modified as[7]

$$m_V = \sqrt{a} f_\pi g \to m_V^* = a f_\pi^* g = \frac{f_\sigma^*}{f_\sigma} m_V \tag{6.92}$$

$$m_N \to m_N^* = \frac{f_\sigma^*}{f_\sigma} m_N. \tag{6.93}$$

- **Dilaton condensate as an IDD**

Here we state the key assumption made in this volume: *If we knew how f_σ^* depends on QCD variables, then its IDD would be known in terms of the density dependence of those QCD variables that enter. It would be encoded in the effective in-medium dilaton potential. The latter is presently unknown due to paucity of precise knowledge on the scale symmetry breaking, both explicit and spontaneous. In the spirit of the FT, however, one can exploit the relations (6.90) and (6.91) at low density linking the σ decay constant to the pion decay constant, since the pion properties can be extracted from deeply bound pionic systems that can be accessed experimentally.* We expect it is reasonable to assume that this holds at least up to n_0 and possibly slightly above. We will apply it in Chapter 9 up to $n_{1/2} \sim 2n_0$.

Now the Lagrangian with χ' describes σ coupling, in a way consistent with scale symmetry, to the vector mesons and the nucleon, with the modified parameters given above. Particularly important to nuclear dynamics as explained below is the σNN coupling

$$\mathcal{L}_{\sigma NN} = -\frac{m_N}{f_\sigma}\bar{\psi}\psi\chi' = g_{\sigma NN}\bar{\psi}\psi\sigma + \cdots \tag{6.94}$$

where the ellipsis stands for higher fluctuating σ fields and the GT_σ relation (6.5) is used in the last equality. Note that the *σNN coupling remains unscaling at the order considered.*

What remains to be determined is the scaling behavior of the dilaton mass. To the LO considered, the property of the dilaton is entirely in the dilaton potential, so how dilaton mass behaves in medium will depend on its characteristic, which is presently unknown.

- **The dilaton mass**

A clear answer to the dilaton mass problem will require a deep understanding of both explicit and spontaneous symmetry breaking of scale invariance. The only thing one can say is that since the dilaton is a pseudo-NG boson, its behavior could be closer to that of the pion than to the vector meson and the nucleon.

[7]These are the scaling obtained in 1991 [Brown and Rho (1991)].

To get some insight, let us proceed with the property of pion in medium, fairly well studied both in experiments and in theory. For the latter, SχPT is sufficiently suitable.

From Eq. (6.91) we have at the mean-field level

$$\frac{m_\pi *}{m_\pi} \approx \sqrt{\frac{f_\pi^*}{f_\pi}}\,. \tag{6.95}$$

It has been established from the study of deeply bound pionic systems, that $(f_\pi^*/f_\pi)^2 \approx 0.64$ at nuclear matter density $n = n_0$ [Kienle and Yamazaki (2004)]. This gives $m_\pi^*/m_\pi \approx 0.9$. However this is a mean-field result. What is observed in nature [Kienle and Yamazaki (2004)] contains also pion fluctuation effects. The latter can be computed reasonably well with the chiral Lagrangian containing "bare" parameters. It comes out to give $\sim$ 10% increase to the mass at $n = n_0$. To compare with the experiment, one has to combine the two. The net effect is $0.9 \times 1.1 \approx 1$. One interpretation of this result — that the pion mass does not scale in density in nuclear mattter — is that the anomalous dimension for the quark mass term in medium is $\gamma(m) \approx 1$.[8] This means that the pion mass should remain unshifted by matter, at least up to nuclear matter density. This plays an important role in the nuclear tensor force which controls the symmetry energy in compact-star matter. We return to this matter in Chapter 9.

It is not feasible at the moment to exploit a similar strategy for the in-medium mass of the dilaton. We can, however, make an indirect argument based on QCD sum-rule analyses of gluon condensates using Eq. (6.6). Ignoring the pion mass relative to the kaon mass, we have in medium

$$m_\sigma^{*\,2} \approx f_\sigma^{*\,-2}\left[-(4+\beta')\left\langle\frac{\beta(\alpha_s)}{4\alpha_s}G^2\right\rangle^* + 3m_K^{*\,2} f_\pi^{*2}\right]. \tag{6.96}$$

In the chiral limit, the second term is absent. From CT analyses, we know that $\left\langle\frac{\beta(\alpha_s)}{4\alpha_s}G^2\right\rangle^* \sim -\beta' v_2 \left\langle\chi^{4+\beta'}\right\rangle^*$. Therefore to the leading order in β',

$$m_\sigma^{*\,2} \approx c\beta' \left(\langle\chi\rangle^*\right)^2 \tag{6.97}$$

so $m_\sigma^* \to 0$ for $\beta' \to 0$ for $f_\sigma^* = \langle\chi\rangle^* \neq 0$. We will see later that in dense medium, scale symmetry emerges with $\langle\chi\rangle^* \to 0$ without requiring that $\beta' \to 0$.

Now from QCD sum rules, one knows that the gluon condensate is little affected by density. This indicates that $f_\sigma^* = \langle\chi\rangle^*$ is also insensitive to density, particularly above some density. In fact this will emerge in

[8] There is an indication for $\gamma(m) \approx 1$ but in "walking" technicolor theories.

phenomenology as discussed in Chapter 9. Also the s-quark condensate $\langle\bar{s}s\rangle$ is expected to be less sensitive to density than the light-quark condensate, so the kaon mass would also be little affected for not too high density. Thus the in-medium σ mass is affected primarily by the σ-decay constant appearing in the denominator. If one assumes the single infrared mass scale in medium, $f_\sigma^* \approx f_\pi^*$, then the σ mass would stay more or less unscaled in density until the DLFP is reached. Thus for densities up to n_0, we can take

$$\frac{m_\sigma^*}{m_\sigma} \approx \frac{f_\sigma^*}{f_\sigma}. \tag{6.98}$$

This relation will be used in the applications discussed in Chapters 8 and 9.

Now we are ready to see what the mean-field approach with bsHLS gives. For this we expand the Lagrangian (6.86) to the lowest order in the fields with only the baryon N, scalar σ and isoscalar vector field ω_μ, appropriate for symmetric nuclear matter retained (the ρ and π fields can be brought in HLS symmetric way for asymmetric nuclear systems if needed) and get

$$\begin{aligned}\mathcal{L} = &\overline{N}[i\gamma_\mu(\partial^\mu + ig_v\omega^\mu) - m_N^* + g_{\sigma NN}\sigma]N\\ &-\frac{1}{4}F_{\mu\nu}^2 + \frac{m_\omega^{*\,2}}{2}\omega^2 + \frac{1}{2}(\partial_\mu\sigma)^2 - \frac{m_\sigma^{*\,2}}{2}\sigma^2.\end{aligned} \tag{6.99}$$

This is precisely the Lagrangian (4.24) discussed in Chapter 4 for deriving Landau Fermi-liquid fixed-point theory using relativistic mean-field approximations. This Lagrangian with the parameters given by the IDDs (supplemented when required by $\mathrm{DD_{induced}}$) gives an appreciable improvement [Song, Brown, Min and Rho (1997)] over Walecka's linear RMF model [Serot and Walecka (1986)] as discussed in Chapter 4. Thus we have derived the Landau Fermi-liquid structure using scale-chiral HLS.

6.6.1.3 *Interplay of vector and scalar mean fields*

What figures importantly in the mean-field treatment is the large cancellation of the scalar mean field energy — which is attractive — and the vector mean field energy — which is repulsive. The binding energy that results, $\sim$ 16 MeV, is tiny compared with the respective mean field energies $\sim$ 300 MeV. We can capture what's happening here in terms of the nucleon self-energies in nuclear matter obtained in in-medium QCD sum rule calculations. The detailed analysis made in [Cohen, Furnstahl, Griegel and Jin (1995)] can be succinctly summarized by the nucleon self-energies in medium at density n

$$\frac{\Sigma_s}{m_N} \approx -\frac{\sigma_{\pi N}}{m_\pi^2 f_\pi^2} n\,, \tag{6.100}$$

$$\frac{\Sigma_v}{m_N} \approx \frac{8m_q}{m_\pi^2 f_\pi^2} n \tag{6.101}$$

where $\sigma_{\pi N}$ is the πN sigma term and m_q is the average light-quark mass. It is found numerically that at normal nuclear matter density, $\Sigma_s/m_N \approx -(0.3-0.4)$ and $\Sigma_v/m_N \approx (0.3-0.4)$. In terms of the scale-invariant baryonic HLS (bsHLS) Lagrangian, these self-energies correspond to two tadpole diagrams, one involving a σ pole giving the scalar attraction and the other an ω exchange leading to the vector repulsion. Both are big but the sum is tiny. We will see in the next chapter that is focused on dense skyrmion matter that this feature emerges in what appears to be a totally different mechanism. It will be shown that this cancellation persists at densities above the normal, however, highly intricately due to the scaling properties of the dilaton and the ω meson at higher densities. This feature plays a crucial role in generating repulsion in the EoS needed to lead to the massive neutron stars observed recently.

To reiterate what was stated before, which counts as one of the principal theses, the simplest way to understand the impressive success of the RMF approach in nuclei and nuclear matter is that the system is at the Landau Fermi-liquid fixed point [Shankar (1994)] and that the RMF approach captures the Landau Fermi-liquid fixed-point theory. What's more important is that the RMF approximation could continue to be valid in dense medium as long as the system is in Fermi liquid. In Chapters 8 and 9, this mean-field structure will be shown to hold in higher density regime $n \gtrsim 2n_0$. It will be seen that it arises in terms of topology due to the emergence of a half-skyrmion phase at $n = n_{1/2} \sim 2n_0$.

6.6.2 *Matching to QCD and vector manifestation*

Up to the normal nuclear matter density and slightly above, what governs the scaling behavior is the dilaton condensate. For densities much above n_0, other condensates, e.g., quark condensates, that enter via matching with QCD could intervene to bring drastic modifications to nuclear dynamics. The most prominent effect, crucial at high density above $\sim 2n_0$, is the VM associated with the ρ meson introduced in Chapter 5, combined with certain topological effects taking place at higher densities.

While the method presented above with (6.86) that can be given both theoretical and phenomenological support near the normal nuclear matter density could be extended to somewhat above the equilibrium density, there are reasons to expect that it will ultimately fail within the hadronic phase at some density not too high above n_0. One reason is that the Fermi-liquid structure could be broken down at high density. This will be discussed later. Another reason has to do with the IDD that the EFT inherits from QCD at the scale where EFT is matched to QCD. This effect is embedded in the parameters of the Lagrangian at the scale the EFT is defined. While the density dependence brought in by the Φ factor is more or less under control, most likely up to nuclear matter density and perhaps slightly above, ignoring the IDD in other parameters of the effective Lagrangian cannot be valid at higher density, particularly as the chiral restoration density is approached. Among others, the most important is the VM property of HLS [Harada and Yamawaki (2003)].

As alluded to above, the ultraviolet completion of the effective Lagrangian, a requirement for an EFT for QCD, is made by matching the correlators of the EFT — the dilaton-implemented HLS (sHLS) — to those of QCD at a matching scale Λ_M slightly below the chiral scale $\Lambda_\chi \sim 4\pi f_\pi$. From it, the parameters of the "bare" sHLS Lagrangian inherit from QCD the dependence on non-perturbative quantities, i.e., the condensates, $\langle\chi\rangle$, $\langle\bar{q}q\rangle$, $\langle G^2\rangle$, etc. defined at the scale Λ_M in the matter-free vacuum. In applying this Lagrangian to in-medium systems at low energy, two operations are required. One is to take into account the vacuum structure caused by density and the other is to RG-decimate from Λ_M to the low-energy scale where nuclear physics is done. The first is captured by the density dependence of the condensates, namely, IDDs, mainly in the quark condensate since the gluon condensate is insensitive to density, and the second is to do quantum (loop) corrections using the density-dependent parameters.

To be precise, one can classify two classes of IDDs. One involves the matter fields, vector mesons and baryons. It comes from matching the isovector vector and axial vector correlators at the scale Λ_M (at $n = 0$). We call this IDD$_m$. The other involves pseudo-NG bosons, σ and π. It is gotten from matching the trace of energy-momentum tensor. We shall call the resulting density dependence IDD$_{pNG}$. The IDD$_{pNG}$ embodies the common IR scale for the scale and chiral symmetries leading to the relations (6.90) and (6.91). Due to the interplay between the two symmetries, IDD$_{pNG}$

affects also the matter fields giving rise to the scalings (6.92) and (6.93). The masses and coupling constants involving the matter fields carry, in addition, the IDD_m.

In doing the mean field approximation with, e.g., (6.99), the IDD_{pNG} effect is included via the dilaton condensate, but the IDD_m effect, which is difficult to calculate except near the VM fixed point, is not taken into account. Fortunately there is a good reason to believe that the latter effect is negligibly far away from the chiral restoration point. Thus for instance, the ρ meson mass is to scale with f_σ^*, but not with the gauge coupling g nor with the parameter a. In fact, it has been verified phenomenologically that the density dependence is fairly well captured by the dilaton condensate, at least, up to nuclear matter density [Brown and Rho (2004)]. But this cannot continue to be valid as the matter approaches the chiral restoration density n_c. Among others, the most compelling indication for the breakdown comes from the VM property of HLS [Harada and Yamawaki (2003)].

As suggested in [Harada and Yamawaki (2003)], when the chiral symmetry gets unbroken — by temperature or density or other mechanisms — so that the quark condensate goes to zero, the ρ mass should approach zero (in the chiral limit) as

$$m_\rho^* \sim g^* \sim \langle \bar{q}q \rangle^* \to 0. \tag{6.102}$$

Thus while the ρ mass drops at low density mainly due to the decrease of f_π^* with the gauge coupling g and a remaining constant, at high density it is the dropping gauge coupling that should take over as indicated in (6.102). It is found also that the pion decay constant does remain more or less constant starting from a density $\sim (2-3)n_0$ up to near the chiral transition. For this, the explanation could be that the dilaton decay constant is controlled at high density by the gluon condensate with the quark condensate strongly suppressed and the locking of the scale and chiral symmetries makes the pion decay constant behave the same as the dilaton decay constant. There is an effective change in the role for precursor for chiral symmetry restoration between f_σ^* (or equivalently f_π^*) and the hidden gauge coupling g^* at $n \sim 2n_0$. This suggests that the ρ mass, not f_π^*, plays the role for order parameter for chiral restoration. A similar phenomenon takes place also for the nucleon mass. This change will be seen in Chapters 8-9 to have an up-to-date unexplored impact on nuclear tensor forces, properties of neutron stars such as sound velocity etc.

6.6.3 *Signal for $\beta' \neq 0$ in nuclear medium?*

The LOSS Lagrangian, applied above and also in the following chapters, is gotten in the case of CT with $\beta' \to 0$ or $c_{1,2} \to 1$ whereas in GS, the invariant potentials $V_I(\sigma - \eta)$ figuring in scale-invariant Lagrangian scale with $\eta = 0$ as $V_I(\sigma) = 1 + O(\Delta)$, so the LOSS follows with no dependence on the anomalous dimension β'. In the next chapter where the skyrmion approach to dense matter is described, it will be shown that the ω-meson contribution to the ground-state energy diverges unless $\beta' \approx 2$ and $c_h \approx 0$ in the homogeneous Wess–Zumino term in the anomalous parity Lagrangian given in the CT formalism as

$$\mathcal{L} \sim \left[c_h + (1 - c_h)\left(\frac{f_\sigma^*}{f_\sigma}\right)^{\beta'}\right]\mathcal{L}_{\text{hWZ}} \tag{6.103}$$

where $\mathcal{L}_{\text{hWZ}}$ is the hWZ Lagrangian which contains π and HLS vector mesons, but no σ field. We will have more to say on what β' of this size implies in the context of emergent scale symmetry. Here we discuss what this would imply vis-à-vis the issue of the quenched g_A in nuclei recounted in Chapters 2 and 4. In the CT approach [Crewther and Tunstall (2015)], the single-particle Gamow–Teller operator in nuclear medium at small momentum transfer is given by

$$\left[c_h + (1 - c_h)\left(\frac{f_\sigma^*}{f_\sigma}\right)^{\beta'}\right]g_A\tau^\pm\boldsymbol{\sigma} \tag{6.104}$$

where c_h is an unknown constant. The LOSS is obtained for $c_h = 1$ or $\beta' \to 0$ and the operator goes to

$$g_A\tau^\pm\boldsymbol{\sigma}. \tag{6.105}$$

This gives what's referred to as Fermi-liquid fixed-point g_A^L at nuclear matter density

$$g_A^L = 0.79 \times 1.2755(11) \approx 1.01 \tag{6.106}$$

in precise agreement with the quoted phenomenological value $g_A^* = 1.0 \pm 0.1$ in simple shell model. Now suppose $\beta' \sim 2$ as indicated in the skrymion matter. This would require that c_h be close to 1. Accurate many-body calculations of this quantity in light nuclei, in such approaches as quantum Monte Carlo Green's function [Pastore *et al.* (2018)], could lead to an estimate of β' and give a quantitative hint at emergence of scale symmetry in nuclei.

Chapter 7

From Skyrmions to Dense Matter

In approaching dense compact-star matter from QCD, we will be invoking the notion of Cheshire Cat (CC) that implements the large N_c limit approximation, thereby trading in the QCD degrees of freedom for topology. Now the question is what one can do with a given topological structure to access nuclear matter and high-density matter. In the preceding chapters, we have simply by-passed this step by putting "by hand" mesons and baryons of the relevant scale along the line of strategy defined by the Folk Theorem. Now the question is how to set up the effective Lagrangian given by the large N_c topological structure discussed above that can directly confront Nature at high density. At present, going directly from the given topological structure to nuclear dynamics is out of feasibility. The possibility, one of the main themes of this volume, is extract as fully and robustly as possible information from topological considerations and translate it into the effective field theory framework constructed with which to do the calculation. This matter is discussed in this chapter.

7.1 Trading-in quarks/gluons for topology in effective field theory

Baryons figured explicitly in the preceding chapters. Although the mass is of the same scale as the chiral scale, hence not soft, the baryon is mostly near its on-shell in nuclear interactions, so "soft theorems" could be applied to nuclear interactions. Furthermore both the vector mesons and the dilaton, though heavy in the matter-free vacuum, can be treated in some sense as light in the environment of baryons in interactions. Therefore the scale-chiral perturbation scheme made sense in the low-density regime treated so far with great success. When density goes high above the normal nuclear matter density n_0 as needed for compact stars, however, it becomes highly doubtful that the "softness approach" à la folk theorem (FT) in nuclear interactions will continue to hold. How to overcome this potential difficulty is currently a big issue in nuclear physics. In this chapter, we will develop the thesis that the topology embedded in nonlinear sigma model introduced in Chapter 2 could be exploited in a way that is not available in the absence of topology to go to high-density regime relevant to compact stars. This is of course related to the fact that the QCD degrees of freedom, quarks and gluons, will become more effective as density, probing shorter distances, increases. The objective is then to trade in the quark-gluon degrees of freedom more effective at high density for the topology encoded in pions, reinforced, it seems, by the vector mesons and dilaton. We will show indeed that this objective works out. Basically, the idea is to "translate" implications of topology in the skyrmion description into the *bs*HLS Lagrangian treatments of baryonic matter at $n \gtrsim 2n_0$.

In Chapter 3, the notion that at low energy at which nuclear dynamics takes place, there is a hadron-quark continuity that can provide a connection between hadrons and quarks/gluons of QCD was developed in terms of the Cheshire Cat principle. There the quarks and gluons are traded in for topological solitons emerging from the pion fields, namely, skyrmions. In nuclear physics, this implies that properties of nuclear matter could in principle be equally described in terms of either quarks/gluons or solitons/mesons or in hybrids of them. The illustration was made using the chiral-bag model that consists of quarks and gluons confined in the bag coupled at the bag boundary to the pions and heavy mesons, subject to the symmetries encoded in QCD. It was shown that the dynamics of baryons should not depend on how the two different degrees of freedom are partitioned. This principle of continuity could be elegantly phrased as a "gauge artifact" as proposed in Chapter 3: The dynamics should not depend on how the partition is made or the gauge is chosen. Except in (1 + 1) dimensions, this "principle" cannot be verified in an exact manner in reality, so it can only be approximate at best in (3+1) dimensions. Nonetheless it is, at least semi-quantitatively — and in some cases even quantitatively — valid. How it works depends of course on the kinematics involved, and our philosophy is that it should be applicable in the kinematic regime of nuclear physics.

In this chapter, we will shrink the bag to a point and describe baryons *entirely* in terms of skyrmions, considered as a particular "gauge choice." Differently from the usual nuclear chiral effective theories including baryons as explicit degrees of freedom, the baryon field here will not figure explicitly in this chapter. Dense baryonic matter will therefore be treated as dense skyrmion matter.

The original soliton model proposed by Skyrme [Skyrme (1962)] consisted of only the pion field, carrying the topology. This model has been — and will continue to be — referred to as "Skyrme model" or "skyrmion$_\pi$" for short.[1] The Skyrme model or "skyrmion$_\pi$" contains a quartic derivative term involving pion fields — referred to as the "Skyrme quartic term" — that assures the topological stability, effectively simulating massive degrees of freedom capturing the higher energy/momentum scale required in nuclear interactions that are integrated out. In our approach, higher energy/momentum scale involved will be represented by the vector fields ρ and ω and a scalar degree of freedom denoted by σ as defined precisely in Chapter 6, embodying, respectively, hidden local (gauge) fields for the vectors and hidden dilaton field for the scalar. These massive degrees of freedom will be introduced, based on symmetries, though invisible, presumed to be encoded in QCD. Those degrees of freedom are found to play a crucial role in defining the effective Lagrangian. It will be explored in a way distinctive from, and arguably more powerful than, standard effective field theory (EFT) typified by chiral perturbation theory (SχPT) discussed in Chapter 4.

We should mention that for a quantitatively more reliable treatment, it could very well be that more massive degrees of freedom than the hidden local vector $V = (\rho, \omega)$ and scalar σ play a significant role. For instance in the instanton description of baryons in holographic QCD model [Hong, Rho, Yee and Yi (2007); Hata, Sakai, Sugimoto and Yamato (2007); Hashimoto, Sakai and Sugimoto (2008)] based on gravity-gauge duality constructed by Sakai and Sugimoto [Sakai and Sugimoto (2005)], the infinite tower of vector mesons $\rho, \rho', \ldots$ and $\omega, \omega', \ldots$ intervene to give rise to the vector meson dominance (VMD) to the EM form factors of both pion and nucleon. While the lowest vector mesons alone can describe surprisingly well the VMD of mesons, they fail badly for the nucleons [Hong, Rho, Yee and Yi (2008)]. This points to the possibility that the infinite tower can be more important in baryons than in mesons. With an astute implementation of gravity-gauge duality borrowed from string theory, one could perhaps access nuclear systems much more quantitatively starting from instantons instead of from

[1] Unless specified otherwise, "skyrmion" will stand generically for the baryon described as a soliton with the winding number 1, independently of what other degrees than the pion are involved.

skyrmions. This is because starting from instantons, one could simulate dense matter more effectively.[2] There is an indication discussed below that flat-space instantons could describe heavy nuclei better, i.e., binding energy, going toward a Bogomol'nyi–Prasad–Sommerfield (BPS) structure. Unfortunately some crucial problems remain unresolved in holographic QCD models, in particular in dealing with chiral symmetry breaking, strangeness flavor, going beyond the leading order in the 't Hooft constant etc., not to mention how to construct higher instanton-number configurations for dense matter. At present little quantitative calculation can be performed in this model for dense baryonic matter.

Throughout the volume, we will be focusing on the Lagrangian in which the lowest vector mesons ρ and ω — and a_1 whenever needed — are incorporated in hidden local symmetry (HLS) and the dilaton σ is implemented as precisely defined in Chapter 6. The resulting Lagrangian is referred to as sHLS where s stands for scale invariance. Unless otherwise specified, "skyrmion" will stand for the solitonic baryon gotten from sHLS.

In this chapter, we give an overview of qualitative structure of dense matter described in terms of skyrmions. A quantitative calculation, properly quantized, is out of reach, and we cannot attempt to compute observables that can be directly confronted with experiments. However, there are certain properties associated with the topological structure of skyrmions that can be taken robust. The strategy is to extract those robust features and implement them in an EFT framework with baryons explicitly present, anchored on a Wilsonian renormalization group approach to nuclear and dense systems. The topology is carried uniquely by the pion, so whatever is robust is closely tied principally with the pion field. The heavier degrees of freedom do intervene at low density but less importantly than at higher density crucially relevant to the interior of compact stars.

7.2 Skyrmions in brief

The basic essential feature of the skyrmion model was briefly introduced in Chapter 2. (We refer for more detailed studies to, among many others in print, [Zahed and Brown (1986); Rho and Zahed (2016); Ma and Rho (2017)].) Here we review the particular properties of skyrmion that will be exploited in describing dense baryonic matter, with a focus on what's essential in strong-interaction many-body systems.

[2] This possibility was suggested, e.g., in [Park, Min, Rho and Vento (2008)], prompted by the idea from [Atiyah and Manton (1989)].

7.2.1 *Baryons*

Consider the Skyrme model, skyrmion$_\pi$ [Skyrme (1962)], with the Lagrangian written as

$$L = \frac{f_\pi}{2e}\int d^3x\left\{-\frac{1}{2}\mathrm{Tr}(L_\mu L^\mu) + \frac{1}{16}\mathrm{Tr}([L_\mu, L_\nu]^2) + \bar{m}^2\mathrm{Tr}(U-1)\right\} \quad (7.1)$$

and the energy of the classical static skyrmion

$$M_{\mathrm{Skyr}} = \frac{f_\pi}{2e}\int d^3x\left\{-\frac{1}{2}\mathrm{Tr}(L_i L_i) + \frac{1}{16}\mathrm{Tr}([L_i, L_i]^2) + \bar{m}^2\mathrm{Tr}(U-1)\right\}, \quad (7.2)$$

with $L_\mu \equiv U^\dagger(\partial_\mu U)$ and $\bar{m} = m_\pi/(f_\pi e)$. Here to simplify the notations, the energy and length units are given, respectively, in terms of $f_\pi/2e$ and $1/2f_\pi$. In (7.1), in addition to the quadratic current algebra term and the quartic Skyrme term written down in Chapter 2, the pion mass term is included. In some applications to nuclei, the pion mass cannot be ignored. In dense matter, it plays an unimportant role and we deal most of the time with an EFT of QCDLite. In terms of an expansion in standard chiral perturbation theory (SχPT), the Skyrme term can be regarded as the linear combination of the L_1 and L_2 terms in the $O(p^4)$ terms — in Table. 2.2 — which capture most of what enters in $\pi - \pi$ scattering at low energy [Diakonov and Petrov (2000)]. However, in the context of the Skyrme model as an effective theory of QCD, it is more appropriate to consider it as an effective higher-order term that captures massive degrees of freedom (e.g., infinite tower in holographic QCD) that are integrated out.

To make the discussion that follows self-contained, let us first recall a few basics from Chapter 2.

Since the unitary field $U(x)$ satisfies $U(x)U(x)^\dagger = U(x)^\dagger U(x) = 1$, for any fixed time, say, t_0, the matrix $U(\boldsymbol{x}, t_0)$ defines a map from R^3 to the manifold S^3 in isospin space. At the low-energy limit, QCD goes to the vacuum, i.e.,

$$U(|\boldsymbol{x}| \to \infty, t_0) = \mathbf{1}, \quad (7.3)$$

therefore, all the points at $|\boldsymbol{x}| \to \infty$ are mapped onto the north pole of S^3 and the energy of the system is finite. We have the nontrivial map

$$U(\boldsymbol{x}, t_0) : R^3 \to S^3, \quad (7.4)$$

for the static configuration $U(\boldsymbol{x}, t_0)$. In the language of topology, these maps constitute the third homotopy group $\pi_3(S^3) \sim Z$ with Z being the additive group of integers which accounts for the times that S^3 is covered by the mapping $U(\boldsymbol{x}, t_0)$, i.e., winding numbers. Because a change of the time coordinate can be regarded as a homotopy transformation which cannot transit between the field configurations in homotopically distinct classes, the winding number is a conserved quantity in the homotopy transformation by the unitary condition of the field $U(x)$ and condition (7.3). The conserved winding number represents the conserved baryon number in QCD. The baryon arises as a topological soliton with the topology lodged in the chiral field $U(x)$ with the expression of the baryon number current

$$B^\mu = \frac{i}{24\pi^2}\epsilon^{\mu\nu\alpha\beta}\mathrm{Tr}[L_\nu L_\alpha L_\beta]. \tag{7.5}$$

Note that even though the Skyrme model does not include this baryon number current, it is an intrinsic quantity in the model.

Since in the N_c counting, $f_\pi \sim O(\sqrt{N_c})$ and $e \sim O(1/\sqrt{N_c})$, the soliton mass $M_{\rm Skyr}$ scales as $M_{\rm Skyr} \propto f_\pi/e \sim N_c$. This N_c counting agrees with that of baryons from QCD per se and the constituent quark model in particular. Suppose one takes the parameters of the effective Lagrangian to be those given in matter-free space, such as for instance $f_\pi \approx 93$ MeV, $g_A \approx 1.27$, $m_\pi \approx 140$ MeV etc., then the Skyrme model would give the nucleon mass to be much too big, say ~ 1500 MeV. This just reflects that the nucleon mass at the leading order in N_c cannot be trustful. One may fiddle with those parameters given that they are "effective quantities," which is what is done in applications of the model to baryon properties. In what follows, throughout the volume, we eschew this practice. There are corrections of subleading order in N_c such as the Casimir energy that appears at $O(N_c^0)$ that can be large. Unfortunately the Casimir effect is extremely hard to calculate in the Skyrme model, but a rough estimate shows that it can be ~ -0.5 GeV [Nowak, Rho and Zahed (1996)]. The difficulty will be greater when the skyrmion description is applied to nuclei and dense nuclear matter. For this reason we will not attempt quantitative comparison with nature with the model.

7.3 Dense baryonic matter

There is no continuum technique available up to date in treating dense matter in terms of multi-skyrmions. Quantizing multiskyrmions is mathematically daunting, if not inaccessible, and hence it is infeasible at present to access the dense matter present inside compact stars. What is feasible, however, is putting skyrmions on crystal lattices and simulate dense matter by squeezing the crystal lattice size. We will resort to this procedure — which has been treated extensively in the literature as well as in books [Rho and Zahed (2016); Rho (2008)]. One cannot, however, directly confront the crystal lattice calculations with nature. This is because putting skyrmions on crystal cannot be reliable in certain density regime involved in compact-star matter. For instance baryonic matter at low density cannot be in crystal. It is most likely that nuclear matter is a liquid, not a solid. It is not at all clear how to arrive, starting from a crystal, at a Fermi-liquid structure at normal matter density n_0, which is known to be the correct picture. On the other hand, as we will argue, at large density and in the large N_c limit, baryons could very well be in the form of skrymion crystals. We will therefore ignore the crystal structure at low density, say, $n \lesssim 2n_0$ and exploit the features anchored on topology that can be extracted as one approaches high density matter $n \gtrsim 2n_0$.

Now what are such high density characteristics and how to implement such features for the physics of normal nuclear matter where there are constraints from nature and go to high density where there are none?

For this, topology is found to figure in a highly intricate and specific way. We list here as concisely as possible the key elements that play important roles in our approach that can be extracted from topology of skyrmions and then be translated into an effective Lagrangian formulation.

7.3.1 *Half-skyrmions*

The notion that nuclear matter can be simulated in the skyrmion approach by putting skyrmions onto the crystal lattice [Klebanov (1985)] will be extended to dense baryonic matter appropriate for massive ($\sim$ 2-solar mass) compact stars. The density involved is of order $\sim$ (5-7) times normal nuclear matter density $n_0 \approx 0.16$ fm^{-3}, which is most likely below the density $n_{\rm deconf}$ at which the deconfinement of quarks/gluon is to set in. What the exact location of $n_{\rm deconf}$ is not important for the discussion. In the application to massive compact stars, $n_{\rm deconf}$ could easily be of order of $10n_0$ or greater.

So far the face-centered-cubic (FCC) crystal is the lowest energy crystal configuration known for nuclear ground state [Kugler and Shtrikman (1988, 1989)]. In the FCC configuration, the baryon number density is given by $4/(2L)^3$ with $2L$ being the crystal size. The size of the crystal lattice that corresponds to the normal nuclear density n_0 is $L \approx 1.43$ fm.[3]

[3] A word of caution is in order here. In the standard notation, n_0 stands for the

Among a variety of properties revealed in the crystal approach, the most important of all is the existence of half-skyrmion configurations at some higher density. As in condensed matter physics, being topological, its presence is a robust prediction, involving in this case maximal symmetry [Goldhaber and Manton (1987)]. It does not, for instance, depend on what degrees of freedom other than pions are involved. What turns out to be highly significant is that it involves a topology change from skyrmions — characterized by integer topological charges Q — to half-skyrmions with half-integer topological charges $2\hat{Q}$ with $\hat{Q} = Q/2$. This topology change turns out to be responsible for a drastic change in the structure of dense matter at a density $n_{1/2} \gtrsim 2n_0$. This feature is absent in other approaches in the literature. In Fig. 7.1 is shown the baryon number density distribution of the skyrmion (left panel) and half-skyrmion states of matter (right panel) in the $z = 0$ plane. In this figure, the white area is the vertex where the baryon number is accumulated. From the plot, one can see that, in the half-skyrmion phase, the vertices where the baryon number density accumulates form the cubic crystal. The space average of the trace

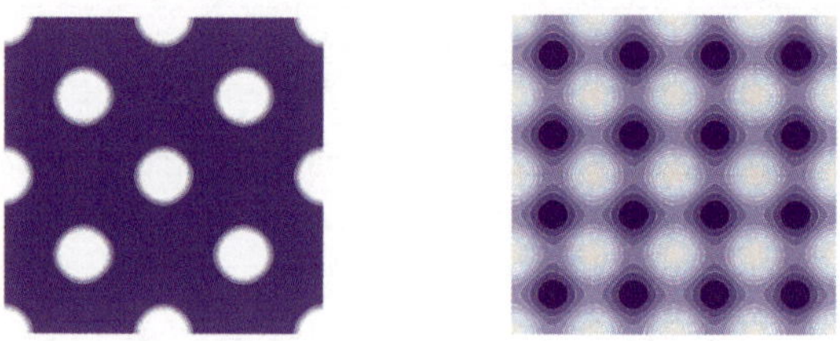

Fig. 7.1 The distribution of the baryon number density in the skyrmion (left panel) phase and half-skyrmion phase (right panel).

of the soliton configuration U_0 defined as

$$\langle \phi_0 \rangle = \frac{1}{(2L)^3} \int_0^{2L} d^3x \, \frac{1}{2} \mathrm{Tr}(U_0), \tag{7.6}$$

which while nonzero in the skyrmion matter, is found to vanish in the half-skyrmion state. This implies that the quark condensate $\langle \bar{q}q \rangle$, when the space is averaged, vanishes. This, as will be seen below, is an important observation, suggesting the possibility of an "emerging" parity-doubling in the nucleon structure.

equilibrium nuclear matter density. In skyrmion crystals, it is not known what the equilibrium density is since the crystal description is valid only at high density. Thus one should not make a direct comparison to n_0.

The topology change brings out certain features that need to be incorporated in dense matter but are not "visible" in non-topological models. First of all, this changeover of state of baryonic matter does not involve phase transitions in the sense of Landau–Ginzburg–Wilson paradigm, since, as will be discussed later, chiral symmetry is not restored although the quark condensate vanishes on average. The pion decay constant f_π is non-vanishing, hence the pion is still present. There is certain resemblance to what's known as "pseudo-gap" in high temperature superconductivity. Even in condensed matter physics, the notion of pseudo-gap is rather confused and controversial [Mueller (2018)], so making comparison between the case we have to the latter may not be meaningful. But it is interesting to to relate the quark condensate (averaged) to the pairing order parameter for superconductivity, $\Delta_0 \sim \langle\psi\psi\rangle$, and the pion decay constant f_π to the spectral gap Δ_s. One interpretation of the pseudo-gap phenomenon in superconductivity with $\Delta_0 = 0$ while $\Delta_s \neq 0$ resembles the half-skyrmion state.

In contrast to the superconductivity case, however, here is no bona-fide local order parameter for this change of states from skyrmion to half-skyrmion. Although it is not a paradigmatic phase change, there is a drastic change in the properties of the matter, so in what follows we will loosely use the term "half-skyrmion phase." In addition, in the inhomogeneous medium, due to the interaction of the fluctuating pions with the medium, the quantities $\mathrm{Tr}(U_0)/2$ and $\mathrm{Tr}(\tau^a U_0)/2$ measure the inhomogeneous quark condensate in the medium [Harada, Lee, Ma and Rho (2015)]. This gives rise to a chiral density fluctuation.

Half-skyrmions and monopoles

Perhaps most relevant physically is that the half-skyrmions that appear at $n = n_{1/2}$ cannot be separated. Thus they are not propagating degrees of freedom. This structure is different from the half-skyrmions that figure at the "deconfined quantum critical point" in the magnetic Néel-to-VBS transition in condensed matter (see Senthil *et al.* in [Rho and Zahed (2016)]) where the half-skyrmions get deconfined in between, whereas in the initial and final phases, the half-skyrmions are confined. This happens because there is a hidden $U(1)$ symmetry in the skyrmion configuration and an emergent monopole that confines the half-skyrmions in the present case, whereas in the condensed matter case, the monopole becomes irrelevant due to a Berry phase. Within the framework adopted in the calculation, there is no way to determine whether the half-skyrmions can be made propagating.

Even though the existence of the half-skyrmion phase is robust depending solely on the pion field, at what density the transition can take place in dense baryonic matter depends on, among others, the other degrees of freedom that intervene. This will be seen in numerical simulations given below. Since in practice a large number of mesons can be involved, e.g., the infinite tower of holographic QCD, and one is limited to drastic approximations, it is difficult to pinpoint the changeover density — denoted $n_{1/2}$. Given that there is no indication for a half-skymrion phase at nuclear matter density,[4] it cannot be lower than the normal density n_0. If it were much higher, say, in the regime where the quark percolation transition can take place, then the hadronic model could not be used. Thus the optimal density for $n_{1/2}$ should be somewhere, but not far, above n_0. The present estimate based on detailed analyses using a semi-realistic model that contains π, ρ, ω and σ indicates that $n_{1/2} \sim 2n_0$.[5] This typical value will be used in the following applications to nuclear matter and denser matter. In the qualitative discussions, the exact position of $n_{1/2}$ will not be specified other than that it is above n_0 and below a putative quark deconfinement transition point $n_{\rm deconf}$. The recent measurement of the tidal deformability in the gravity waves coming from coalescing neutron stars [Abbott *et al.* (2017)] could provide a bound on $n_{1/2}$. This matter will be discussed in Chapter 9.

It should be stressed before we enter into detailed discussions that the skyrmion crystal picture would make sense only if N_c were very large. Indeed if $N_c \to \infty$, the baryonic matter would be a crystal. In reality $N_c = 3$ is not so "large" and normal nuclear matter is known, as mentioned, to be a liquid and not a crystal. One might hope to arrive at a structure compatible with nuclear matter by doing proper quantum-mechanical calculations taking into account $1/N_c$ corrections starting from the crystal structure, but no such calculations have been properly performed so far. Even within the large N_c scheme, indications from holographic QCD calculations are that at high density, a variety of different forms, such as dyonic salt (resembling half-skyrmions), layers of "popcorns," salty or sugared etc, can be

[4]There is, however, an intriguing, albeit indirect, hint that half-skyrmions, not explicitly visible, are hidden in light nuclei — such as alpha particle. This is discussed by Battye *et al.* in [Rho and Zahed (2016)] and more recently by Manton [Manton (2017)].

[5]It will be pointed out later that the possible smooth cross-over from hadrons to strong-coupled quarks which takes place at a comparable density could be related to this topology change.

present [Rho, Sin and Zahed (2010); Gorsky, Kopnin and Krikun (2014); Kaplunovsky, Melnikov and Sonnenschein (2012); Elliot-Ripley, Sutcliffe and Zamaklar (2016)]. It may be that all or some of these exotic configurations could be artifacts of the large N_c models, but at present none of them can be ruled out.

Although what's presented up to here is robust and its qualitative structure is not sensitively dependent on other than the pion, the other degrees of freedom that intervene in the lattice crystal structure have highly important influences on the dynamics for dense medium. For this we will need the results of what will be developed in the later chapters involving the dilaton and the vector mesons. For illustration, we can take the leading order ($\mathcal{O}(p^2)$) HLS containing the dilaton and a homogeneous Wess–Zumino (hWZ) term that carries the anomalous dimension β' that encodes the QCD trace anomaly as explained in Chapters 2 and 6. What is important for this discussion is the role of the dilaton mass for the interplay as density increases between the quark condensate $\langle\bar{q}q\rangle$ — which is equivalent to ϕ_0 in (7.6) — and the dilaton condensate $\langle\chi\rangle$. Consider the light dilaton mass $m_\sigma^{\rm light} \ll \Lambda$ and the heavy dilaton mass $m_\sigma^{\rm heavy} \gg \Lambda$ where Λ is an appropriate scale that will be matched to QCD as described in Chapter 8. One can take $\Lambda \sim 1$ GeV, to be specific. The interplay between the two condensates is very sensitive to the dilaton mass. For $m_\sigma^{\rm light}$, both condensates fall with density and then go to zero simultaneously at the point where skyrmions transform to half-skyrmions. For $m_\sigma^{\rm heavy}$, the quark condensate drops to zero, first forming the half-skyrmion phase, while the dilaton condensate drops slowly and then goes to zero after traversing the half-skyrmion phase. This feature is schematically shown in Fig. 7.2. What this result illustrates is that for a dilaton mass greater than that of a pion but less than that of vector mesons, there could be a phase in which half-skyrmions populate at above certain density that we identify as $n_{1/2}$ with non-zero dilaton condensate. It also illustrates that for a dilaton mass that is tweaked downward near to that of the pion mass, there can be a phase transition that brings both the pion and dilaton to the same zero-mass multiplet in the chiral limit. Another remarkable feature that we will encounter below in a more refined treatment is that at the topology change from skyrmions to half-skyrmions, the dilaton condensate comes to a substantially large constant, which will be made precise below, independent of density instead of decreasing continuously. This will turn out to be proportional to the nucleon mass that remains non-zero after chiral restoration but still confined.

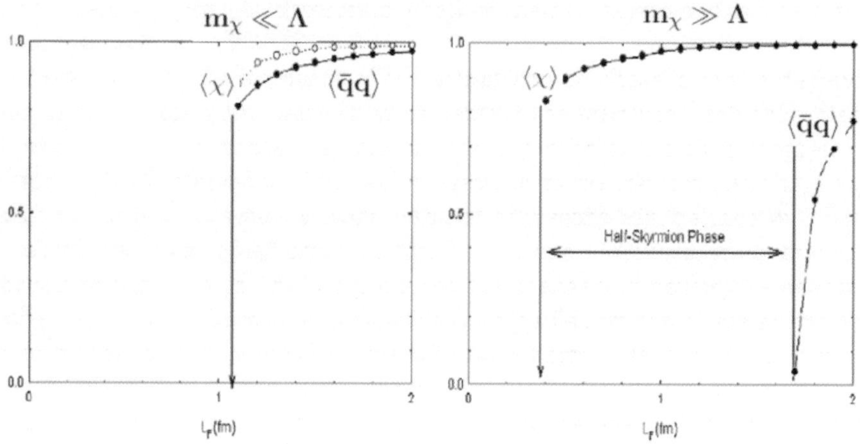

Fig. 7.2 Schematic tendency of dilaton and quark condensates vs. L. Increasing density corresponds to decreasing L.

7.3.1.1 *Half-skyrmion phase as a Fermi liquid*

Remarkably the half-skyrmion phase possesses a feature that is characteristic of a Fermi liquid at its fixed point. This can be illustrated explicitly.

While it is most likely that the dilute matter treated in crystal lattice cannot be given as the mean-field approximation of the Lagrangian, which is equivalent to saying that it is not in Fermi liquid, the same analysis shows, however, that once the system is in a half-skyrmion phase, the matter does seem to satisfy the mean-field criterion, suggesting that the state is in Fermi liquid. Here we show that the half-skyrmion phase is in a state that can be described entirely by noninteracting quasiparticles, more or less undistorted by strong interactions. This resembles Landau Fermi-liquid fixed-point theory where the β function for the quasiparticle interactions is suppressed. This striking feature was first found in the Skyrme model with the Atiyah–Manton ansatz in [Park, Min, Rho and Vento (2008)]. Here we will show the phenomenon using the HLS Lagrangian [Ma *et al.* (2012)].[6]

We write the chiral field U as $U(\vec{x}) = \phi_0(x, y, z) + i\phi_\pi^j(x, y, z)\tau^j$ with the Pauli matrix τ^j and $j = 1, 2, 3$. Including ρ and ω, we write the fields

[6]Since what matters as in the structure of the tensor forces is the topology and symmetry involved, largely independent of strong interactions mediated by non-topological fields, the same argument should apply to the dilaton in the half-skrymion phase in sHLS.

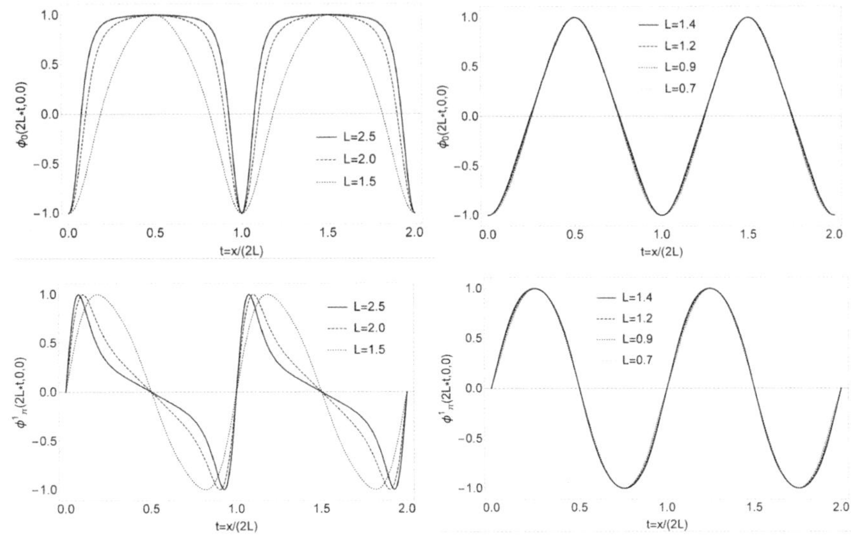

Fig. 7.3 The field configurations ϕ_0 and ϕ^1_π as a function of $t = x/L$ along the y = z = 0 line. The maximum values for $\eta = 0, \pi$ are $\phi_{0,\,L,\,\mathrm{max}} = \phi_{\pi,\,L,\,\mathrm{max}} = 1$. The half-skyrmion phase sets in when $L = L_{1/2} \lesssim 1.45\,\mathrm{fm}$.

placed in the lattice size L as $\phi_{\eta,\,L}(\vec{x}\,)$ with $\eta = 0,\ \pi,\ \rho,\ \omega$ and normalize them with respect to their maximum values denoted $\phi_{\eta,L,\mathrm{max}}$ for given L. It can be shown, as in [Park, Min, Rho and Vento (2008)], with HLS that in the half-skyrmion phase[7] with $L \lesssim L_{1/2}$ where $L_{1/2} \simeq 1.45$ fm, the field configurations are invariant under scaling in density as the lattice is scaled from L_1 to L_2

$$\frac{\phi_{\eta,\,L_1}(L_1\vec{t}\,)}{\phi_{\eta,\,L_1,\,\mathrm{max}}} = \frac{\phi_{\eta,\,L_2}(L_2\vec{t}\,)}{\phi_{\eta,\,L_2,\,\mathrm{max}}}. \tag{7.7}$$

Since other fields are quite similar, we only show in Fig. 7.3 the case of $\phi_{0,\pi}$ for $\phi_{0,\pi}(t,0,0)$ vs. t with $t \equiv x/L$. What is seen there is that density-scale invariance sets in for $L \lesssim L_{1/2}$ (corresponding to $n > n_{1/2}$). One can see that the field is independent of density in the half-skyrmion phase with $L \lesssim L_{1/2}$, whereas for the skyrmion phase with lower density with $L > L_{1/2}$, it is appreciably dependent on density.

What does this imply for the energy density?

[7]The precise value of the half-skyrmion density which depends on the parameters is not important for our discussions.

The energy density for the skyrmion matter put on the lattice of size L can be written as

$$\epsilon = E/A/V(= (2L)^3) = \frac{1}{(2L)^3}\int_0^{2L} d^3x \sum_{n,\,m} c_{n,m}\, f_{n,m}\left(\vec{\nabla}_x,\, \phi_{\eta,\,L}(\vec{x}\,)\right)\,,$$

where $c_{n,m}$ is the coefficient of $f_{n,m}$ which is the function of $\vec{\nabla}_x$ and $\phi_{\eta,\,L}(\vec{x}\,)$ having nth power of $\vec{\nabla}_x$ and mth power $\phi_{\eta,\,L}(\vec{x}\,)$ with $\nabla_{x,\,j} = \frac{\partial}{\partial x^j}$. One can reduce it to

$$\begin{aligned}
\epsilon &= \sum_{n,\,m}\left(\frac{1}{2L}\right)^n (\phi_{\eta,\,L,\,\max})^m \int_0^{2L}\frac{d^3x}{(2L)^3}\, c_{n,m}\, f_{n,m}\left(L\vec{\nabla}_x,\, \frac{\phi_{\eta,\,L}(\vec{x}\,)}{\phi_{\eta,\,L,\,\max}}\right)\\
&= \sum_{n,\,m}\left(\frac{1}{2L}\right)^n (\phi_{\eta,\,L,\,\max})^m \int_0^{1} d^3t\, c_{n,m}\, f_{n,m}\left(\vec{\nabla}_t,\, \frac{\phi_{\eta,\,L}(2L\vec{t}\,)}{\phi_{\eta,\,L,\,\max}}\right)\\
&= \sum_{n,\,m}\left(\frac{1}{2L}\right)^n (\phi_{\eta,\,L,\,\max})^m A_{n,\,m}\,,
\end{aligned} \tag{7.8}$$

where $A_{n,\,m}$ is a constant independent of the lattice size L.

Calculating the energy density (7.8) in skyrmion-crystal simulations involves field configurations satisfying their equations of motion. Hence (7.8) can be considered as a system of non-interacting quasiparticles. It captures all essential dynamics in terms of the mean fields of each degrees of freedom involved, with residual interactions suppressed. The density dependence lies, apart from the $(1/L)^n$ factor, in the maximum field configuration $(\phi_{\eta,\,L,\,\max})^m$. This implies that in the half-skyrmion phase, considered to set in at high density, the mean-field structure dominates. This agrees with the lore that at high density — and in the large N_c limit, the skyrmion crystal picture becomes valid in QCD. In clear contrast, however, as one can see in Fig. 7.3, the mean-field structure breaks down in the lower-density phase with $L > L_{1/2}$. Taking the relativistic mean-field approximation (RMF) to be more or less equivalent to Fermi-liquid fixed-point theory at high density, one can take the breakdown of the mean-field structure as a signal for non-Fermi liquid structure. This also agrees with the understanding that the property of low-density baryonic matter — including nuclear matter — may be poorly if any captured in crystal.

This indicates that while low density may not be accessible by skyrmion crystals, the high density regime can be. An equally significant observation is that the skyrmion-half-skyrmion transition can be reproduced in terms of a chiral-quark model, described in Chapter 3 where the transition is described as a delocalization of the baryon number density leading to half a baryon-number structure [Mantovani Sarti and Vento (2014)]. This may be interpreted in terms of the Cheshire Cat mechanism described in Chapter 3. It may also be connected to the quarkyonic phase [McLerran (2010); McLerran and Pisarski (2007)] to which we will return in Chapter 9 in connection with the structure of ~ 2 solar mass stars.

7.3.2 *Pion decay constant in the half-skyrmion phase*

The medium-modified hadron properties can be studied without involved calculations in the skyrmion crystal approach. For describing hadron properties in dense medium, one can consider the skrymion crystal as a background that provides "warping" of the space in which fluctuations take place. In Chapter 8, this effect is identified with the intrinsic density dependence (IDD) inherited from QCD. One of the most interesting quantities in dense medium is the effective in-medium pion decay constant f_π^* which can be taken as the order parameter of chiral symmetry in medium.

The procedure for calculating f_π^* in the skyrmion matter is quite simple. We take the skyrmion crystal solution as the background classical field and interpret the fluctuating fields on top of it as the corresponding mesons in dense baryonic matter. Following [Ma, Harada, Lee, Oh, Park and Rho (2013)], we write the skyrmion crystal solution as $U_{(0)}$ and introduce the fluctuating field as

$$U = u_{(0)}\check{U}u_{(0)}, \tag{7.9}$$

where $\check{U} = \exp(2i\tau_a\check{\pi}_a/f_\pi)$ stands for the corresponding fluctuating field and $u_{(0)}^2 = U_{(0)}$. Since for each crystal size, there is a solution for $U_{(0)}$, $U_{(0)}$ is dependent on the crystal size, or equivalently, the baryon density. The decomposition in Eq. (7.9) guarantees that to each order of fluctuations the chiral invariance of the model is preserved. By substituting the fields in Eq. (7.9) into the Skyrme model Lagrangian one obtains the medium-modified Lagrangian for the pion. Consequently the parameters in the pion Lagrangian become density-dependent through their dependence on $U_{(0)}$.

To define the pion decay constant in the skyrmion matter, we write the axial-vector current correlator

$$iG_{\mu\nu}^{ab}(p) = i\int d^4x e^{ip\cdot x}\langle 0|TJ_{5\mu}^a(x)J_{5\nu}^b(0)|0\rangle \tag{7.10}$$

which is decomposed in medium as [Kapusta (1989)]

$$G_{\mu\nu}^{ab}(p) = \delta^{ab}\left[P_{T\mu\nu}G_T(p) + P_{L\mu\nu}G_L(p)\right], \tag{7.11}$$

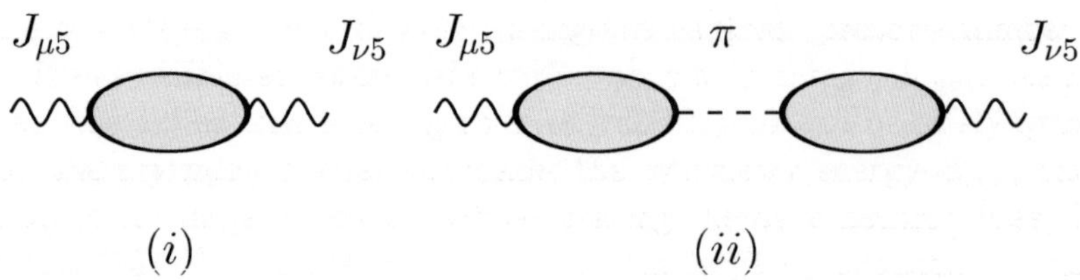

Fig. 7.4 Two types of contributions to the correlator Eq. (7.10): (i) the contact diagram and (ii) the pion exchange diagram. Shaded blobs stand for interaction vertices in the skyrmion matter.

where the polarization tensors $P_{L,T}$ are defined as

$$P_{T\mu\nu} = g_{\mu i}\left(\delta_{ij} - \frac{p_i p_j}{|\boldsymbol{p}|^2}\right) g_{j\nu},$$

$$P_{L\mu\nu} = -\left(g_{\mu\nu} - \frac{p_\mu p_\nu}{p^2}\right) - P_{T\mu\nu}. \tag{7.12}$$

Then the medium-modified (time component of) pion decay constant is given by the longitudinal component in the low-energy limit

$$f_\pi^{*2} \equiv -\lim_{p_0\to 0} G_L(p_0, \boldsymbol{p}=0). \tag{7.13}$$

Now for the calculation, we shall ignore the contributions from the loop diagrams of the fluctuation fields. Therefore, there are only two types of diagrams that contribute as illustrated in Fig. 7.4: (i) the contact diagram and (ii) the pion exchange diagram. They are given by

$$\text{(i)}: i f_\pi^2 g_{\mu\nu}\delta^{ab}\left(1 - \frac{2}{3}\langle\boldsymbol{\phi}_\pi^2\rangle\right),$$

$$\text{(ii)}: -i f_\pi^2 \frac{p_\mu p_\nu}{p^2}\delta^{ab}\left(1 - \frac{2}{3}\langle\boldsymbol{\phi}_\pi^2\rangle\right). \tag{7.14}$$

Summing the two contributions, one obtains

$$f_\pi^{*2} = f_\pi^2\left[1 - \frac{2}{3}\left(1 - \langle\phi_0^2\rangle\right)\right], \tag{7.15}$$

where what corresponds to the IDD is brought in by the minimal energy solution ϕ_0^2. To arrive at (7.15), one uses the relation $\phi_0^2 + \boldsymbol{\phi}_\pi^2 = 1$. Equation (7.15) shows the direct relation between the "vacuum"-modified pion decay constant f_π^* and the parameter $\langle\phi_0\rangle$ which signals the topology change.

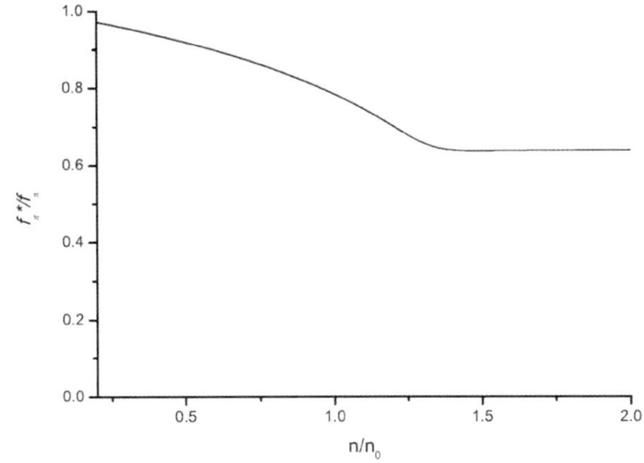

Fig. 7.5 f_π^*/f_π in the Skyrme model as a function of nuclear matter density normalized by the normal nuclear density.

In Fig. 7.5 is plotted as an illustration the nuclear matter density dependence of f_π^*/f_π. We see that f_π^* decreases to $\sim 0.65 f_\pi$ at the density $n_{1/2}$ at which skyrmions fractionize to half-skyrmions and then stays constant for $n > n_{1/2}$. In terms of the quark condensate $\langle\bar{q}q\rangle$, $\langle\phi_0\rangle$ corresponds to the space average of the latter. Note that $\langle\phi_0\rangle \to 0$ does not signal the restoration of chiral symmetry since the pions are still around with the non-zero pion decay constant. This means that the quark condensate is locally non-zero in the half-skyrmion phase, confirming that there is chiral-density wave [Harada, Lee, Ma and Rho (2015)]. This feature will remain unmodified when heavy mesons are included.

7.3.3 *The cusp in the nuclear symmetry energy*

The key observable that links topology to the effective Lagrangian that we shall use is the nuclear symmetry energy that controls the equation of state (EoS) for asymmetric dense matter. Topology via its change from skyrmions to half-skyrmions brings an aspect that is robust into a highly complex system and is found to play the principal role in the development made in this volume.

Proceeding, we return to the energy per particle $E(n, \alpha)$ of a dense system with density n and asymmetry α given by

$$E(n, \alpha) = E_0(n, x = 0) + \alpha^2 E_{sym}(n) + O(\alpha^4) + \cdots , \tag{7.16}$$

where $\alpha = (N-P)/A$ with $N(P)$ being the neutron (proton) number and $A = N + P$. This equation defines the symmetry energy E_{sym}. As mentioned, the corrections entering at $O(\alpha^2)$ and higher can be safely ignored.

In the skyrmion crystal approach, the symmetry energy E_{sym} can be calculated for large N_c by collective-rotating the pure neutron [Lee, Park and Rho (2011)]. It comes at $O(1/N_c)$ in the large N_c expansion. Though sub-leading in N_c, it is primarily controlled by topology and hence more or less independent of other degrees of freedom than the pion field.

Doing the collective rotation, well described in the literature — introduced in Chapter 2 above, one finds the energy of the A-nucleus as

$$E_{\rm A} = AE_{\rm Sol} + \frac{1}{2A\mathcal{I}} I^{\rm Tot}\left(I^{\rm Tot}+1\right), \tag{7.17}$$

where $E_{\rm Sol}$ is the single soliton energy in the nucleus and $A\mathcal{I}$ is the iso-spin moment of inertia of the $A = N$ system. $I^{\rm Tot}$ is the total isospin of the whole system which could be the same as the third component of the isospin I_3 for pure neutron matter, i.e., $I^{\rm Tot} = A/2$. For $\alpha \equiv (N-P)/(N+P) \lesssim 1$ we have

$$I^{\rm Tot} = \frac{1}{2}(N-P) = \frac{1}{2}(N+P)\frac{(N-P)}{(N+P)} = \frac{1}{2}A\alpha. \tag{7.18}$$

The energy per nucleon can then be expressed as

$$E_{\rm Nucleon} = \frac{E_{\rm Nuclei}}{A} \simeq E_{\rm Sol} + \frac{1}{8\mathcal{I}}\alpha^2, \tag{7.19}$$

for a large mass number A. Consequently, the symmetry energy takes the simple form

$$E_{\rm Sym} = \frac{1}{8\mathcal{I}}. \tag{7.20}$$

The moment of inertia $\mathcal{I}$ is given by the integral over the single cell of a quantity that contains U_0 and other degrees of freedom such (as the dilaton configuration) warped by the background. As noted before, the warping by the background encodes the IDDs, representing the sliding vacuum structure in the presence of density. The symmetry energy calculated with the Skyrme model — skyrmion$_\pi$ — implemented with scale invariance [Lee, Park and Rho (2011)] is given in the left panel of Fig. 7.6. This is qualitative, with no fine tuning of the parameters with the vacuum scalar mass of $\sim$ 700 MeV. What's noticeable is the cusp at the density $n_{1/2}$. The symmetry energy first decreases with density up to the density $n_{1/2}$ at which skyrmions fractionize to half-skyrmions and then increases as density increases. Of high significance is that the presence of this (perhaps) strange-looking cusp does not depend on what massive degrees of freedom intervene. It depends only on the pion field. The cusp shows up precisely at the density at which skyrmions fractionalize into half-skyrmions, and this is insensitive to the dilaton mass. Its robustness is due to topology. This cusp will be found to have a drastic impact on the structure of the nuclear tensor force, which dominates the symmetry energy in dense matter $n \gtrsim n_{1/2}$. This will be taken up in Subsection 8.2.4.

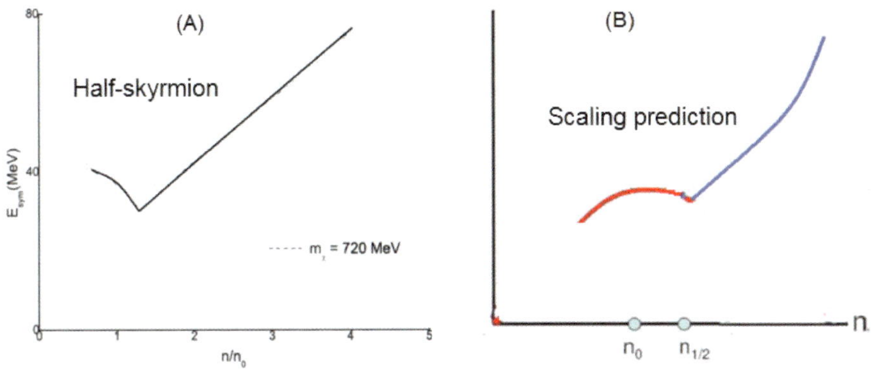

Fig. 7.6 (A) Symmetry energy in the skyrmion crystal model for the dilaton mass $\sim$ 720 MeV as a function of density calculated in Ref. [Lee, Park and Rho (2011)]. The cusp is located at $n_{1/2}$. (B) Symmetry energy expected from the nuclear tensor force affected by the topology change described in Chapter 8.

7.3.4 *Effects of massive mesons on skyrmion matter*

We have seen that the minimal skyrmion crystal approach using the skyrmion$_\pi$ model, with or without a scalar degree of freedom, simulates, semi-quantitatively correctly, certain characteristic properties of nuclear matter and gives certain new predictions. It is the long wave-length pionic fluctuation on the solitonic background that is responsible for it. However, it is well known that the scalar and vector resonances with masses much greater than that of pions are important in nuclear dynamics involving short-range interactions. We now turn to their effects in the structure of both elementary baryons and baryonic matter, based on the scale-invariant HLS (sHLS) approach and the holographic model of QCD with infinite tower of vector mesons based on gauge-gravity duality. Without adequate quantization, too difficult to perform, the results cannot be taken quantitatively reliable. They however illustrate the essential qualitative features that characterize certain important aspects of nuclear dynamics at normal as well as high densities.

7.3.5 *The effect of the vector mesons ρ and ω*

It was stated in Chapters 2 and 5 that both the vector mesons and the scalar meson can be brought in as hidden symmetries, the former as HLS and the latter hidden scale symmetry. We first discuss HLS, to which scale symmetry will be implemented later. At high density, the flavor $U(2)$ symmetry for $V = (\rho, \omega)$ will be found broken, so we will work in the flavor space of $SU(2) \times U(1)$. However for convenience of notation, in writing down Lagrangians, we will employ $U(2)$. Whenever necessary, we will exhibit the $SU(2) \times U(1)$ structure.

We recall from Chapter 2 the most general HLS Lagrangian written to $O(p^4)$ and in the large N_c limit (i.e., $O(N_c)$)

$$\mathcal{L}_{\rm HLS} = \mathcal{L}_{(2)} + \mathcal{L}_{(4)} + \mathcal{L}_{\rm anom}, \tag{7.21}$$

with

$$\mathcal{L}_{(2)} = f_\pi^2 \,{\rm Tr}\, (\hat{\alpha}_{\perp\mu}\hat{\alpha}_\perp^\mu) + a f_\pi^2 \,{\rm Tr}\, \left(\hat{\alpha}_{\parallel\mu}\hat{\alpha}_\parallel^\mu\right) + \mathcal{L}_{\rm kin}, \tag{7.22}$$

where $\hat{\alpha}_{\parallel\mu}$ and $\hat{\alpha}_{\perp\mu}$ are the Maurer–Cartan 1-forms that transforms covariantly under hidden gauge transformation.

The full expressions to $O(p^4)$, given in Chapter 2, are not repeated here.

The calculations for the structure, both elementary and in-medium, of skyrmions from this Lagrangian are quite involved depending on a variety of undetermined parameters, and are not in a position with which we can confront nature directly. For a qualitative study to bring out general property of the skyrmion description for both elementary baryons and dense matter, it is found to be illuminating to resort to a holographic QCD model, such as the Sakai-Sugimoto model [Sakai and Sugimoto (2005)] that has correct symmetries of QCD. Decomposing the 5D model to 4D leading to infinite tower of vector mesons with hidden local symmetry and integrating out all higher-mass vector mesons other than the lowest $V = (\rho, \omega)$, it has been shown that one can write an HLS Lagrangian to $O(p^4)$ with only two parameters f_π and m_ρ [Harada, Matsuzaki and Yamawaki (2010)]. This comes about because the holographic QCD model has only two parameters (for $U(2)$ symmetry for the vectors) and there is a "master formula" that connects all the parameters of the HLS Lagrangian so obtained, including a

in (7.21), to the two parameters. This Lagrangian is of course limited to the applicability of the holographic model, which as noted, has its own limits and defects with respect to nature. However, it has the merit of singling out the role of the vector mesons, in particular, the role of ω meson that is lodged in the Chern–Simons (CS) term. We summarize what comes out for the properties of both elementary baryons and dense baryonic matter. For qualitative considerations, we simply take the free-space parameters assuming $U(2)$ symmetry.

$$f_\pi = 92.4 \text{ MeV}, \quad m_\omega = m_\rho = 775.5 \text{ MeV}. \tag{7.23}$$

For dense matter, as noted, the IDD that is crucial in field theory approach described in Chapter 8 does not enter in the approach used here, the skyrmion crystal. This is because the skyrmion background does the role of IDDs.

7.3.5.1 *Elementary baryons*

To see how the degrees of freedom affect the skyrmion structure, it is feasible to turn off one by one the ρ and ω mesons. The results for the soliton mass, the winding number radius $\sqrt{\langle r^2\rangle_W}$ and energy radius $\sqrt{\langle r^2\rangle_E}$ are [Ma, Yang, Oh and Harada (2013)]:

- π, ρ, ω:

$$\begin{aligned} M(\pi,\rho,\omega) &\approx 1184 \text{ MeV}, \\ \sqrt{\langle r^2\rangle_W} &\approx 0.433 \text{ fm}, \qquad \sqrt{\langle r^2\rangle_E} \approx 0.608 \text{ fm}. \end{aligned} \tag{7.24}$$

- π, ρ:

$$\begin{aligned} M(\pi,\rho) &\approx 835 \text{ MeV}, \\ \sqrt{\langle r^2\rangle_W} &\approx 0.247 \text{ fm}, \qquad \sqrt{\langle r^2\rangle_E} \approx 0.371 \text{ fm}. \end{aligned} \tag{7.25}$$

- π:

$$\begin{aligned} M(\pi) &\approx 922 \text{ MeV}, \\ \sqrt{\langle r^2\rangle_W} &\approx 0.309 \text{ fm}, \qquad \sqrt{\langle r^2\rangle_E} \approx 0.417 \text{ fm}. \end{aligned} \tag{7.26}$$

It follows that putting the ρ in the skyrmion brings in attraction, and compresses the size of the skyrmion, whereas the ω meson does the opposite. This feature is consistent with the conventional nuclear physics notion of the role of the vector mesons. Note also that with the parameters of the $O(p^4)$ terms in the nonlinear sigma model Lagrangian that results from integrating out all vector mesons from the holographic QCD model, one obtains the soliton mass comparable to the observed nucleon mass. This is different from that of the Skyrme model — skyrmion$_\pi$ — which with the parameters of the vacuum values, gives typically ~ 0.5 GeV higher. Nature is of course with all vectors, and the role of ω makes the resulting soliton mass somewhat higher than the physical nucleon mass. The attraction that is brought in by the ρ meson will be found later to be crucial for the binding energies of heavy nuclei in skyrmion models. It leads to the BPS structure that seems to surprisingly agree with experiments in heavy nuclei.

7.3.5.2 *Baryonic matter*

When the HLS Lagrangian of [Ma, Yang, Oh and Harada (2013)] is put on crystal lattice (in FCC), there result some remarkable observations on the in-medium properties of the soliton (effectively nucleon) and the mesons.

Here we focus on the behavior of the soliton mass (representing the in-medium nucleon mass) and the pion decay constant f_π^* as a function of density. For this, the hWZ term in the HLS Lagrangian plays an important role. As written in Eq. (2.95), there are three terms that satisfy the homogeneous part of the Wess–Zumino anomaly equation, $\mathcal{L}_{\rm anom}$,

$$\Gamma_{\rm hWZ} = \int d^4x \mathcal{L}_{\rm anom} = \frac{N_c}{16\pi^2}\int_{M^4}\sum_{i=1}^{3} c_i \mathcal{L}_i, \tag{7.27}$$

where M^4 stands for the 4-dimensional Minkowski space.[8] In the literature, one finds the approximation that reduces the hWZ Lagrangian to only one term

$$\mathcal{L}_{\rm anom}^{\rm min} = \frac{3}{2} g\omega_\mu B^\mu \tag{7.30}$$

where B_μ is the topological baryon current. This is obtained if one takes $c_1 = -c_2 = 2/3$ and $c_3 = 0$ in the $\mathcal{L}_{\rm anom}$ term. It is equivalent to substituting the equation of motion for the ρ field into (7.27) and then taking

[8] Explicitly

$$\begin{aligned}\mathcal{L}_1 &= i\,{\rm Tr}\left[\hat{\alpha}_{\rm L}^3\hat{\alpha}_{\rm R} - \hat{\alpha}_{\rm R}^3\hat{\alpha}_{\rm L}\right], &(7.28a)\\ \mathcal{L}_2 &= i\,{\rm Tr}\left[\hat{\alpha}_{\rm L}\hat{\alpha}_{\rm R}\hat{\alpha}_{\rm L}\hat{\alpha}_{\rm R}\right], &(7.28b)\\ \mathcal{L}_3 &= {\rm Tr}\left[F_{\rm V}\left(\hat{\alpha}_{\rm L}\hat{\alpha}_{\rm R} - \hat{\alpha}_{\rm R}\hat{\alpha}_{\rm L}\right)\right], &(7.28c)\end{aligned}$$

in the 1-form and 2-form notations with

$$\begin{aligned}\hat{\alpha}_L &= \hat{\alpha}_{\parallel} - \hat{\alpha}_{\perp},\\ \hat{\alpha}_R &= \hat{\alpha}_{\parallel} + \hat{\alpha}_{\perp},\\ F_V &= dV - iV^2. \end{aligned}\tag{7.29}$$

the mass $m_\rho \to \infty$. We will consider, for comparison, what is called "minimal model," $\mathrm{HLS_{min}}(\pi, \rho, \omega)$ [Meissner, Rakhimov and Yakhshiev (2000)], which is obtained from (7.21) by switching off $\mathcal{L}_{(4)}$ and taking (7.30) for the hWZ term. Of course taking the vector-meson mass going to infinity is incompatible with the possible presence of the vector manifestation (VM) fixed point at which the vector mass is to go to zero and hence cannot be reliable for dense matter. Indeed we will see below that the properties predicted by this model are unacceptable. This hints at that it is important to have all three terms in the hWZ term taken into account. Below, however, given its simplicity, we will use this model in discussing the generic role of the hWZ terms.

A physical quantity closely tied to the effective in-medium nucleon mass m_N^* is the effective pion decay constant f_π^*. Note that m_N^* and f_π^* are closely related because of the large N_c relation $m_N^* \sim \kappa f_\pi^*$ where κ is a density independent constant of $O(\sqrt{N_c})$. This relation turns out to show up strikingly in the half-skyrmion phase.

The calculation of f_π^* in HLS can be performed straightforwardly by including the ρ-meson contribution to the axial correlator (7.10). They are depicted in Fig. 7.7.

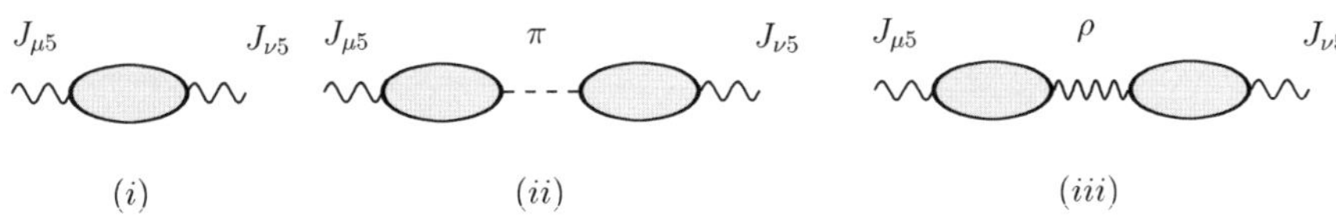

Fig. 7.7 Three types of contributions to the correlator of Eq. (7.10): (i) the contact diagram, (ii) the pion exchange diagram, and (iii) the rho exchange diagram. Shaded blobs stand for the skyrmion matter interaction vertices.

The three graphs contribute

$$(i): i f_\pi^2 g_{\mu\nu}\delta^{ab}\left\langle 1 + \frac{1-a}{2}\left[\left(1-\frac{2}{3}\phi_\pi^2\right)-1\right]\right\rangle$$

$$(ii): \; -i f_\pi^2 \frac{p_\mu p_\nu}{p^2}\delta^{ab}\left\langle\left(1-\frac{2}{3}\phi_\pi^2\right)-1\right\rangle$$

$$(iii): i\delta^{ab}\left\langle\frac{a^2 g^2 f_\pi^4}{p^2-m_\rho^2}\left(g_{\mu\nu}-\frac{p_\mu p_\nu}{m_\rho^2}\right)\left[\left(1-\frac{2}{3}\phi_\pi^2\right)-1\right]\right\rangle. \quad (7.31)$$

To the order of the expansion considered, the sum of the three terms is Lorentz invariant and has the same expression of f_π^* as that from the Skyrme model, Eq. (7.15).

Now let us look at m_N^* and f_π^* juxtaposed to each other. Their behavior at the skyrmion-half-skyrmion transition density $n_{1/2}$ is of the primary importance in the development that follows. In Fig. 7.8 are given m_N^* and f_π^* vs. the lattice size L [Ma, Harada, Lee, Oh, Park and Rho (2013)].

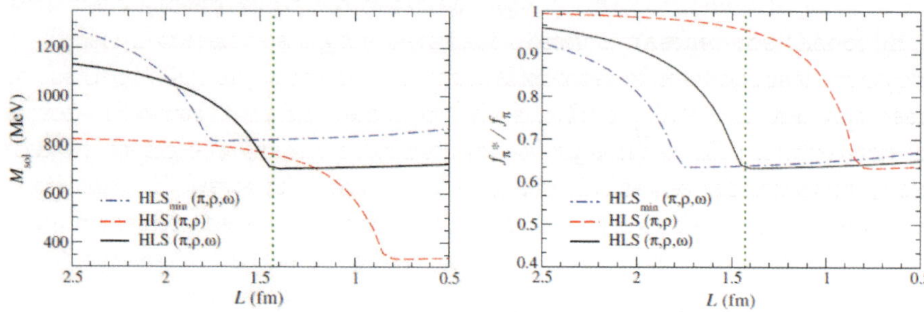

Fig. 7.8 $m_{\rm sol}^*$ (left) and f_π^* (right) vs. lattice size L.

There are several remarkable features in these results that will figure importantly in the EoS of compact stars.

- While the presence of the topology change is robust, the location of the skyrmion-half-skyrmion transition density $n_{1/2}$ depends, as mentioned before, sensitively on the degrees of freedom taken into account. This can be seen in Fig. 7.8 (right panel). The role of the ω meson is particularly striking. This can be seen in how the hWZ terms influence it. This reflects that the ω couples to other fields via the hWZ terms. This feature was noted in the holographic QCD model where the CS term that is subleading in the 't Hooft constant brings in the ω degree of freedom.
- The most striking and highly significant feature is that both the soliton mass $M_{\rm sol}$ and the pion decay constant f_π become independent of density in the half-skyrmion phase. The similarity in the density dependence of the two quantities reflects the applicability of the large N_c behavior mentioned above. Note the striking insensitivity of f_π^* to the degrees of freedom involved, coming to $f_\pi^*/f_\pi \approx 0.6 - 0.7$.
- Particularly notable is that the value of the pion decay constant is independent of massive degrees of freedom in the half-skyrmion phase although the location of $n_{1/2}$ is extremely sensitive to them. This comes about because terms dependent on the space-averaged chiral conden-

sate (7.6) affected by the massive degrees of freedom vanish in the half-skyrmion phase. This accounts for the insensitivity to ω and hence to the hWZ terms. *This robust property of the pion decay constant in the density regime $n \geq n_{1/2}$, independent of the degrees of freedom included as one can see in Figs. 7.5 and 7.8, is one of the key features of the approach anchored on the topology change. The topology is dictated by the pion field and the vector mesons do not carry any influence on the topology. This feature will lead in Chapter 9 to the "conformal sound velocity" in compact stars, $v_s^2 = 1/3$, at a precocious density slightly above the onset of half-skyrmions, i.e., $n \sim 3n_0$.*

- In the low-density regime $n < n_{1/2}$, in a stark contrast to the half-skyrmion phase, the pion decay constant — and the soliton mass — is quite sensitive to the massive degrees of freedom. Note specially the minimal model $\text{HLS}_{\text{min}}(\pi, \rho, \omega)$ with its truncated hWZ term. This behavior can enable one to sort out which degrees of freedom are indispensable and which are not. Deeply bound pionic systems provide data on density dependence of f_π^* up to the normal nuclear matter density [Kienle and Yamazaki (2004)].

7.3.6 HLS_{reduced}

So far we have been considering HLS Lagrangian written up to $O(p^4)$ including the hWZ terms and found that the $O(p^4)$ terms are important and hence need to be taken into account for both elementary baryons and dense baryonic matter. But this Lagrangian contains much too many terms to confront Nature with. Up to date there is no systematic effort to reduce the number of terms involved here. In this subsection, we will attempt to make a numerical exercise to reduce the complicated Lagrangian to a manageable form. In this connection, it is worth reiterating that there is something magical about the HLS limited to $O(p^2)$, namely the Lagrangian (7.22).

As already suggested, the notion of HLS makes sense only if the vector meson mass m_V is considered to be "light" in the sense that NG boson mass is light [Harada and Yamawaki (2003); Komargodski (2011)]. Chiral perturbation expansion can then be made, both in the matter-free space and in dense medium, in terms of π and V, with the mass of V treated on the same footing as the NG boson mass [Harada and Yamawaki (2003)]. The notion of the VM fixed point in approaching chiral restoration as one dials the chiral quark condensate to zero does make sense then. Even in the vacuum, we recall, the famous KSRF formula for the ρ mass, $m_\rho^2 = 2f_\pi^2 g^2$ with $g = g_{\rho\pi\pi}$, turns out to be exact to all orders of loop terms [Harada, Kugo and Yamawaki (1993)]. The vector dominance works fairly well for mesons and also for the nucleon if the infinite tower of holographic model is included. There are other surprising properties with the notion of HLS with the vector-meson mass taken as "light."

It seems therefore a highly pertinent question whether one cannot stick to the $O(p^2)$ Lagrangian with the minimal number of corrections from $O(p^4)$ terms. This issue was addressed in [Ma and Rho (2017)]. It has not been feasible to make a drastic simplification of high-order HLS Lagrangian in continuum. However it is found in the soliton lattice description of baryonic matter that to the accuracy of 97.28% of the total of the soliton mass, the HLS Lagrangian can be simplified to

$$\mathcal{L}_{\rm SHLS} = f_\pi^2 {\rm Tr}[\hat{a}_{\perp\mu}\hat{a}_\perp^\mu] + af_\pi^2{\rm Tr}[\hat{a}_{\|\mu}\hat{a}_\|^\mu] - \frac{1}{2g^2}{\rm Tr}[V_{\mu\nu}V^{\mu\nu}] \\ + iz_4\,{\rm Tr}\,[V_{\mu\nu}\hat{\alpha}_\perp^\mu\,\hat{\alpha}_\perp^\nu] + \mathcal{L}_{\rm hWZ}. \tag{7.32}$$

The model (7.32) — that we dub as $HLS_{\rm reduced}$ — is found to capture the dominant physics of other quantities predicted by HLS (7.21). One can understand this as follows. The $O(p^2)$ terms encode the current algebra of hadron physics, this to all orders of loops as in the KSRF relations, so they are dominant in low-energy nuclear interactions. The ω meson — which is indispensable for stability of nuclear matter — figures through the hWZ terms. Among all the $O(p^4)$ terms — apart from the hWZ terms — the z_4 term, involving the strong ρ-π-π interaction, contributes most importantly, whereas the z_5 term, describing ρ-ρ-ρ interactions, is suppressed. In fact it is this simplified Lagrangian that through the effect of explicit scale symmetry breaking removes the "ω disaster" in dense skyrmion matter discussed below.

7.3.7 *Scale symmetry and homogeneous Wess–Zumino term in dense matter*

What is strikingly clear from the results discussed above is that the presence of the ω field substantially modifies, and mostly worsens, the skyrmion description of elementary baryons as well as baryonic matter in comparison with nature. Here we show that a correct treatment of the ω meson in the skyrmion crystal description of dense matter may require the presence of a scalar field, in our formulation the dilaton described in detail in Chapter

6. This interplay between the vector and scalar is well-known in standard nuclear many-body approaches such as in Walecka-type RMF theory. In the framework of the scale-chiral symmetric Lagrangian approach treated in Chapter 6, what enters crucially is the explicit scale symmetry breaking in QCD attributed to the trace anomaly. How to introduce the dilaton, a NG boson of spontaneously broken scale symmetry, is detailed in Chapter 6. For the discussion that follows, what is important is that scale symmetry is *hidden* in generic sigma model that encompasses EFT in hadronic physics that models QCD at low energy [Yamawaki (2017)] and we will have it "exposed" to see the interplay between the ω and the dilaton in the skyrmion matter.

Keeping to $O(p^2)$ in the leading chiral order and incorporating the dilaton field via the conformal compensator $\chi = f_\sigma e^{\sigma/f_\sigma}$ as described in Chapter 6, we write the pertinent Lagrangian as

$$\mathcal{L}_{\rm inv} = \frac{f_\pi^2}{4}\left(\frac{\chi}{f_\sigma}\right)^2 {\rm Tr}\left(\partial_\mu U \partial^\mu U^\dagger\right) + \frac{1}{2}\partial_\mu\chi\partial^\mu\chi + O(p^4). \tag{7.33}$$

Hidden-local-symmetrizing this, we have [Ma and Rho (2017)]

$$\begin{aligned}\mathcal{L}_{sHLS} = {} & f_\pi^2\left(\frac{\chi}{f_\sigma}\right)^2 {\rm Tr}\left[\hat{a}_{\perp\mu}\hat{a}_\perp^\mu\right] + a f_\pi^2\left(\frac{\chi}{f_\sigma}\right)^2 {\rm Tr}\left[\hat{a}_{\parallel\mu}\hat{a}_\parallel^\mu\right] \\ & - \frac{1}{2g^2}{\rm Tr}\left[V_{\mu\nu}V^{\mu\nu}\right] + \frac{1}{2}\partial_\mu\chi\partial^\mu\chi + O(p^4).\end{aligned} \tag{7.34}$$

There is a dilaton potential that we have left out above since it does not figure directly in the discussion. (7.34) is essentially what we have referred to as "LOSS."

In the presence of vector fields, as noted above, there contributes the anomalous-parity hWZ term $\mathcal{L}_{\rm anom}$ that contains three terms. For an approximate calculation we again take only one term in the form $g\omega_\mu B^\mu$. It is straightforward to treat all three terms involved therein at the cost of more parameters. We do not expect qualitative differences from treating all of them. By itself, this term is of scale dimension 4, hence scale-invariant in the scaling defined in Chapter 6. Therefore it has no coupling to the dilaton field χ. This is, however, not consistent with scale-chiral symmetry à la Crewther–Tunstall [Crewther and Tunstall (2015)]. The correct expression should be [Ma and Rho (2017)]

$$\mathcal{L}_{hWZ\chi} = \left(c_h + (1-c_h)\left(\frac{\chi}{f_\sigma}\right)^{\beta'}\right)\mathcal{L}_{\rm hWZ}. \tag{7.35}$$

where β' is the anomalous dimension of $(G_{\mu\nu})^2$ and signals the explicit breaking of scale symmetry. Setting $\beta' = 0$ or $c_h = 1$ amounts to ignoring scale symmetry explicit breaking. Thus $\beta' \neq 0$ with $c_h \neq 1$ would be signaling explicit scale symmetry breaking.

The question is: Is there an evidence for this in the skyrmion crystal? In [Park, Rho and Vento (2004)], where skyrmions were put on crystal lattice with $\beta' = 0$ and $c_h = 1$ with the Lagrangians (7.34) to simulate dense matter, the contribution to the energy of the system from the hWZ term diverged at high density unless the pion decay constant (or equivalently the ω mass) went to ∞. This defect can be eliminated if the explicit breaking of scale symmetry dampens the strength of the hidden gauge coupling g. In [Park, Rho and Vento (2004)], the problem was resolved when the factor $(\frac{\chi}{f_\sigma})^n$ with $1 < n \lesssim 3$ was *arbitrarily* multiplied to $\mathcal{L}_{hWZ}$. This happens because in medium f_σ^*/f_σ decreases monotonically at increasing density up to $n_{1/2}$, therefore the coupling gets sufficiently suppressed. This means that the solution to the problem found in [Park, Rho and Vento (2004)] corresponds to (7.35) with [Ma and Rho (2017)]

$$1 < \beta' \lesssim 3, \quad c_h \approx 0. \tag{7.36}$$

The message one received from this subsection is that to correctly describe baryonic matter at high density, the scalar degree of freedom with *scale symmetry explicitly broken* seems to be absolutely necessary. They could perhaps be generated at high (chiral) orders in perturbative scheme and could come in as a dilaton in the broken scale symmetry. For this, it seems essential, if the analysis described in this subsection is correct, that how the scale symmetry is explicitly broken has to be understood. What we have observed above regarding the disastrous role of the ω meson in dense matter in the presence of a dilaton indicates the necessity of the dilaton coupling to the ω in the hWZ term in 4D skyrmion matter or the CS term in 5D instanton matter.

We should point out here a serious caveat to the solution to the "ω catastrophe" that required an important role of β'. In Chapters 2 and 6, it was seen that the LOSS approximation worked well without any intervention of β'. The same approximation will be found to fare well in compact-star phenomenology in Chapters 8 and 9. In the CT scheme, the LOSS corresponds to setting coefficients $c_{1,2} = 1$ in Eqs. (6.14a) and (6.14b) which then does away with the dependence on β' except in the dilaton potential. This is drastically different from what's needed for the ω problem (7.36). What makes a crucial difference between the two cases is how the ω meson intervenes in the description. If the given solution to the ω problem is viable, then how the explicit scale symmetry breaking manifests in dense medium remains a totally open issue.

7.3.8 *Infinite tower of vector mesons in holographic QCD*

We made references in the preceding chapters to the infinite tower of vector mesons in connection with the vector meson dominance (VMD) for the nucleon EM form factors and the two-parameter HLS model resulting from integrating out the infinite tower in the holographic QCD models. The infinite tower arises by Kaluza–Klein reduction "top-down" from 5D to 4D of holographic dual QCD. As alluded to, the infinite tower can also arise "bottom-up" from 4D by resorting to infinite redundancies present in the chiral field $U = \xi_L^\dagger \xi_R$. It can lead to deconstructing 5D Yang–Mills Lagrangian with a CS term (encoding chiral anomalies) [Son and Stephanov (2004)].

In this subsection, we discuss the influence of the infinite tower on the structure of dense matter, with focus on the "apparent" approach to the BPS structure associated with the binding energies of heavy nuclei. It indicates how dense matter could be accessed by packing the infinite tower. For this we use the "top-down" approach based on the Sakai–Sugimoto (SS) model [Sakai and Sugimoto (2005)] which has the properties closest to QCD.

In holographic models, the generic action has the structure

$$S_5 = S_5^{\rm DBI} + S_5^{\rm CS}, \tag{7.37}$$

where $S_5^{\rm DBI}$ is the 5D Dirac-Born-Infeld (DBI) part and $S_5^{\rm CS}$ is the 5D CS part. For different models the expressions of $S_5^{\rm DBI}$ could be different.

In the SS model, the DBI part is written in terms of the pure bulk gauge fields as

$$S_5^{\rm DBI} \approx S_{\rm YM} = -\kappa \int d^4x dz \frac{1}{2e(z)^2}\,{\rm Tr}\,\mathcal{F}^2 \tag{7.38}$$

with $\kappa = \frac{\lambda N_c}{216\pi^3}$. $e(z)$ is the effective YM coupling that depends on the holographic direction z and is proportional to the KK mass as $M_{KK}^{-1/2}$. The 5D CS term is

$$S_5^{\rm CS} = \frac{N_c}{24\pi^2}\int d^4x dz {\rm Tr}\left(\mathcal{A}\mathcal{F}^2 + \frac{i}{2}\mathcal{A}^3\mathcal{F} - \frac{1}{10}\mathcal{A}^5\right). \tag{7.39}$$

We use the index $m = (\mu, z)$ with $\mu = 0, 1, 2, 3$. The gravity enters in the z dependence of the YM coupling, giving rise to the warping of the space. $\mathcal{A} = \mathcal{A}_\mu dx^\mu + \mathcal{A}_z dz$ is the five-dimensional $\mathrm{U}(N_f)$ gauge field and $\mathcal{F} = d\mathcal{A} + i\mathcal{A}\mathcal{A}$ is its field strength. In the case of $N_f = 2$ flavors,

$$\mathcal{A} = A_{\mathrm{SU}(2)} + \frac{1}{2}\tilde{A}_{\mathrm{U}(1)}. \tag{7.40}$$

The resulting YM action is

$$S_{\mathrm{YM}} = -\kappa \int d^4x dz \frac{1}{2e^2(z)} \left(\mathrm{Tr}\, F_{mn}^2 + \frac{1}{2}\tilde{F}_{mn}^2 \right), \tag{7.41}$$

and the CS term is

$$S_{\mathrm{CS}} = \frac{N_c}{16\pi^2} \int \tilde{A} \wedge \mathrm{Tr} F^2 + \frac{N_c}{96\pi^2} \int \tilde{A} \wedge \tilde{F}^2. \tag{7.42}$$

In Eqs. (7.41) and (7.42) F_{mn} is the field strength for the SU(2) gauge field and $\tilde{F}_{mn}$ stands for the field strength of the U(1) gauge field.

We should stress that the SS model — which is holographically dual to QCD in the large N_c and λ limit (and the chiral limit) — has no free parameters. To the leading order in λ, that is, to $O(\lambda)$, $e(z)$ is a constant, so the 5D YM action can be taken to be in flat space. The CS term, which comes at $O(N_c\lambda^0)$, does not contribute. Thus to the leading order, i.e., $O(N_c\lambda)$, the static baryon is given by the instanton solution that is self-dual [Kaplunovsky, Melnikov and Sonnenschein (2012)].

Let us first look at the instanton given by the SS action in the leading order, say, $O(N_c\lambda)$. At this order only the S_5^{DBI} in flat space contributes and the $U(1)$ degrees of freedom decouple from $SU(2)$ degrees of freedom. The resulting 5D Yang–Mills theory, in unit of an arbitrary mass dimension, is

$$S = -\frac{1}{2} \int \mathrm{Tr} F_{mn}^2 d^4x\, dz, \tag{7.43}$$

where

$$F_{mn} = \partial_m A_n - \partial_n A_m + [A_m, A_n] \tag{7.44}$$

with $A_m = T^a A_m^a$ normalized as $\mathrm{tr}(T^a T^b) = \frac{1}{2}\delta_{ab}$. The gauge field transforms

$$A_m \to g(A_m + \partial_m) g^{-1} \tag{7.45}$$

with $g \in SU(2)$.

The static energy coming from the action (7.43), known as BPS action, has a well-known bound, the Bogomol'nyi bound,

$$E \geq 8\pi^2 B, \tag{7.46}$$

where B is the instanton number representing baryon number

$$B = \frac{1}{16\pi^2} \int \mathrm{Tr}(F_{MN}{}^* F_{MN}) d^3x\, dz, \tag{7.47}$$

in which $M = 1, 2, 3, z$ and ${}^*F_{MN} = \frac{1}{2}\epsilon_{MNAB} F_{AB}$ is the dual field strength. Now the bound is satisfied if F_{MN} is self-dual, i.e.,

$$F_{MN} = {}^* F_{MN}. \tag{7.48}$$

This means that the energy of the system cannot be lower than the bound. In other words, a system of instanton number A describing mass-number A nucleus has the binding energy equal to zero. This can be referred to as "BPS skyrmion" as will be explained below.

The 5D action (7.38), when KK-reduced to 4D, can be written as the sum of an infinite number of vector and axial-vector mesons possessing HLS [Sakai and Sugimoto (2005)]. In the leading order, the gauge coupling constant will be a constant independent of metric warping with no coupling to the infinite tower of iso-scalar vector mesons. The skyrmion constructed with this infinite tower will then be equivalent to the instanton discussed above. A highly illustrative study was made by Sutcliffe [Sutcliffe (2010, 2011)] which showed how the vector mesons entered in the tower.

Start with the pion only with all the vector mesons set equal to zero. This is the Skyrme model, skyrmion$_\pi$. The soliton static energy is ~ 1.24 times the Bogomol'nyi bound (7.63)[9]

$$E^{(0)} = 1.235\,(8\pi^2 B). \tag{7.49}$$

Next when the lowest-lying vector meson ρ is included, the soliton mass is drastically reduced to

$$E^{(1)} = 1.071(8\pi^2 B). \tag{7.50}$$

The next-lying axial-vector meson a_1 brings this further down to

$$E^{(2)} = 1.048(8\pi^2 B). \tag{7.51}$$

Since the full tower of isovector vector mesons brings this to the bound $E^{(\infty)} = 8\pi^2 B$, it follows that the high-lying vector mesons make the theory flow to a conformal theory. And, the lowest-lying vector meson ρ does nearly all the work in flowing to the conformality. This feature is most likely related to the near complete saturation, as in the case of the pion, of the EM form factors of the nucleon form factors [Hong, Rho, Yee and Yi (2008); Hashimoto, Sakai and Sugimoto (2008)]. *An important point to*

[9]This corresponds in the case of the Skyrme Lagrangian to the "Faddeev bound" $12\pi^2 B$.

underline here is that as far as the nuclear binding energy is concerned the BPS limit — with zero binding energy — arrived at by the infinite tower of isovector vector mesons in flat space is close to nature. We return to this point below in connection with BPS models.

Now suppose that the same packing strategy is applied to multi-skyrmion or nuclear systems. The above result implies that the nuclear binding energy, $\sim 24\%$ of the nucleon mass with the pion field only, will reduce to near zero as the higher-lying vector mesons are packed in. This tendency is beautifully illustrated in Fig. 7.9.

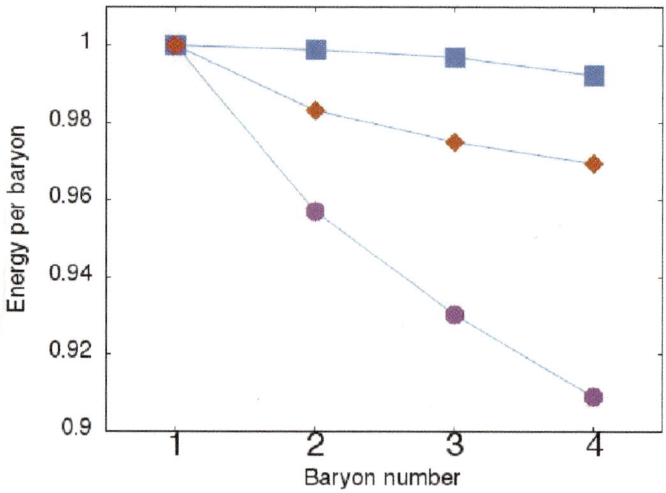

Fig. 7.9 The energy per baryon, in units of the single baryon energy vs. baryon number up to $A = 4$. Squares are the experimental data, circles are the skyrmion energies with only pions, and diamonds are the skyrmion energies in the extended theory with pions, ρ and a_1 mesons. This figure is taken from [Sutcliffe (2011)].

In the SS holographic QCD model, the ω meson figures at the next order in λ. It figures via the CS term. In standard paradigm of nuclear physics, as stressed above, the ω degree of freedom is indispensable. In chiral perturbation theory, its effect, at least partially, is captured in short-range three-body and higher-body forces brought in by high-order derivative counter terms in the chiral Lagrangian. To confront the predictions of the holographic model on nuclear physics with the empirical values, therefore, one should consider the warped space in the YM action and the CS term.

Here we analyze how the ω, entering at $O(N_c\lambda^0)$, influences the flow to conformality seen above with the isovector vector mesons. At the same order, the metric warping also enters but its effect is less significant. To explore the effect of the $U(1)$ degrees of freedom, one should include the effect of the CS term. The CS term is background-independent and hence should be independent of the warping. Using our energy unit, we have the BPS mass $M_{\rm BPS} \approx \frac{\lambda N_c}{27\pi} M_{KK} \approx 559$ MeV [Hong, Rho, Yee and Yi (2007, 2008)],[10] in agreement with [Sutcliffe (2011)]. When the CS term contribution is added, we get $M_{\rm BPS-CS} = M_{\rm BPS} + \sqrt{\frac{2}{15}} N_c M_{KK} \approx 1038$ MeV which shows that the contribution from the CS term to the soliton mass is significant. Along the same procedure applied before, we found that, for the given $\mathcal{M}$, using the HLS given by Eq. (7.21) in the following and determining the low energy constants by the 5D YM theory (7.43), the results are [Ma *et al.* (2012); Ma, Yang, Oh and Harada (2013)]:

- π, ρ, ω:

$$M_{\text{``BPS''}}(\pi, \rho, \omega) \approx 1162 \text{ MeV}. \tag{7.52}$$

- π, ρ:

$$M_{\text{``BPS''}}(\pi, \rho) \approx 577 \text{ MeV}. \tag{7.53}$$

- π:

$$M_{\text{``BPS''}}(\pi) \approx 673 \text{ MeV}. \tag{7.54}$$

The ω meson is seen to block the flow to the conformal fixed point. In Eqs. (7.52), (7.53) and (7.54), the subindex "BPS" means the parameters are calculated from the 5D YM theory (7.43).

We next examine the effect of the tower of the infinite vector mesons from the SS model in which the warp effect is included. A direct calculation of the mass of the topological object, here, the *approximate* instanton, yields the mass [Hong, Rho, Yee and Yi (2007, 2008); Hashimoto, Sakai and Sugimoto (2008); Hata, Sakai, Sugimoto and Yamato (2007)]

$$M_{\rm (approx)instanton} \simeq 1800 \text{ MeV}, \tag{7.55}$$

with $f_\pi = 92.4$ MeV and $\lambda = 17$ fixed in the meson sector [Sakai and Sugimoto (2005)]. Note that the mass (7.55) corresponds to the mass of a skyrmion with the effect of the infinite tower of vector mesons in a warped space and the CS term taken into account. The effect of the tower of vector mesons are reported above.

[10]Here we used $\lambda = 16.66, M_{KK} = 948$ MeV determined from our inputs.

In sum, what is found is the following: In the holographic QCD of SS, to the leading λ order, the more isovector vector mesons are included, the lighter soliton mass becomes and the sum of the contributions from the infinite tower reduces the soliton mass to the BPS instanton limit. The residual interaction between the skyrmions (as quasiparticles) gets weaker and the size becomes smaller, all going in the right direction. However, this tendency gets blocked at the next order in λ, namely at $O(\lambda^0)$, primarily by the presence of the ω meson present in the CS term with the effect of metric warping less significant. This situation is changed by the scalar meson effect as reported above.

7.4 The BPS skyrmions

We have seen above that the instanton structure equivalent to the infinite tower of isovector vector mesons in the flat space (in the absence of the iso-scalar vector degrees of freedom that are mediated by the CS term that is sub-leading in the 't Hooft constant) — BPS structure — gives the binding energies of heavy nuclei closer to nature — with very small binding energy — than the skyrmion models described above. This description contains the current algebra term $\propto \mathrm{Tr}(L_\mu L^\mu)$ as well as the quartic term that represents the L_i's that dominate π-π scattering, so the low-energy theorems can be assured. However, as shown, once the iso-scalar ω is present, the good agreement with nature is ruined. The presence of dilaton χ ameliorates the situation, but how χ enters in the infinite-tower structure is not known.

7.4.1 *Hidden BPS structure*

An extremely intriguing possibility, suggested by Adam, Sanchez-Guillen and Wereszcynski (ASGW) [Adam, Sanchez-Guillen and Wereszczynski (2017)], is that a potentially realistic effective theory for nuclear matter may comprise of hidden BPS substructures. Imagine there exists a chiral Lagrangian — with pion fields — that is given by an infinite derivative terms representing low-energy nonperturbative QCD. Following ASGW, we write

$$\mathcal{L} = \sum_{n=0}^{\infty} \mathcal{L}_{2n} \tag{7.56}$$

where n is the number of derivatives. They truncate at 6 derivatives as

$$\mathcal{L}_0 = -\mu^2 F(\mathrm{Tr}U), \quad \mathcal{L}_2 = \frac{1}{2}\mathrm{Tr}(L_\mu L^\mu),$$
$$\mathcal{L}_4 = \frac{1}{16}\mathrm{Tr}[L_\mu, L_\nu]^2, \quad \mathcal{L}_6 = \lambda^2\pi^4 B_\mu B^\mu \tag{7.57}$$

where B_μ is the topological baryonic current. Here units of energy and length are taken and λ and μ are dimensionless coupling constants. One can identify $\mathcal{L}_0$ as the pion mass term, $\mathcal{L}_2$ the current algebra term and $\mathcal{L}_4$ as the Skyrme quartic term. In the infinite tower of vector mesons, the quartic term can be thought of arising when the isovetor vector mesons are integrated out. Similarly the sextet term $\mathcal{L}_6$ can be considered as arising when the infinite tower of iso-scalar vector mesons are integrated out. Now the hedgehog configuration for the chiral field U is written in terms of a real scalar ξ and complex scalar u as

$$U = e^{i\xi\vec{\tau}\cdot\vec{n}}, \quad \vec{n} = \frac{1}{1+|u|^2}\Big((u+u^*), -i(u-u^*), 1-|u|^2\Big). \tag{7.58}$$

A remarkable observation made in [Adam, Sanchez-Guillen and Wereszczynski (2017)] is that one can write $\mathcal{L}_2$ and $\mathcal{L}_4$ in two terms in the new target space coordinates ξ and u,

$$\mathcal{L}_2 = \mathcal{L}_2^{(1)} + \mathcal{L}_2^{(2)} \tag{7.59}$$

and similarly for $\mathcal{L}_4$ in such a way that $\mathcal{L}^{(m)} \equiv \mathcal{L}_2^{(m)} + \mathcal{L}_4^{(m)}$ for $m = 1, 2$, separately, satisfies a Bogomol'nyi equation. As will be detailed below, the sum $\mathcal{L}^{(3)} \equiv \mathcal{L}_0 + \mathcal{L}_6$ also satisfies a Bogomol'nyi equation. Thus there are three separate combinations that give rise to BPS structure. *This implies that (7.56), non-BPS as a whole, has hidden BPS substructures.*

Now the question is: Which set of BPS substructure is relevant for given physical processes?

This question can be addressed using the Sakai–Sugimoto holographic gauge-gravity dual model for baryons. Sutcliffe's derivation of the BPS structure of nuclei was based on the large N_c and large λ ('t Hooft constant) limit. In those two limits, the holographic space flattens and the CS term is absent. Suppose one derives the infinite derivative Lagrangian (7.56) from the SS model. There are only two unknown constants in the SS action, the series (7.56) can in principle be written down containing the two constants N_c and λ by integrating out the infinite tower of vector mesons. This has been done with certain approximations up to six derivatives in [Bartolini, Bolognesi and Proto (2017)]. It is shown there that if one considers "small 't Hooft constant (λ) regime" instead of the large one that gives the low-energy dynamics of the SS model, there are two possibilities that differ from the current algebra regime. The authors argue that in the limit $\lambda \to 0$, one gets $\mathcal{L}_2 + \mathcal{L}_6$ if the pion is massless and $\mathcal{L}_0 + \mathcal{L}_6$ if the pion is massive. It is the latter that gives the BPS model of Adam *et al.* [Adam, Sanchez-Guillen and Wereszczyn (2013)]. This suggests that one can interpolate from the current algebra domain to the BPS domain by tuning λ and the pion mass. This is another illustration that one theory can be tuned to another theory by varying "conditions," resembling the interpolation of linear sigma model from nonlinear sigma model with scale symmetry to scale invariant theory tuned by a parameter. Again *the key question would be: What are the pertinent conditions for given physical processes and what are the parameters to "dial?"* In this connection, recall that one can go from a linear sigma model to nonlinear sigma model and to scale-invariant model by dialing a parameter. This is an important aspect for the structure of compact stars discussed in what follows.

7.4.2 *BPS skyrmion for medium-heavy nuclei*

Let us examine the BPS structure, $\mathcal{L}^{(3)} = \mathcal{L}_0 + \mathcal{L}_6$. In the "derivation" of the small binding energy from a flat-space YM Lagrangian discussed above, the relevant degrees of freedom in nuclear physics, namely, the pion and the lowest-lying isovector vector mesons, are augmented with the infinite tower leading to the BPS structure. Suppose one simply posits a BPS structure in 4D. Here the philosophy would then be, instead of starting from a chiral Lagrangian with pions only and building up the tower to go to heavy nuclei, why not start with the simplest structure that could work for heavy nuclei with nearly vanishing binding energy and then go from there making appropriate corrections? This is the point of view taken by Adam *et al.* [Adam, Sanchez-Guillen and Wereszczynski (2010); Adam *et al.* (2016)] in picking $\mathcal{L}^{(3)}$. This option eschews contact with low-energy QCD, e.g., low-energy theorems. In fact the parameters entering in this approach will have no direct connection with QCD proper, in a spirit totally different from what is implicit in *all* nuclear EFTs.

The justifications offered for this drastically unorthodox approach are [Adam *et al.* (2016)]

(1) The binding energy is zero at the *leading order* even though highly nonlinear strong interactions may be involved. One might think of this as an extreme quasiparticle description with no residual interactions. This uncannily resembles the half-skyrmion structure encountered in Section 7.3.1.1.
(2) The model effectively describes a perfect fluid with its energy-momentum tensor in the perfect fluid form, and the static energy functional is invariant under volume-preserving diffeomorphisms.
(3) While the standard quasiparticle description of baryonic matter,

namely, Landau Fermi-liquid theory, can be approximately given in the mean field of a relativistic field theory, here the quasiparticle picture is to capture the *exact* result of the theory. The mean-field result of the relativistic field theory can be an approximation that is obtained by "averaging" the BPS theory.

In the BPS skyrmion model of [Adam, Sanchez-Guillen and Wereszczynski (2010)], the basic quantity is the $SU(2)$ valued field $U(\boldsymbol{x}, t)$ in which topology is lodged. The Lagrangian picked is $\mathcal{L}^{(3)}$ given above,

$$\mathcal{L}_{\text{BPS}} = -\lambda^2\pi^4 B_\mu B^\mu - \mu^2 V(\text{tr}(U)), \tag{7.60}$$

where λ and μ are parameters that are not related to QCD parameters. They will be fit to nuclear binding energies that are focused on. The potential $V(\text{tr}(U))$ is suitably chosen, involving nonderivative terms. It is taken in the forms

$$\begin{aligned} V(\text{tr}(U)) &= V_\pi = \frac{1}{2}\text{tr}(1-U), \qquad \text{model I;} \\ V(\text{tr}(U)) &= (V_\pi)^2, \qquad\qquad\qquad \text{model II.} \end{aligned} \tag{7.61}$$

The baryon number B_μ takes the familiar expression

$$B_\mu = \frac{1}{24\pi^2}\epsilon^{\mu\nu\alpha\beta}\text{tr}\left(U^\dagger\partial_\nu U U^\dagger\partial_\alpha U U^\dagger\partial_\beta U\right). \tag{7.62}$$

First to see that the BPS skyrmion has zero binding energy, one shows that it satisfies the Bogomol'nyi equation. The static energy given by Eq. (7.60) has the bound

$$\begin{aligned} E &= \int d^3x\left(\pi^2\lambda B_0 \pm \mu\sqrt{V}\right)^2 \mp 2\pi^2\lambda\mu\int d^3x B_0\sqrt{V} \\ &\geq \mp 2\pi^2\lambda\mu\int d^3x B_0\sqrt{V} \\ &= 2\pi^2\lambda\mu|B|\left\langle\sqrt{V}\right\rangle_{\mathbb{S}^3}, \end{aligned} \tag{7.63}$$

where $\left\langle\sqrt{V}\right\rangle_{\mathbb{S}^3}$ is the average value of $\sqrt{V}$ on the target space S^3. The last equality follows from the condition that $B > 0$. The bound is saturated by the solutions of the BPS equation of motion from (7.60)

$$\pi^2\lambda B_0 \pm \mu\sqrt{V} = 0. \tag{7.64}$$

By coupling gravity to (7.60), one can write the action of the BPS model in curved space as

$$S_{\text{BPS}} = \int d^4x|g|^{\frac{1}{2}}\left(-\lambda^2\pi^4|g|^{-1}g_{\rho\sigma}B^\rho B^\sigma - \mu^2 V\right). \tag{7.65}$$

From this action, one can show that the energy-momentum tensor has the perfect fluid form

$$T^{\rho\sigma} = (p+\varepsilon)u^\rho u^\sigma - p g^{\rho\sigma} \tag{7.66}$$

with the four-velocity u^ρ, energy density ε and pressure p given by

$$\begin{aligned} u^\rho &= B^\rho/\sqrt{g_{\sigma\pi}B^\sigma B^\pi}, \\ \varepsilon &= \lambda^2\pi^4|g|^{-1}g_{\rho\sigma}B^\rho B^\sigma + \mu^2 V, \\ p &= \lambda^2\pi^4|g|^{-1}g_{\rho\sigma}B^\rho B^\sigma - \mu^2 V. \end{aligned} \tag{7.67}$$

Going further, it can be shown that the BPS skyrmion is equivalent to a relativistic perfect fluid in the Eulerian formulation.

The energy of the BPS system described by the Lagrangian (7.60)

$$E = \int d^3x \left\{\pi^4\lambda^2 B_0^2 + \mu^2 V(\mathrm{Tr}U)\right\} \tag{7.68}$$

has, as stated above, a Bogomol'nyi bound. In fact it has infinitely many BPS solutions saturating the bound. The topology is lodged in the field U which is of the pion field type as in the usual chiral Lagrangian. In this approach, the pion field is a highly correlated nuclear collective excitation of pionic quantum number that supports a solitonic structure that has no fluctuation component that enters in the low-energy theorems of strong interactions. This would naturally exclude the possibility that the pion field that enters in this description be directly related, if any, to the asymptotic pions observed in nature. Assuming that it is of the soltion structure similar to that of large N_c QCD model, i.e., the Skyrme model, one takes the axial symmetric ansatz $U(r) = e^{i\vec{\tau}\cdot\vec{r}\theta(r)}$. With this it is straightforward to calculate the soliton energy for the BPS Lagrangian (7.60)

$$E_{\rm sol} = \frac{64\sqrt{2}\pi}{15}\mu\lambda N, \tag{7.69}$$

where $N = B$ is the winding number, i.e., baryon number. Note that the soliton energy per particle then is just a constant $\propto \mu\lambda$ given by the combination of the potential and the constant multiplying the topological term, so the energy of N baryon number system is a multiple of baryon number and hence zero binding energy at the level of the soliton.

To go beyond the classical soliton structure, one needs to perform the collective quantization. The resulting rotational spectra are given by

$$E_{rot} = \frac{\hbar^2}{2}\left(\frac{j(j+1)}{\mathcal{J}_1} + \frac{i(i+1)}{\mathcal{I}_1} + \left(\frac{1}{\mathcal{I}_3} - \frac{1}{\mathcal{I}_1} - \frac{B^2}{\mathcal{J}_1}\right)k_3^2\right), \tag{7.70}$$

where $\mathcal{J}_i$ and $\mathcal{I}_i$ are, respectively, the angular momentum moment of inertia and the isospin moment of inertia. For even B nuclei, $k_3 = 0$. The resulting binding energy including the Coulomb energy and isospin violation is given in Fig. 7.10. The theoretical prediction with only 3 fit-parameters agrees surprisingly well with the experimental data and the Weizsacker's empirical formula.

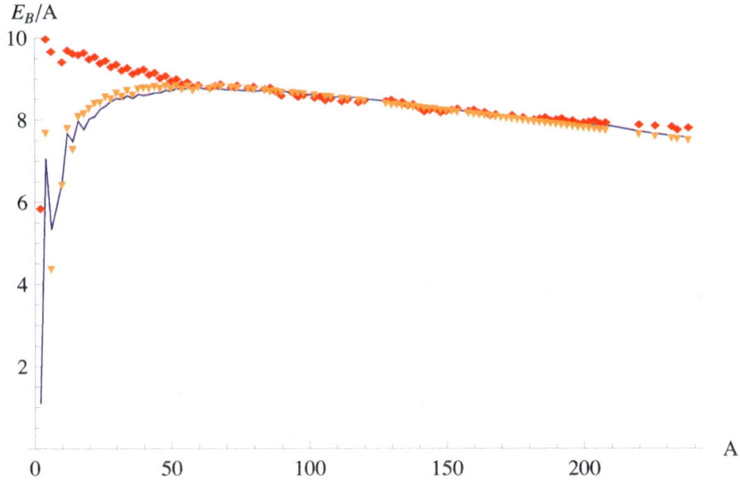

Fig. 7.10 Binding energies per nucleon in MeV. The diamond is from the present BPS model. The triangle is from the Weizsacker's formula. And, the solid line is from experimental data. The figure is taken from [Adam, Sanchez-Guillen and Wereszczyn (2013)].

A comment on the contrast of the BPS approach to the current lore of nuclear physics is in order here. There are two key ingredients in the currently predominant attitude among nuclear theorists: The first is that the small binding energy observed in nuclei is an interplay between a big attraction characterized by scalar degrees of freedom and a big repulsion provided by vector degrees of freedom, and the second the theory has to be able to accurately satisfy the conditions met both experimentally and theoretically at around nuclear matter density.

The BPS model is contrary to the first in that the BPS "nucleon" is a topological blob ignorant of the intricate interactions that lead to its property. As it stands, the BPS model cannot account for low-energy pionic processes that are described by low-energy theorems that are the ground to effective chiral theories.

As for the second, it cannot access the accurately measured nuclear response functions to the electroweak fields. For instance, while it may provide bulk properties of nuclear ground states, it cannot account for the important role of meson-exchange currents that are accurately described by effective chiral Lagrangian approaches, offered in Chapter 2, as an example of a nuclear physics "proof" of the FT.

The departure from the soft-pion theorems notwithstanding, one should not dismiss the BPS model. It may be possible to start from the BPS matter and develop a theory that can access those processes given by the low-energy theorems. This could be in the same spirit as starting from the Landau Fermi-liquid fixed point applicable to the regime of nuclear equilibrium density and arriving at processes involving soft pions. This is exemplified by relativistic mean-field approaches that are, as argued in this volume, connected to Landau Fermi-liquid theory. There pionic interactions can be — and are — readily incorporated. *That the "exact" effective Lagrangian (7.56) may have a hidden BPS structure of the type described above, together with hidden symmetries of QCD explored in dense matter, suggests that different parts of the EFT Lagrangian could be manifested in different conditions.* We will return to this point in connection with what are referred to as "constraints" extracted at low density, say, at n_0 to be satisfied by the EoS at high density, say, at $\sim (6-7)n_0$ appropriate for dense compact stars.

7.5 Cheshire Cat and duality

It was argued in Chapter 3 that the Cheshire Cat provided a duality between quarks/gluons and skyrmions — nucleons and nuclear matter — in terms of the chiral bag. The half-skyrmion structure described above, which will play a bridge from low to high density, can also be given a Cheshire Cat interpretation [Mantovani Sarti and Vento (2014)]. It will be seen in the next chapter that the behavior of the skyrmion-half-skyrmion changeover can be translated into a changeover of the behaviors in the parameters in the *bs*HLS. In Chapter 9, an analogy is made between this changeover and hadron-quarkyonic crossover in the equation of state of compact-star matter.

In condensed matter physics, there exist exotic phase transitions that involve a duality between two EFTs. A well-known example is the putative quantum critical phenomenon mentioned above, namely the Néel-valence bond solid (VBS) transition between two different breaking phases of a spin-1/2 system on the square lattice [Janssen and He (2017)]. Here half-skyrmions appear between the two phases in which they are confined. The appropriate theoretical description, the non-compact CP^1 model, is considered to be dual to (2+1)-dimensional QED. Unlike this condensed-matter case where a phase transition, though not of Ginzburg–Landau–Wilsonian paradigm, does take place with the half-skyrmions "deconfined" at the boundary between the two phases, in our case, the half-skyrmions are, as noted above, seem to be confined. No phase transition is involved, so it could be put in the class of Cheshire Cat.

Chapter 8

Compressed Baryonic Matter

In the preceding chapters, specially in Chapter 6, we constructed the baryonic scale-invariant HLS (bsHLS) Lagrangian built on as precise a notion as possible of hidden symmetries in QCD — in particular, scale symmetry and local flavor symmetry — coupled to pseudo-scalar Nambu–Goldstone (NG) bosons and baryons. The Lagrangian so constructed purports to cover the range of densities from that of normal nuclear matter to high-density compact-star matter. In this chapter we confront nuclear matter and a selected aspect of dense compact star matter with the bsHLS approach so formulated. For confrontation with Nature, it remains yet to specify and determine the "bare parameters" sliding with the vacuum change as a function of density, i.e., intrinsic density dependence (IDD), so as to do calculations to confront experiments. Already in various places in the preceding chapters, the bsHLS Lagrangian in the leading scale-chiral order was used to bring out certain features in a qualitative way, e.g., Fermi-liquid structure of dense nuclear matter. In doing so, sliding bare parameters were employed, however, without explaining how they were determined. In this chapter, we first give as precise a definition as possible, as to how the IDDs figure in the parameters of the Lagrangian and then apply the Lagrangian to dense matter using the V_{lowk} renormalization group (V_{lowk}RG) formulation.

As already stated and will be explained in more precise terms below, essentially two strategies are involved for the calculation. First, we match the effective field theory (EFT) Lagrangian to QCD at a matching scale Λ_M, given typically by the chiral scale $\Lambda_\chi \approx 1$ GeV. This procedure, more or less standard and fully explained in [Harada and Yamawaki (2003)], enables the "bare parameters" of the Lagrangian to capture certain non-perturbative characteristics of QCD. Second we extract certain (robust) topological properties of skyrmion matter simulated on crystal lattice as described in Chapter 7 and translate them into the behavior in medium of the parameters of the Lagrangian. This strategy is foreign to what we refer to as "standard nuclear effective field theory (SnEFT) approaches" anchored on standard chiral perturbation theory (χPT), currently highly popular in nuclear community. In this chapter, this formalism will be applied to nuclear matter at the equilibrium nuclear matter density $n_0 \simeq 0.16$ fm^{-3} and dense matter appropriate to compact-star matter $n \sim (5-7)n_0$. Application to the phenomenology of massive compact stars will be given in Chapter 9.

There is no reason why this formalism could not be applied also to finite nuclei. The details remain still to be worked out and we won't not go into that matter in this volume.

For convenience — and also strongly suggested by topology, it is profitable to separate the density regime into two, "regime I" (R-I) and "regime II" (R-II). There could be a third which is to be identified with deconfined quarks/gluons. In our approach exploiting hadron degrees of freedom only (say, à la Cheshire Cat), this third regime will not figure in the dense baryonic matter. We will pick the separation density to be $n_{1/2}$ given by the topology change introduced in Chapter 7. In consistency with the Cheshire Cat philosophy presented in Chapter 3 — and argued further in Chapter 9 — the possible presence of strongly-coupled quark-gluon phase within the range of densities involved in massive compact stars, such as for example what could be identified as "quarkyonic phase," will be treated in terms of hadronic degrees of freedom. What intervenes in the changeover is identified with a topology change at $n_{1/2}$. This topology change is taken to correspond to the mechanism of trading-in of quark-gluon degrees of freedom for hadronic description. As mentioned, up to date, the precise value for $n_{1/2}$ could not be pinned down either within the theoretical framework or phenomenologically. An exciting possibility, discussed in the next chapter, is that the tidal deformability from gravity waves coming from coalescing neutron stars could provide the means to determine it. The recent LIGO/Virgo observation [Abbott *et al.* (2017)] indicates that $n_{1/2} \gtrsim 2n_0$. Fortunately the numerical results turn out not to depend on the precise value of $n_{1/2}$ around $\sim 2n_0$. We will find, based on both models and phenomenology, that the most likely value will be $n_{1/2} \sim (2-3)n_0$. Unless otherwise stated — and for specificity, we will pick $n_{1/2} = 2.0n_0$ for numerical analyses.

8.1 What's *not* in the standard nuclear effective field theory approach

To start with, we point out what makes the approach adopted in this volume different from SnEFT approaches anchored on standard chiral Lagrangian.[1] The best way is to list what's manifestly *not* in SχPT but figures importantly in bsHLS.

As described in Chapter 2, the SnEFT starts with chiral Lagrangian of nonlinear sigma model type with nucleons and pions only and treats nuclear systems in systematic chiral expansion along the line first proposed by Weinberg [Weinberg (1990)] and then developed in early 1990's both for nuclear forces [Ordónez and van Kolck (1992); Weinberg (1991); Ordónez, Ray and van Kolck (1994)] and nuclear electro-weak currents [Rho (1991)]. Since then there have been extensive developments along that line going to high orders, currently up to n-th order to the leading order (N^{n}LO) for $n \leq 4$ in the chiral counting. There are in the literature a large number of reviews of the current developments applying SnEFT to nuclear systems, both finite and infinite, detailing the results that provide an indication of a convincing, or what one might even say the most successful, proof of Weinberg's folk theorem (FT) on EFT. We won't go into the details here — they are too numerous to list them all. The approach taken in this chapter agrees more or less with what's obtained in SnEFT up to normal nuclear matter density and possibly up to the half-skyrmion density $n_{1/2} \approx 2n_0$. It is for the density regime R-II with $n \gtrsim n_{1/2}$ that our description differs from SnEFT.

In SnEFT, excitations of the quantum numbers of massive vector mesons V_μ and dilaton scalar χ with their energy scale "much higher" than that of pions are to be generated — in principle — by high-order corrections. Since the cut-off employed in nuclear calculations is typically $\sim (400 - 500)$ MeV, the SnEFT Lagrangian is to contain in the higher derivative "counter terms" effects due to those massive mesons that are integrated out. In contrast, in our bsHLS approach, the V_μ and χ are explicitly included so that their effects figure at the tree level in the effective scale-chiral expansion

[1] By "SnEFT," a term to be used only in this chapter to avoid the plethora of abbreviations, we mean standard EFT approaches to nuclear physics currently in vogue. It includes not only SχEFT based on pion-only baryonic chiral Lagrangian treated in various sophisticated numerical techniques (such as Greens function Monte Carlo etc.) but also a variety of energy-density functionals using SχPT inputs and hybrid approaches using SχPT for low density and strongly-coupled quark matter for high density. Some of them will be described in Chapter 9.

(see Chapter 6). What makes this approach particularly more powerful in dense-matter physics is that the possible decrease in medium of the masses of those excitations sliding with the vacuum change as described in Chapter 6, and further elaborated below, can be naturally accounted for at low order of scale-chiral expansion.

An equally important feature absent in SnEFT is a possible intervention of other degrees of freedom at $n \gtrsim n_0$. To explain the observed $\sim$ 2-solar mass stars, the SnEFT needs to invoke either strong-coupling quark degrees of freedom or quarkyonic phase for $n \sim (2-4)n_0$. The transition, assumed to be continuous and hence not a phase transition, takes place from the normal nuclear matter to a quark-hadron state with the normal matter treated either in realistic phenomenological nuclear models [Masuda, Hatsuda and Takatsuka (2016); Kojo, Powell, Song and Baym (2016)] or in high-order χPT [Hell and Weise (2014)]. Both are essentially equivalent up to n_0 in the sense of Weinberg's FT.[2] What is required however is the intervention of new degrees of freedom at $n \gtrsim 2n_0$, be that a strongly coupled quark hadron phase or a quarkyonic phase implemented with color-superconductivity [Fukushima and Kojo (2016)].

In the treatment given in this chapter, the quark degrees of freedom do not figure directly. They are traded in for topology in the sense argued in Chapter 3. It involves only hadronic degrees of freedom with instead a topology change from a state of skyrmions to a state of half-skyrmions at a density $n = n_{1/2} \sim 2n_0$ taken into account. No other degrees of freedom intervene in the structure of compact-star dynamics. No strangeness figures in the process either as discussed in Chapter 10.

8.2 Scale-invariant HLS Lagrangian in medium

In this section, we address the problem of precisely defining and determining the IDDs in the bare parameters of the in-medium *bs*HLS Lagrangian. IDDs have been invoked in various places in the preceding chapters but without precise explanations except that they are inherited from QCD at the matching scale $\sim$ 1 GeV. For practical applications of the theory, one can only work at present with the leading-order terms of *bs*HLS, which seems fully justified. Certain caveats to this will be mentioned. To determine the sliding parameters of *bs*HLS for $n \gtrsim 2n_0$, we rely on the information provided by the topology change and for this we need to consider

[2]The former is a part of MEEFT in the language of [Rho (2008)].

baryons generated as skyrmions, not incorporated *ab initio* as with *bs*HLS, for which the appropriate Lagrangian is scale-invariant HLS (*s*HLS) as given in Chapter 6. It will be argued that what is equivalent to the IDD bare parameters of *bs*HLS, defined below, will be generated by the background of the topological objects, not carried in the parameters of *s*HLS.

8.2.1 *Intrinsic density dependence*

Let us go back to the leading-order Lagrangian (6.86) in *bs*HLS (i.e., LOSS), given in Chapter 6

$$\mathcal{L} = \mathcal{L}^M_{\chi PT_\sigma}(\pi, \chi, V_\mu) + \mathcal{L}^B_{\chi PT_\sigma}(\psi, \pi, \chi, V_\mu) - V(\chi) \tag{8.1}$$

with the *s*HLS Lagrangian

$$\begin{aligned}\mathcal{L}^M_{\chi PT_\sigma}(\pi, \chi, V_\mu) = f_\pi^2 \left(\frac{\chi}{f_\sigma}\right)^2 \mathrm{Tr}[\hat{a}_{\perp\mu}\hat{a}^\mu_\perp] + a f_\pi^2 \left(\frac{\chi}{f_\sigma}\right)^2 \mathrm{Tr}[\hat{a}_{\|\mu}\hat{a}^\mu_\|] \\ + \frac{1}{2g^2}\mathrm{Tr}[V_{\mu\nu}V^{\mu\nu}] + \frac{1}{2}\partial_\mu\chi\partial^\mu\chi\end{aligned} \tag{8.2}$$

and the baryonic coupling to *s*HLS fields

$$\mathcal{L}^B_{\chi PT_\sigma}(\psi, \pi, \chi, V_\mu) = \mathrm{Tr}(\bar{B} i\gamma_\mu D^\mu B) - \frac{\chi}{f_\sigma}\mathrm{Tr}(\bar{B}B) + \cdots \tag{8.3}$$

For numerical analysis, we take the dilaton potential of the standard form

$$V(\chi) \approx \frac{m_\sigma^2 f_\sigma^2}{4}\left(\frac{\chi}{f_\sigma}\right)^4 \left[\ln\left(\frac{\chi}{f_\sigma}\right) - \frac{1}{4}\right]. \tag{8.4}$$

This potential is valid in the CT formalism in the limit that $\beta' << 1$. We will have more to say whether this makes sense when applied to nuclear and dense matter.

The Lagrangian (8.1), linearized to the form (8.7) given below, was used in Chapter 4 to derive Landau Fermi-liquid fixed-point theory. It was also analyzed in Chapter 6 for certain in-medium low-density parameters including the dilaton mass. The IDDs used there are equivalent to what will be derived below in the density regime R-I.

How one arrives at the Lagrangian (8.1) in Crewther–Tunstall (CT) theory is described in Chapter 6. As stated clearly, the Lagrangian is valid in the CT scheme at the order the scale symmetry explicit breaking β' is ignored in the matter sector, with its influence lodged entirely in the dilaton potential. This is consistent with the notion that scale invariance is hidden in the linear sigma model — and in the standard Higgs Lagrangian. As shown in Chapter 6, in the CT scheme, this comes about when one naively sets $\beta' \approx 0$. However as mentioned in Chapter 6, LOSS can be obtained without setting β' equal to zero. In fact it was observed in dense skyrmion matter [Ma and Rho (2017)] that $|\beta'|$ is ~ 2 and scheme-independent analyses indicate that the anomalous dimension for the gluon field tensor squared $G^2_{\mu\nu}$ in gauge theory [Ryttov and Shrock (2016)] seem also to give a similar bound. Nevertheless there are several indications that (8.1) is quite reasonable as we have seen before and also in compact stars in Chapter 9.

In order to confront Nature (nuclear and dense matter), there are two indispensable ingredients to take into account in the *bs*HLS Lagrangian (8.1). The first is what was mentioned before: The "bare" parameters of the effective Lagrangian need to be matched to QCD. Secondly, strong correlations between nucleons, including possible "phase changes," have to be included as one goes up in density. Since high density, $n \sim (5-7)n_0$, involves going toward chiral transition, the "bare" parameters need to have contact with QCD parameters. This is done in our treatment by matching various current correlators of the *bs*HLS Lagrangian to those of QCD at an appropriate matching scale Λ_M lying below the typical chiral scale $\Lambda_\chi \sim 1$ GeV. This matching, explained more in detail below, automatically endows the "bare" parameters of the *bs*HLS Lagrangian with dependence on the quark condensate $\langle\bar{q}q\rangle$, the gluon condensate $\langle G^2\rangle$ etc., generically referred to as "condensates." Since those condensates, VeV's of QCD field operators, depend on the "vacuum," they will of course depend on density, which modifies the vacuum if the Lagrangian is embedded into a medium. The principal point in the development that follows is that the density dependence involved is *intrinsic* of QCD [Lee, Paeng and Rho (2015); Paeng, Kuo, Lee, Ma and Rho (2017)], to be distinguished from the density dependence coming from nuclear many-body correlations including multi-body forces that are integrated out of the effective Lagrangian. The latter due to nuclear correlations, entering in renormalization group (RG) decimations in medium, enters in the induced density dependence ($\mathrm{DD_{induced}}$).

In numerical calculations performed and compared with empirical data, it is sometimes difficult to attribute the observed density dependence *solely* to the IDD because one picks the scale from which the RG decimation is made at a lower point than Λ_M. If the decimation is made from the energy/momentum scale at which the matching is made Λ_M, then the IDD will be entirely given by the matching. However in practical calculations, what is done is to start the RG decimation in medium from a scale — call it Λ_{EFT} — lower than the scale Λ_M, typically, even lower than the vector-meson mass scale, roughly ~ 700 MeV. This is because in practice precise experimental data in nucleon interactions to which the first decimation can be done are available up to $\sim 2\ \mathrm{fm}^{-1} \sim 400$ MeV. In this case, the effective scaling of the parameters involved is not totally of the Lorentz-invariant IDD — more or less unchanged from Λ_M — but can contain nuclear correlations that are in the class of $\mathrm{DD}_{\mathrm{induced}}$, i.e., nuclear many-body dynamics, e.g., short-range three-body forces, that are of higher power counting in the given perturbation scheme. A highly illustrative example will be the C14 dating matrix element that we will describe more in detail below.

Thus in all practical calculations, the scaling parameters that figure in *bs*HLS Lagrangian will be given by IDD* which is the sum of IDD and $\mathrm{DD}_{\mathrm{induced}}$ although it is not specifically mentioned.

At Λ_{EFT}, the "bare" parameters of (8.1) are the pion decay constant f_π, the dilaton decay constant $f_\sigma \sim \langle\chi\rangle$, the dilaton mass m_σ, the vector-meson mass m_V, the nucleon mass m_N, and the coupling constants, specially, the hidden gauge coupling constant g, all of which are taken to be unchanged from Λ_M. Except for the dilaton condensate (and the pion decay constant), as explained below, this is not a bad assumption at low densities, say, $n \lesssim n_0$. This is because the degrees of freedom (except for the NG bosons) involved are massive, lying higher than an effective cutoff, say, Λ_{EFT}, and hence can be considered as integrated out, with the effect appearing in the constants multiplying higher-dimension field operators, e.g., quartic fermion fields (as explained in Chapter 4). We will see that this is not valid when, e.g., the hidden gauge coupling drops toward the vector manifestation (VM) fixed point as described below.

The "bare" constants are known phenomenologically in the medium-free vacuum denoted as $|0\rangle$. In our scheme, they depend on density as the vacuum is modified, with the density dependence denoted by superscript with $*$. We denote such vacuum by $|0^*\rangle$. For contrast, note that in SnEFT, the standard chiral expansion usually takes those parameters fixed at zero density. The IDD effects are missing although the parameter-fitting compounds them in the numerical calculations.

In medium, Lorentz invariance is broken so the time and space derivatives bring different components into the Lagrangian. An example is the time and space components of the axial coupling constant g_A discussed in Chapter 2.[3]

[3]The scaling known as "BR scaling" introduced in 1991 [Brown and Rho (1991)] belongs to the effective one that contains both IDD and induced scaling $\mathrm{DD}_{\mathrm{induced}}$. This aspect is illustrated in connection with the anomalous orbital gyromagnetic ratio δg_l and nuclear

8.2.2 *Two density regimes delineated by topology change*

At present, there is no way to calculate in a controlled systematic way the IDD in the parameters of the *bs*HLS Lagrangian. It is plausible that lattice QCD could ultimately do so, but that's for the far future. As we will show, one can, however, resort to experimental data up to the equilibrium density n_0, and possibly slightly above n_0, but this is not feasible due to the lack of experiments — and no model-independent theoretical tools — for $n \gtrsim 2n_0$. For that density regime, we shall therefore rely on what can be taken as robust from the topological structure provided by the skyrmion matter obtained with *s*HLS. Thus it makes a natural sense to make the separation of density regimes into "Region-I" (or R-I for short) and "Region-II" (R-II) delineated by the topology change at $n_{1/2}$:

$$\text{Region} - \text{I}; \quad n < n_{1/2}, \tag{8.5}$$

$$\text{Region} - \text{II}; \quad n \geq n_{1/2}. \tag{8.6}$$

It was briefly stated in Chapter 4 that the appropriate approach to many-nucleon problems is the V_{lowk} RG with *bs*HLS in two steps. First, the decimation is done from $\Lambda_{\rm EFT}$ down to just above the Fermi sea and the second decimation takes into account high-order correlations that take place above the Fermi sea. Being a Wilsonian-type RG procedure, as density increases, the Fermi-liquid fixed-point approximation becomes better as $1/N \equiv (\Lambda_{\rm EFT} - k_F)/k_F \to 0$ [Shankar (1994)]. This means that the approximation will be more reliable in R-II. The latter could of course break down if the Fermi-liquid structure does not hold any more as perhaps in percolation transition. We will assume that this breakdown does not occur within R-II relevant to massive compact stars.

8.2.2.1 *IDD in R-I*

Up to n_0, the IDDs can be deduced from experimental data and we assume that they can be extrapolated to $n_{1/2} \sim 2n_0$. In medium we need to take into account the change in the condensates. We therefore shift the dilaton field around the in-medium vacuum value, $\chi \to \langle\chi\rangle^* + \chi'$. We then expand and linearize the *bs*HLS Eq. (8.1)

$$\mathcal{L}_I = \overline{N}[i\gamma_\mu(\partial^\mu + ig^* V^\mu) - m_N^* + g_\sigma^*\sigma]N - \frac{1}{4}V_{\mu\nu}^2 + \frac{{m_V^*}^2}{2}V^2 + \frac{1}{2}(\partial_\mu\sigma)^2 - \frac{{m_\sigma^*}^2}{2}\sigma^2 + \frac{1}{2}\partial^\mu\vec{\pi}\cdot\partial_\mu\vec{\pi} - \frac{1}{2}{m_\pi^*}^2\vec{\pi}^2 + \mathcal{L}_{\pi\rm matter} + \cdots \tag{8.7}$$

weak transitions in heavy nuclei described in Chapters 2 and 4. This point is reiterated here because there continues to be a confusion on it in the nuclear community.

where the ellipsis stands for possible terms that are of higher order in chiral-scale counting and of higher fields and $V_\mu = \vec{\tau}\cdot\vec{\rho}_\mu + \omega_\mu$ assumed to be flavor-$U(2)$ symmetric and $\mathcal{L}_{\pi\mathrm{matter}}$ is for the pion coupling to other fields. Here the exponential of the conformal compensator is expanded in terms of the dilaton field σ and likewise for the pion field which is redefined so as to give the canonical kinetic energy term in (8.7). Note that (8.1) encodes by construction all the scale-chiral low-energy theorems, and (8.7) inherits them from it. This fact will figure importantly below.

To express the parameters of (8.7) with IDDs suitably incorporated, let us define the scaling parameter

$$\Phi_\sigma(n) \equiv f_\sigma^*/f_\sigma \tag{8.8}$$

where $f_\sigma^* = \langle\chi\rangle^*$. The Wilsoninan matching via the vector and axial-vector correlators at the matching scale $\Lambda_M \sim (1.0-1.2)$ GeV [Harada and Yamawaki (2003)] with $\Lambda_{QCD} \sim (0.3-0.45)$ GeV shows that the mass parameters and coupling constants involving both mesons — and baryons — depend very weakly on the quark and gluon condensates $\langle\bar{q}q\rangle, \langle G^2\rangle, \ldots$. This corresponds to what was referred to in Chapter 6 as the IDD$_m$. It is a good approximation to take them independent of density in R-I. Therefore it is principally the dilaton condensate Φ_σ, (8.8), giving the IDD$_{pNG}$ that controls the IDDs in R-I. This observation is supported by nuclear phenomenology [Brown and Rho (2004)]. This means that the bare parameters are given, assuming $U(2)$ symmetry for the vectors $V = (\rho, \omega)$, by[4]

$$m_N^* = \Phi_\sigma m_N, \quad m_V^* = \Phi_\sigma m_V, \quad m_\sigma^* = \Phi_\sigma m_\sigma \tag{8.9}$$

and

$$g_V^* = g_V, \quad g_\sigma^* = g_\sigma. \tag{8.10}$$

We will now show how the unique scaling factor can be determined from experiments up to n_0. For this we need to look at the pion mass term[5]

$$\mathcal{L}_{\mathrm{mass}} = \frac{f_\pi^2}{4}\left(\frac{\chi}{f_\sigma}\right)^3 \mathrm{Tr}(M^\dagger U + h.c.) \tag{8.11}$$

[4]In the framework of $\chi\mathrm{PT}_\sigma$ discussed in Chapter 6, the dilaton mass has dependence on the anomalous dimension β' as $m_\sigma^*/m_\sigma = \Phi^{(\frac{\beta'}{2}+1)}$. Since β' is unknown, although there is a hint for its presence in skyrmion crystal, we shall ignore it in the numerical applications made in this and next chapters. This leaves a certain uncertainty in the scaling of the dilaton mass with respect to the other degrees of freedom. This aspect differs from what has been done in the past with the scaling of [Brown and Rho (1991)].

[5]Note that in the *exact* chiral limit this procedure does not allow to determine the scaling factor Φ_σ.

where M is the mass matrix $M = m_\pi^2 \mathrm{diag}(1,1)$ (for $N_f = 2$). From this, we deduce that[6]

$$f_\pi^* = \Phi_\sigma f_\pi, \tag{8.12}$$

$$m_\pi^* = \sqrt{\Phi_\sigma} m_\pi. \tag{8.13}$$

Equation (8.12) determines Φ_σ as a function of density

$$\Phi_\sigma = f_\pi^*/f_\pi \tag{8.14}$$

which reflects *the locking in medium of the chiral condensate with the dilaton condensate in scale-chiral symmetry* [Crewther and Tunstall (2015); Catà, Crewther and Tunstall (2018)]. f_π^* is known up to normal nuclear matter density n_0 from deeply bound pionic nuclei [Kienle and Yamazaki (2004)]. At n_0 it is

$$\Phi_\sigma(n_0) \approx 0.8. \tag{8.15}$$

To go beyond to the density $n_{1/2}$, we will simply extrapolate it assuming that there is no phase change along the way.

It is not surprising that the density scaling (d-scaling) by IDD of the σ mass is controlled by the dilaton condensate. It will be seen below why the d-scaling of the nucleon mass follows the same scaling factor.

Since the $U(2)$ is a fairly good symmetry for V in the vacuum and also most plausibly at low density — although there is no proof why, one can focus on the ρ meson. (The reasoning given here will apply to higher density in R-II although the scaling factor will not be Φ_σ as in R-I but the VM in the gauge coupling g_ρ^*.)

At the Wilsonian matching scale Λ_M, the bare parameter for the ρ meson is given by the famous KSRF relation

$$m_\rho^2 = a f_\pi^2 g_\rho^2. \tag{8.16}$$

In the matter-free vacuum, with the phenomenological values for f_π and g_ρ, $a \simeq 2$ in (8.16) gives the ρ mass in good agreement with the experimental value. It also gives the vector meson dominance for the pion electromagnetic form factors.[7] This observation brings out a puzzle as to how and why hidden local symmetry for the vector mesons works so well.[8] The only

[6]We are ignoring the anomalous dimension of the quark mass term γ_m which is not very well determined. As noted before, were we to take $\gamma_m \approx 1$ just for illustration, we would find $m_\pi^* = m_\pi$. This uncertainty does not affect the phenomenology given below.

[7]It also works well with the nucleon EM form factors but with an infinite tower of hidden gauge fields as mentioned in Chapter 7.

[8]A potentially important new development on the power of hidden local symmetry combined with scale symmetry is in Yamawaki's discussion in terms of a Grassmannian model $G/H \simeq O(N)_{\mathrm{global}} \times [[O(N-3) \times O(3)]_{\mathrm{local}}$ [Yamawaki (2018)].

way to understand this, as pointed out before, is that the vector mesons are taken to be "light" on par with the pseudo-Goldstone mesons. We recall that there are at least two known cases where the "lightness" of V_μ is realized. One is the presence of the VM fixed point at which the vector meson mass goes to zero as does the pion mass (in the chiral limit) [Harada and Yamawaki (2003)]. The other is supersymmetric QCD in certain parameter space [Komargodski (2011)]. Remarkably — and most importantly in dense matter, the low-energy theorem (8.16) is shown to be exact to all orders of loop corrections [Harada, Kugo and Yamawaki (1993)] with an $O(p^2)$ HLS. This result holds also with $O(p^4)$ HLS, modulo corrections of $O(m_\rho^2/\Lambda_M^2)$.[9] Now if we assume that this low-energy theorem holds well in R-I, then both a and g_ρ do not scale, so only the pion decay constant will scale. It then follows from (8.12) that the ρ mass scales with Φ_σ. The same will hold for the ω meson.

8.2.2.2 *Topology change*

We will now exploit the observation made in Chapter 7 that when A skyrmions are put on crystal lattice, as density increases, it is energetically more favorable for the A skyrmions to change over to $2A$ half-skyrmions. This topology change turns out to be quite robust and figures, though indirectly, even in small nuclei as in the α particle as pointed out in Chapter 7. This changeover occurs in the crystal simulation with the quark condensate $\Sigma \equiv \langle \bar{q}q \rangle$ averaged over space $\overline{\Sigma}$ going to zero whereas it is not zero locally, giving rise to a chiral density wave [Harada, Lee, Ma and Rho (2015)]. Since pions are still present in the system with non-zero pion decay constant, the half-skyrmion phase is not in the bona-fide Wigner–Weyl phase. However it has certain characteristics of chirally restored phase. One feature is the emergence of parity-doubling in the baryon mass structure as noted in Chapter 7, with a chirally invariant component of the skyrmion mass that closely resembles m_0 in the parity-doubling nucleon model [Detar and Kunihiro (1989)]

$$m_N = m_0 + \Delta(\Sigma) \tag{8.17}$$

where $\Delta(\Sigma) \to 0$ as $\Sigma \to 0$. In the crystal model, what corresponds to m_N is found to go in the half-skyrmion phase as

$$m_{\text{skyrmion}} \propto \langle \chi \rangle \tag{8.18}$$

[9] In anticipation of what's to come, when the vector mass drops rapidly as in R-II, the KSRF relation would work lot more accurately.

which is a chiral singlet, as the m_0 is in the parity-doubling scenario. In fact the mass splitting of the heavy-light mesons with chiral partner structure in nuclear matter, described by putting skyrmions on FCC crystal lattice, goes to zero in the half-skyrmion phase [Suenaga, He, Ma and Harada (2015)] in the same way as it does in the continuum description signaling chiral symmetry restoration [Nowak, Rho and Zahed (1993); Bardeen and Hill (1994)]. This suggests that chiral symmetry is in some sense restored, resembling perhaps a pseudo-gap phase, with parity-doubling.

The most striking outcome of this parity-doubling in the crystal model [Lee, Park and Rho (2011)] is the cusp structure discussed in Chapter 7 in the nuclear symmetry energy $E_{sym}(n)$ at $n = n_{1/2}$ which figures in the energy per particle of $A = N + Z$-nucleon system

$$E(A, n) = E_0(n, \alpha = 0) + \alpha^2 E_{sym}(n) + O(\alpha^4)\,. \tag{8.19}$$

As we will see in Chapter 9, the symmetry energy plays a particularly important role in the EoS for compact stars for which α is near 1 and also figures importantly in asymmetric nuclei away from stability being explored in RIB-type physics. Given a reliable phenomenological model with a large number of parameters or high-order standard chiral perturbative treatments, a fairly good fit can be achieved up to normal nuclear matter density $n \sim n_0$. The quantities of interest are not only the symmetry energy E_{sym}, but also the first and second derivatives of E_{sym}, L and K_{sym}, at $n = n_0$. The major problem in nuclear theory relevant to compact stars, however, is that beyond the equilibrium density n_0, the symmetry energy is totally unknown: It can be anything with parameters totally uncontrolled. This "wilderness" situation is illustrated in Fig. 8.1 [Olson *et al.* (2016)]. While E_{sym} is constrained by the available data (represented by the shaded blue area) up to $n \sim n_0$, there are no constraints beyond n_0, experimental or theoretical. What seems to be needed for the observed massive neutron stars is stiffening of the E_{sym} at a density $> n_0$. Some are stiff enough and some are not. In fact the E_{sym}'s that become softer beyond n_0 cannot support the massive stars unless the standard gravity theory itself is drastically modified.

As stressed in Chapter 7, the striking prediction of skyrmion models derived from the topology involved is that the whole skyrmion system of neutrons, collectively quantized, exhibits a cusp in the symmetry energy coming in at the $1/N_c$ order [Lee, Park and Rho (2011)]. In Fig. 8.2, the result of the Skyrme model is shown, implemented with scale symmetry, simulated on crystal lattice [Lee, Park and Rho (2011)]. It is given here

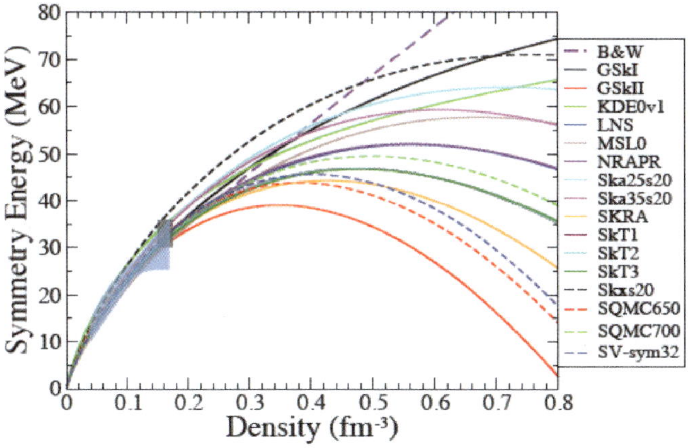

Fig. 8.1 The predicted E_{sym} by models that are fit to the matter properties up to n_0 and extrapolated to higher densities, showing the total wilderness at high densities. The shaded blue area represents experimental constraints.

again, this time for two different masses for the dilaton. It shows while the behavior at density $n > n_{1/2}$ is sensitive to the mass, the location of $n_{1/2}$ is not. As noted, the existence of the cusp is topological with the topology carried by the pion with other degrees of freedom affecting inappreciably the properties of the cusp.

It may be suspected that this cusp is an artifact of the crystal lattice in the description of the topology change and hence should be distrusted. But most fortunately for what comes below, that is not the case. We shall show (in Subsection 8.2.4) that this can be simply explained from a specific structure of the nuclear tensor force in *bs*HLS. This connection provides a support to our procedure anchored on topological information of QCD. *We cannot overemphasize this point, since it is the most crucial element in the development and if it turned to be wrong, could invalidate the whole premise of this volume.*

8.2.2.3 *IDD in R-II*

In Subsection 8.2.2.1, we noted that it is mainly the dilaton condensate that controls the IDD in R-I. The quark condensate $\langle\bar{q}q\rangle$ plays no direct role — apart from the fact that the pion decay constant encoded with the quark condensate is locked to the dilaton condensate. What happens above $n_{1/2}$ in R-II is a different story because of the topology change. The most

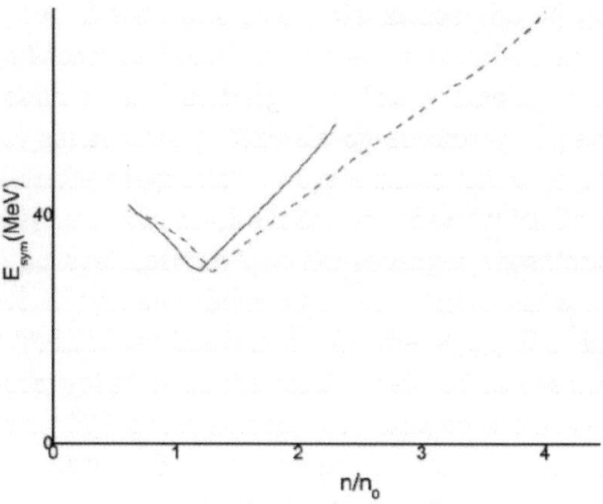

Fig. 8.2 Effect of dilaton mass on the symmetry-energy cusp in the skyrmion crystal model. The solid (dotted) line is for $m_\sigma = 600(720)$ MeV.

important is the behavior of the ρ meson mass as density passes $n_{1/2}$ into R-II. As mentioned, the KSRF relation, being a low-energy theorem valid to all orders of loop corrections [Harada, Kugo and Yamawaki (1993)], should be more valid in R-II with corrections of $\mathcal{O}((m_\rho^*/\Lambda_\chi)^2)$

$$m_\rho^* = \sqrt{a^*} f_\pi^* g_\rho^*. \tag{8.20}$$

f_π^* drops with density in R-I as observed in deeply bound pionic matter but remains constant $\sim f_\pi^*(n_{1/2})$ in R-II. As for g_ρ^*, the matching to QCD indicates it will remain nearly independent of density up to $n_{1/2}$, but beyond $n_{1/2}$ the vector manifestation (VM) dictates that it should fall rapidly to zero approaching the VM fixed point. This then gives the VM behavior of the ρ mass

$$m_\rho^* \sim g_\rho^* \to 0 \tag{8.21}$$

as $n \to n_{\rm VM}$ with the VM fixed point being the same as or near to the chiral restoration density. Note that it is the gauge coupling g_ρ that goes to zero that signals a phase change, not the bilinear quark condensate that is conventionally associated with chiral restoration. This feature confirms Gerry Brown's early argument that for chiral symmetry behavior, be on the look out for the ρ mass going to zero, say, in relativistic heavy ion collisions, instead of observables explicitly and directly linked to quark condensate — such as the pion decay constant — for monitoring the QCD transition [Adami and Brown (1992)].

Another striking feature is the behavior of the "bare" nucleon mass in the Lagrangian in R-II. While it drops in R-I at increasing density because f_π^* does, the skyrmion crystal model predicts that f_π^* connected to the dilaton condensate $\langle\chi\rangle^*$ as (8.18) stops decreasing and stays constant up to the point where possible phase transition(s) takes (take) place. This is associated with the emergence mentioned above of parity-doubling in R-II. Thus assuming the equivalence of the skyrmion crystal in R-II to the mean-field of bsHLS, we deduce that

$$m_N^* \propto \langle\chi\rangle^* \to m_0. \tag{8.22}$$

Therefore we have

$$\frac{m_N^*}{m_N} \approx \langle\chi\rangle^*/f_\sigma \equiv \kappa \sim \text{constant}. \tag{8.23}$$

As argued before, this can be taken as the quasiparticle mass at its Fermi-liquid fixed point.

As for the ω meson, the situation is less clear. The flavor $U(2)$ symmetry for ρ and ω holds well in the vacuum and it is reasonable to assume that it holds as well in dilute systems up to n_0. We will assume that it is valid for $n \lesssim n_{1/2}$. There is, however, a strong indication that the $U(2)$ symmetry breaks down in R-II [Paeng, Lee, Rho and Sasaki (2013)]. The ω meson does not join the ρ meson in the VM. This will become clear in applications to compact stars discussed in Chapter 9. But one can already see this in the mean field of bsHLS. Analogously to the nucleon mass, the ω mass is not directly connected to chiral symmetry breaking but dominated by the dilaton condensate.

8.2.3 *Nuclear tensor forces*

We discussed in Chapter 4 how the mean-field approach with an effective Lagrangian, with IDDs (plus $\mathrm{DD_{induced}}$) suitably incorporated, can be made equivalent to single-decimation RG procedure. There the method was applied to the properties of nuclear matter at $n \sim n_0$. In addressing dense matter, one can also work without potentials either in chiral perturbation [Holt, Rho and Weise (2016)] or standard relativistic mean-field energy functional approach. But these approaches are limited to low density and cannot be used way beyond $\sim n_0$. The reason is that for the former which relies on expansion in k_F, one cannot push the treatment to high density typically involved in compact stars, $(5-7)n_0$, and for the latter, without IDD effects, mean-field approximation cannot be reliable in R-II. To bypass these difficulties, we adopt the V_{lowk} RG procedure. The V_{lowk} RG procedure relies on nuclear potentials as in the chiral perturbation approach à la Weinberg. In the Weinberg strategy [Weinberg (1991)], one first constructs potentials by summing n-particle — typically $n \leq 3$ — "irreducible diagrams" in systematic chiral expansion to as high orders as feasible, i.e., mth-to-the leading order "N^{m}LO" with $m \gg 1$, and then the potentials are inserted into Schrödinger equations for bound-state problems or into Lippmann–Schwinger equations for scattering problems to account for "reducible" diagrams. In the V_{lowk} RG approach, instead, the potentials, computed with the *bs*HLS defined at the matching scale, are then used for doing RG decimations, resorting to the scale-chiral expansion as explained in Chapter 6. As explained, this allows one to do double-decimation Fermi-liquid fixed-point theory, going beyond the mean-field approximation.

We focus on the tensor forces in this subsection as they play the most important role in the EoS of compact-star matter.

In *bs*HLS, two-body tensor forces are given by one-pion exchange and one-ρ exchange with the IDDs determined as described above. Here the ω meson turns out to affect little of the phenomena considered, where the leading order is considered, so it can be ignored. The one-π- and one-ρ-exchange tensor forces, in the non-relativistic approximation, which is valid for the nucleon in all relevant ranges of density as described above, have the generic form [Holt, Brown, Kuo, Holt and Machleidt (2007); Machleidt (1989)]

$$V_M^T(r) = S_M \frac{f_{NM}^{*\,2}}{4\pi} \boldsymbol{\tau}_1 \cdot \boldsymbol{\tau}_2 \, S_{12} \mathcal{I}(m_M^* r), \tag{8.24}$$

$$\mathcal{I}(m_M^* r) \equiv m_M^* \left(\left[\frac{1}{(m_M^* r)^3} + \frac{1}{(m_M^* r)^2} + \frac{1}{3 m_M^* r} \right] e^{-m_M^* r} \right), \tag{8.25}$$

where $M = \pi$, ρ, $S_{\rho(\pi)} = +1(-1)$ and

$$S_{12} = 3 \frac{(\vec{\sigma}_1 \cdot \vec{r})\,(\vec{\sigma}_2 \cdot \vec{r})}{r^2} - \vec{\sigma}_1 \cdot \vec{\sigma}_2 \tag{8.26}$$

with the Pauli matrices τ^i and σ^i for the isospin and spin of the nucleons with $i = 1, 2, 3$. A qualitatively important quantity is the ratio f_{NM}^*/f_{NM}

$$\mathcal{R}_M \equiv \frac{f_{NM}^*}{f_{NM}} = \frac{g_{MNN}^*}{g_{MNN}} \frac{m_N}{m_N^*} \frac{m_M^*}{m_M} \tag{8.27}$$

where g_{MNN} are the effective meson (M)-nucleon couplings. Their relations to the coupling constants in *bs*HLS Lagrangian will be specified below.

What is the most notable in (8.24) is that given the same radial dependence, the two forces (through the pion and ρ meson exchanges) come with an opposite sign.

First, we discuss the d(ensity)-scalings of the tensor forces in medium given by IDDs and predict how the tensor forces scale in density. For the π tensor force, we have

$$\mathcal{R}_\pi = \frac{g^*_{\pi NN}}{g_{\pi NN}} \frac{m_N}{m^*_N} \frac{m^*_\pi}{m_\pi} \approx \frac{m^*_\pi}{m_\pi} \quad \text{in both R-I and R-II.} \tag{8.28}$$

Thus the π-tensor force principally depends only on the d-scaling of the pion mass.

8.2.3.1 *Pion-exchange tensor force in medium*

How the pion mass scales in density is a highly intricate matter. A comment is in order regarding its property with respect to the tensor force structure in dense matter. In arriving at the density scaling of the pion mass (8.13) we have simply ignored the possible contribution from the anomalous dimension for the quark mass term γ_m which figures in the chiral symmetry breaking term in χPT_σ (Eq. (6.14c) in Chapter 6)

$$\mathcal{L}_{\text{mass}} = \frac{f_\pi^2}{4} \left(\frac{\chi}{f_\sigma}\right)^{3-\gamma_m} \text{Tr}(M^\dagger U + h.c.). \tag{8.29}$$

Together with the anomalous dimension β' for $G^2_{\mu\nu}$, the γ_m is a fundamental quantity in the evolution of an asymptotically free gauge theory, i.e., QCD, from the ultraviolet (UV) to infrared (IR). The issue has some resemblance to Higgs physics with a possible conformal window although they most likely involve basically different physics. At present both β' and γ_m are not known except (perhaps) for large $N_f \gtrsim 8$. Only the bounds are known $\gamma_m < 2$ and scheme-independent calculations indicate that it goes to zero for $N_f \gtrsim 16$ [Ryttov and Shrock (2016)] and increases to ~ 1 for $N_f \sim 8$. Nothing is known for QCD with $N_f \sim 3$. But suppose, for illustration, $\gamma_m \approx 1$ in compressed matter. In this case, we would have

$$\mathcal{R}_\pi \approx \frac{m^*_\pi}{m_\pi} \approx 1. \tag{8.30}$$

It turns out that even without assuming the anomalous dimension $\gamma_m \approx 1$, a simple calculation assures that the pion tensor is unmodified in medium to the accuracy we need. A numerical result is shown in Fig. 8.3 [Paeng, Kuo, Lee and Rho (2015)].

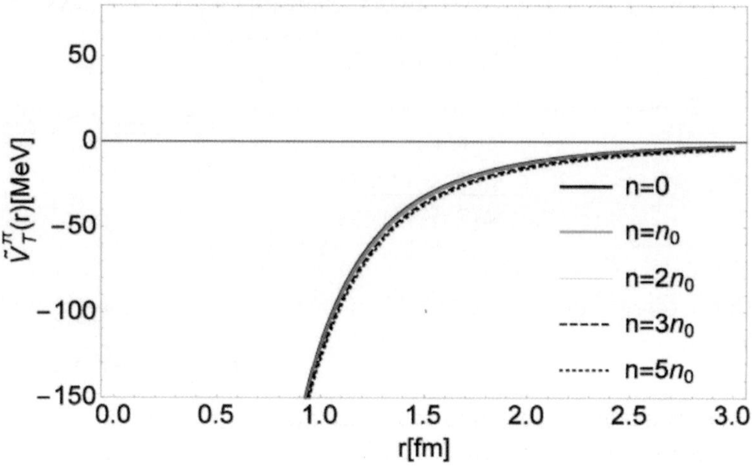

Fig. 8.3 $\tilde{V}_T^\pi(r) \equiv V_T^\pi(r)\,(\tau_1 \cdot \tau_2\, S_{12})^{-1}$ with $n_{1/2} = 2n_0$.

This result is obtained as follows. The pion tensor depends on density only via m_π^*. In R-I, for numerical estimate, we take

$$\frac{m_\pi^*}{m_\pi} = \Phi_\sigma^{1/2}(n) \approx \left(\frac{1}{1 + 0.13 * n/n_0}\right)^{1/2}. \tag{8.31}$$

In R-II, the situation is subtler due to the parity-doubling in the nucleon structure. While R_π is to scale proportionally to the in-medium pion mass, the pion mass will not d-scale proportionally to $\sqrt{\langle\bar{q}q\rangle^*}$ since the bilinear quark condensate tends to zero for $n \gtrsim n_{1/2}$. The chiral symmetry is still spontaneously broken in R-II, hence we expect the Gell-Mann–Oakes–Renner (GMOR) relation, expected to hold in the tree (or mean-field) order in medium, to be modified to

$$f_\pi^{*\,2} m_\pi^{*\,2} = m_q \langle\bar{q}q\rangle^* + \sum_{n>1} c_n \left\langle(\bar{q}q)^n\right\rangle^* \tag{8.32}$$

$$\Rightarrow \kappa^2 f_\pi^2 m_\pi^{*\,2} = \sum_{n>1} c_n \left\langle(\bar{q}q)^n\right\rangle^*. \tag{8.33}$$

We have indicated by the multiquark (or higher dimension field) condensates possible non-vanishing contribution to the GMOR mass formula in which the quark condensate $\langle\bar{q}q\rangle$ is vanishing. There are several proposals for specific form for the order parameter(s). Whatever the precise form may be, multiquark condensates are expected to be quite suppressed in R-II as shown in [Ma, Harada, Lee, Oh, Park and Rho (2013)]. Since the bilinear quark condensate is to tend to zero, it is reasonable to take it to decease rapidly, going toward — but not reaching on top of — the VM fixed point. Thus we take

$$\frac{m_\pi^{*\,2}}{m_\pi^2} \approx \frac{1}{\kappa^2}\sum_{n>1} c_n\langle(\bar{q}q)^n\rangle^* \sim \left(1 - 0.15 * \frac{n}{n_0}\right)^2 . \tag{8.34}$$

The result in Fig. 8.3, showing high insensitivity to density, verifies the long-standing assumption used in dealing with the pionic tensor force in nuclear dynamics and also in compact star matter. One may interpret this near independence on density of the pion tensor as protection by chiral symmetry.

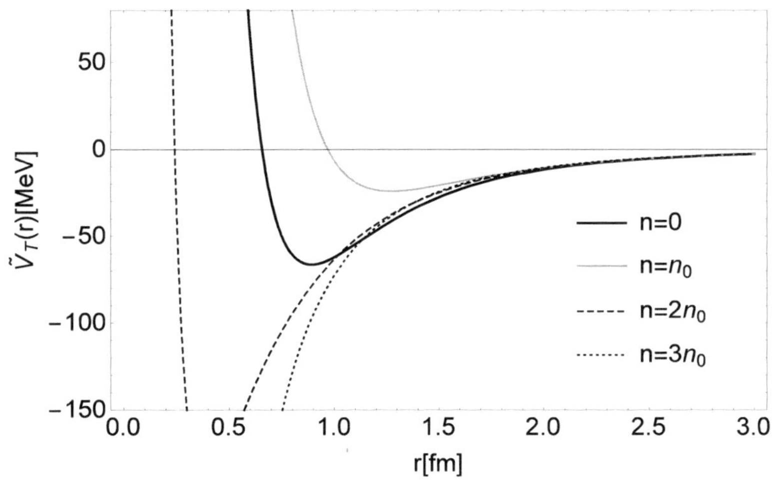

Fig. 8.4 $\tilde{V}_T(r) \equiv V_T(r)(\tau_1 \cdot \tau_2 S_{12})^{-1}$. For illustration, we take $n_{1/2} = 2n_0$, $\Phi_\sigma = \Phi_\rho = 1 - 0.15\frac{n}{n_0}$ and $\kappa = 1$.

8.2.3.2 *ρ-exchange tensor force in medium*

Now we turn to the d-scaling of the ρ-tensor force. As argued above, the KSRF could be taken to be exact in both R-I and R-II in giving the mass parameter in bsHLS. In R-I, we have $m_\rho^*/m_\rho \approx \Phi(n)$ from (8.9). In R-II, the ratio $\sqrt{\frac{a^*}{a}\frac{f_\pi^*}{f_\pi}} \approx \kappa$ is constant independent of density with $\kappa < 1$ while

$$\frac{g_\rho^*}{g_\rho} \equiv \Phi_\rho \to 0 \tag{8.35}$$

as $n \to n_{\rm VM}$, the VM fixed-point density. Thus

$$\frac{m_\rho^*}{m_\rho} \approx \kappa\Phi_\rho \quad \text{in} \quad \text{R-II}. \tag{8.36}$$

The ratio $\mathcal{R}_\rho$ simplifies to

$$\begin{aligned} \mathcal{R}_\rho &= \frac{g^*_{\rho NN}}{g_{\rho NN}} \frac{m_N}{m^*_N} \frac{m^*_M}{m_M} \\ &\approx 1 \quad \text{in} \quad \text{R-I} \qquad (8.37) \\ &\approx \Phi_\rho^2 \quad \text{in} \quad \text{R-II}. \qquad (8.38) \end{aligned}$$

8.2.3.3 *Net tensor force*

The π and ρ tensor forces have the same spin-isospin operator and radial form, but differ in the overall coefficients with strongly different masses, and more importantly come with opposite signs. With the sign difference, the pion tensor gets reduced by the ρ tensor in the intermediate and long ranges in which the potentials act. That the tensor force is quenched in HLS due to the interplay between the π tensor and the ρ tensor at the leading chiral order is an important point that differentiates HLS from standard chiral perturbation with the pion only. In fact there is a clear indication for this kind of quenching in the tensor force in QCD lattice calculations, as one can see in Fig. 8.5.

In SχPT, this quenching could arise mostly, if not all, from high-order pion-loop terms in chiral expansion. But in HLS, this effect is taken into account already at tree order. This difference becomes greatly more important when interactions take place in dense medium, with the simplicity of HLS over SχPT.[10]

Now since the ρ mass scales appreciably with density because of the density dependence of f^*_π, whereas the pion mass remains more or less unscaled, the ρ tensor, becoming stronger at increasing density up to $n_{1/2}$, cancels the pion tensor further at increasing density, the cancelation taking place almost completely near $n_{1/2}$. However precisely at $n_{1/2}$, due to the topology change, the ρ tensor gets strongly suppressed because the hidden gauge coupling drops rapidly due to the vector manifestation. For $n \geq 2n_0$, the pion tensor effectively takes over and the net tensor increases again. This can be seen in Fig. 8.4 [Lee, Park and Rho (2011); Paeng, Kuo, Lee, Ma and Rho (2017)].

[10] This is the point at which there has been profound disagreement between the aficionados of HLS and those of SχPT.

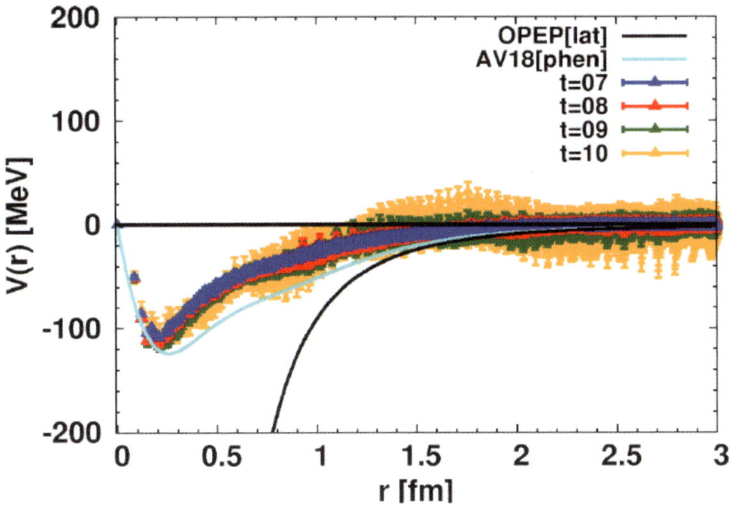

Fig. 8.5 Tensor force obtained in lattice QCD [Doi *et al.* (2017)] is compared with the phenomenological potential (AV18) and the one-pion exchange potential (OPEP).

8.2.3.4 *"Induced density dependence"*

In arriving at Eqs. (8.36) and (8.38), we have written

$$g_{\rho NN} = g_\rho, \tag{8.39}$$

where g_ρ is the hidden gauge coupling, common to $\rho\pi\pi$ and ρNN couplings. If one were to apply mean-field approximation to bsHLS in a single decimation, then it should have an induced factor or renormalization factor $F_\rho \equiv (g_{V\rho} - 1)$ that multiplies the hidden gauge coupling. This kind of effect will arise in decimating from a higher scale, e.g. the matching scale $\Lambda_{\rm M}$ to the effective cutoff scale from which the V_{lowk}RG decimation is performed. This can arise also when many-body forces involving higher scales than the effective cutoff are integrated out, i.e., $\rm DD_{induced}$. In the double-decimation V_{lowk}RG procedure adopted in this volume (described precisely below), this factor should be naturally generated in the first decimation RG procedure. This accounts, e.g., for the fact that g_A in the space component of the weak axial current is different from the one in the time component. This matter was treated in Chapter 2 as a case of nuclear physics proof of the FT. It also accounts for the difference between the Brown–Rho scaling effective in the anomalous orbital gyromagnetic ratio δg_l.

8.2.4 The cusp in E_{sym} from the tensor force

The structure of the tensor force described above gives in a mean-field approximation a very simple explanation for the "exotic" cusp observed at the topology change density $n_{1/2}$ given in the skyrmion crystal simulation.

Assuming, as accepted in the field and confirmed in our approach, that the tensor force does dominate the symmetry energy (e.g. [Carbone *et al.* (2014)]) — and that it is a good approximation to have the excitations produced by the tensor force saturated by the states in the vicinity of $\bar{E} \sim 200$ MeV, the symmetry energy can then be approximately given by the closure approximation [Brown and Machleidt (1994)][11]

$$E_{sym} \propto \frac{\left\langle |V^T|^2 \right\rangle}{\bar{E}} \tag{8.40}$$

where V^T is the radial part of the total tensor force with the scaling properties of the parameters taken into account. It can be seen from Fig. 8.4 that this description would give E_{sym} decreasing as density approaches from below the topology change density $n_{1/2}$ and then increasing above $n_{1/2}$. This structure was schematically depicted in the right panel of Fig. 7.6. This is precisely the cusp structure predicted by the skyrmion model. Formulated more precisely, as seen in Fig. 9.3 (of Chapter 9), this cusp, smoothed by

[11]We give here in a footnote, as a side remark, the apparent link between the cusps seen in the skyrmion crystal picture, the closure-sum approximation and the V_{lowk} RG given below and in Chapter 9. This illustrates the web of possible dualities intricately involved in nuclear dynamics in extreme conditions and points at the utmost importance of strict adherence to one unified formulation of an effective field theory as is emphasized in this volume. For instance, the statements made in the heavy-ion communities based on various different models used for analyzing dilepton data that "the BR scaling (alias IDD*) is ruled out" are totally unfounded.

The cusp seen in the skyrmion crystal matter, as mentioned, consists in the large N_c limit of QCD of a possible pseudo-gap phase with the vanishing of space-averaged quark condensates associated with a topology change. On the other hand, when seen from standard nuclear theory implemented with IDD*s sliding with the vacuum change, it reflects effects of iterated tensor forces, with the tensor force exciting states of high excitation energies comparable to the excitation of Δ resonances. And when seen from V_{lowk}RG, it reflects the change in the EoS from soft to hard smoothed by high-order correlations. In the latter two cases, what plays an important role is the feature of vector manifestation which leads to the vanishing ρ mass at large density or temperature, an effect diletpon experiments purported to reveal. In a given unified formulation, all these are intimately connected although manifested in entirely different ways far away from the vector manifestation fixed point. It would make no sense if one were to add a part of what's given by the model "A" to what's given by the model "B" for what's considered to be the same physics, with A and B formulated in a manifestly unrelated way. This is what happened in the analyses of the dilepton experiments.

higher order correlations in renormalization-group (RG) decimations, will give rise to a dramatic changeover from soft to stiff in $E_{\rm sym}$ and hence in the EoS. This feature, reproduced in the EFT, will be taken as a support for the mapping of topology, with its implication of sliding-vacuum property, to the parameters of the Lagrangian. This could figure crucially for the observed massive compact stars discussed in the next chapter, most interestingly in the tidal deformability recently detected in the gravity waves generated from coalescing neutron stars. More on this in Chapter 9.

8.2.4.1 *Evidence for IDD*: C14 dating beta decay*

It turns out that the long life time of Carbon-14 bears a crucial evidence of the behavior of the nuclear tensor force affected by the IDD, somewhat modified by higher order nuclear correlation effects. Explaining this phenomenon illustrates how the $\rm DD_{induced}$ gets added to the IDD lodged in the two-body tensor forces when short-range three-body forces are integrated out as discussed above.

Up to density $n \sim n_{1/2}$, $\mathcal{R}_\rho$ is constant. Therefore, density affects only the mass that figures in the function $\mathcal{I}(m_M^* r)$ in Eq. (8.25) for $M = \rho$. It is easy to see that the dropping ρ mass would simply increase the strength of the ρ tensor, effectively reducing the strength of the tensor force attraction as shown in Fig. 8.4. This reduction in the tensor was probed in the beta decay transition from the initial state $(J^\pi, T) = (0^+, 1)$ of ^{14}C to the final state $(J^\pi, T) = (1^+, 0)$ of ^{14}N which turns out to be singularly retarded, say, 5730 years despite that the allowed Gamow–Teller transition is involved. It was shown [Holt, Brown, Kuo, Holt and Machleidt (2007)] that the wave functions involved in the transition probe the density of $\sim 0.75 n_0$, and it is at this density that the Gamow–Teller matrix element nearly vanishes. The cancellation in the matrix element is extremely sensitive to the decrease in the tensor force, and it is at that density near n_0 that the overlap of the wave functions peaks. The interplay between the cancellation in the Gamow–Teller matrix element and the splitting between the 1_1^+ and 0_1^+ in ^{14}N shown in Fig. 8.6 provides a strong support to the validity of the interpretation.

The approach of Holt *et al.* [Holt, Brown, Kuo, Holt and Machleidt (2007)] takes the Gamow–Teller operator given solely by the single-particle operator. That the Gamow–Teller matrix element comes out correctly in that process is consistent with the chiral filter hypothesis (discussed in Chapter 4) that says that the meson exchange currents in the space

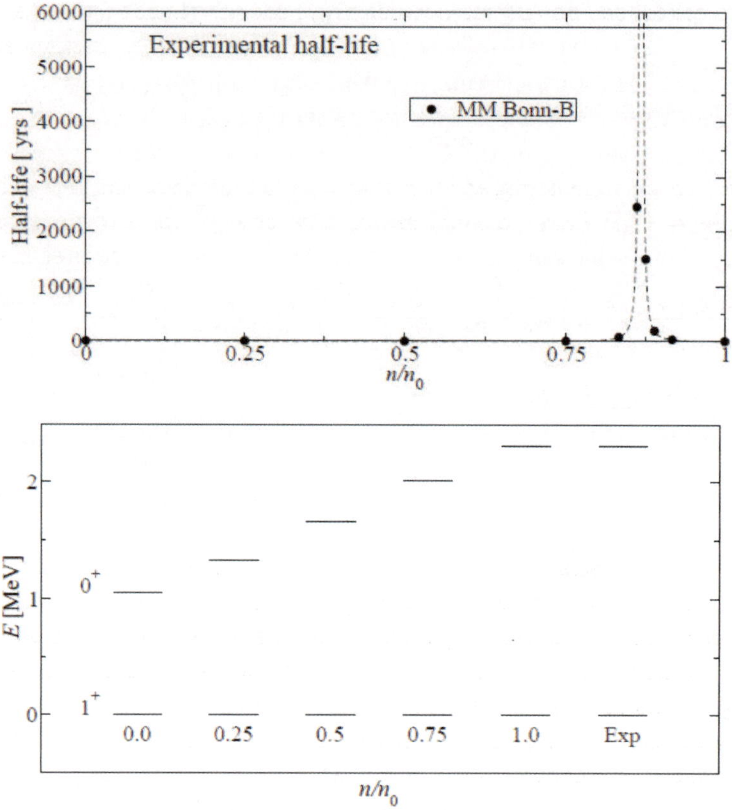

Fig. 8.6 The C-14 dating beta decay half-life and the splitting between the 1_1^+ and 0_1^+ in ^{14}N as a function of density. The half-life and the splitting of the states at the same density near n_0 give a strong consistency condition of the theory.

component of the axial current should be suppressed. This stands for another support of hidden scale symmetry in medium [Li, Ma and Rho (2018)].

It has been shown that with the introduction of chiral three-body forces, the suppressed rate can be more or less explained [Holt, Kaiser and Weise (2009)] although the accuracy is not as high as in Holt *et al.* [Holt, Brown, Kuo, Holt and Machleidt (2007)]. What is significant in this treatment is that the principal mechanism is provided by the short-range three-body force. In terms of the *bs*HLS Lagrangian, such short-range interactions can be generated by the vector-meson exchanges, in particular, one and two ω exchanges. It will require a detailed analysis, yet to be performed, to settle the issue, but one can say that there is no disagreement between the two ways of explaining the process. In the V_{lowk} RG approach of [Holt, Brown, Kuo, Holt and Machleidt (2007)], the decimation is done from a cutoff slightly below the vector-meson mass. The effect of three-body interactions exchanging massive mesons that give rise to the contact interactions are to be integrated out and hence will be effectively lodged in the parameter for the tensor forces as an induced density-dependence correction to the IDD discussed above. Thus the "BR scaling" (or IDD*) figuring in [Holt, Brown, Kuo, Holt and Machleidt (2007)] could contain the effect of contact three-body interactions. It does not contain, however, long-ranged three-body force effects. Such long-range three-body effects could be included in the V_{lowk} approach as a higher-order correction.

Suppose one chose to include three-nucleon interactions in the V_{lowk} calculation. The three-body interactions exchanging pions, vector mesons and dilaton, would then consist of long-ranged interaction involving pion exchanges and short-range ones involving vectors with appropriate IDD's incorporated. It is likely that the short-range three body force generated in this approach with no parameters will have its strength much reduced from what is needed in [Holt, Kaiser and Weise (2009)] without IDD's which is just an unconstrained parameter.

8.2.4.2 *The "master formula" for scaling bare masses*

The "bare" mass parameters in the bsHLS Lagrangian sliding with density applying to both R-I and R-II can be summarized by [Paeng, Kuo, Lee, Ma and Rho (2017)][12]

$$\frac{m_N^*}{m_N} \approx \frac{g_{\rho,\omega}}{g_{\rho,\omega}^*}\frac{m_{\rho,\omega}^*}{m_{\rho,\omega}} \approx \frac{m_\sigma^*}{m_\sigma} \approx \left(\frac{m_\pi^*}{m_\pi}\right)^2 \approx \frac{f_\pi^*}{f_\pi} \approx \frac{\langle\chi\rangle^*}{f_\sigma} \equiv \Phi\,, \tag{8.41}$$

where

$$m_\sigma^{*\,2} \equiv \left.\frac{\partial^2 V(\sigma)}{\partial\,\sigma^2}\right|_{\sigma=\langle\sigma\rangle^*} \tag{8.42}$$

where V is the dilaton potential. The density-scaling functions that we need to fix are $\Phi_V = g_V^*/g_V$ for $V = (\rho,\omega)$ and Φ for R-I and R-II. With these functions determined as described below, this formula will be applied to both normal nuclear matter and high dense matter with only small fine tuning.

[12]The scaling m_σ^*/m_σ as mentioned above should read $\Phi^{\frac{\beta'}{2}+1}$ but the β' dependence will be ignored.

8.3 Renormalization group with V_{lowk}

The basic strategy of V_{lowk} renormalization-group (V_{lowk}RG) was briefly sketched in Chapter 4. Starting with the *bs*HLS Lagrangian with the IDDs incorporated, one can proceed in applying RG technique: Single-decimation RG and double-decimation RG [Brown and Rho (2004)]. Here we elaborate on the procedures in preparation for its application to compact-star matter treated in Chapter 9.

8.3.1 *Single-decimation RG*

Here one takes the effective Lagrangian *bs*HLS and does the mean-field treatment of the Lagrangian along the line of Walecka mean field theory [Serot and Walecka (1986)]. This belongs to the class of what we call "standard relativistic mean field approaches" that are highly popular — and successful — in nuclear physics. This consists of doing the decimation from the cutoff scale at which the "bare Lagrangian" is defined to the top of the Fermi sea. In principle one could start at the scale where the matching to QCD is made, i.e., the chiral scale. In practice, it can be a lot lower, typically in the range of vector-meson mass m_V. In the current activities with the standard relativistic mean-field approaches described in Chapter 4, higher-dimension non-linear meson fields are included in consistency with the symmetries involved, i.e., Lorentz invariance, chiral symmetry etc., in order to improve on certain thermodynamic properties of matter, such as the compression modulus. This procedure has the potential of capturing multi-body forces that are missing in the linear models. With *bs*HLS one limits oneself to leading order in derivatives (thus exploiting low-energy theorems, e.g., KSRF relations) with the IDD *and, what is important,* the $\mathrm{DD_{induced}}$ that is to capture the effects not included in the matching, such as n-body forces for $n > 2$ that are integrated out. That the single-decimation procedure involves both IDD and $\mathrm{DD_{induced}}$ is reflected in the difference in the scaling factor Φ_σ effective in the single decimation from what is needed in the double decimation procedure.[13] This (single-decimation) approach has been verified to work well up to density $\sim n_0$ as explained in Chapter

[13]The precise theoretical prediction of the anomalous orbital gyromagnetic ratio δg_l requires $\Phi_\sigma(n_0) = 0.78$ close to $f_\pi^*/f_\pi \approx 0.8$ [Friman and Rho (1996)] whereas for the double-decimation procedure, $\Phi(n_0)$ is required to be typically larger — that is with a slower drop with density — as in the C14 dating case [Holt, Brown, Kuo, Holt and Machleidt (2007)] and also in compact-star matter discussed in the next section. This seems quite reasonable.

4. As discussed in Chapter 7, there is an indication from crystal lattice simulation that the dense matter in R-II can also be well described in mean field and hence related to a Fermi liquid near its fixed point. We will return to this issue in the confrontation with compact-star physics.

8.3.2 *Double-decimation RG*

For the double-decimation procedure using V_{lowk} potential that we will apply in confronting Nature,[14] one starts with the NN potential constructed by the one-boson potential with bsHLS Lagrangian. Three- or more-body potentials will, for the moment, be ignored. First one decimates from the scale near the matching scale where the Lagrangian is endowed with IDDs to $\Lambda_{\rm FS}$ just above the Fermi-liquid scale $\sim k_F$. The potential $V_{\rm NN}(n)$ will have dependence — IDD — inherited from the matching with QCD, encoding the degrees of freedom integrated out. In practice the decimation cutoff will be lower than the matching scale Λ_M. We do not need the precise value for this. A reasonable choice could be near the lowest-lying vector meson mass, m_V.

In the matter-free vacuum, one can phrase the decimation in terms of the T matrix to arrive at $V_{lowk}(n=0)$ as [Bogner, Kuo and Schwenk (2003)]

$$T(k',k,k^2) = V_{\rm NN}(k',k) + \frac{2}{\pi}\mathcal{P}\int_0^\infty \frac{V_{\rm NN}(k',q)T(q,k,k^2)}{k^2-q^2}q^2dq, \quad (8.43)$$

$$T_{\rm lowk}(k',k,k^2) = V_{\rm lowk}(k',k) + \frac{2}{\pi}\mathcal{P}\int_0^{\Lambda_m} \frac{V_{\rm lowk}(k',q)T_{\rm lowk}(q,k,k^2)}{k^2-q^2}q^2dq, \quad (8.44)$$

$$T(k',k,k^2) = T_{lowk}(k',k,k^2); \quad (k',k) \le \Lambda_m, \quad (8.45)$$

where $\mathcal{P}$ denotes principal-value integration and the intermediate state momentum q is integrated from 0 to ∞ for the whole-space T and from 0 to Λ_m — to be specified below — for $T_{\rm lowk}$. Closely analogous to SχPT, V_{lowk} contains counter terms to regularize the series. How this can be done in the presence of vector mesons has been discussed [Harada and Yamawaki (2003)]. Incorporating the dilaton in the potential has not been worked out up to date: It is being formulated in terms of scale-chiral expansion [Li, Ma and Rho (2017)].

[14] There are a large variety of RG techniques developed for nuclear physics. The V_{lowk} approach is a convenient approach that allows straightforward incorporation of IDDs.

Once V_{lowk} is computed in the vacuum ($n = 0$), then one can do the same decimation in the presence of matter, $n > 0$. There are no density-dependent counter terms — the free-space counter terms will do. Now implementing the IDDs in the RG and arriving at density-dependent $V_{lowk}(n)$ amounts to the first decimation.

Given the intrinsic density-dependent $V_{lowk}(n)$, one can then do the Fermi-liquid calculation using this as the potential to obtain the Fermi-liquid interactions. This has been done [Holt, Brown, Holt and Kuo (2007)]. The second decimation consists of arriving, via V_{lowk}RG, at the Fermi-liquid fixed point and then computing fluctuations around the Fermi-liquid fixed point.

In the numerical results that will be given below, the V_{NN} will be picked to be the realistic BonnS [Machleidt (1993)] NN interaction with the IDD encoded in the "bare" parameters taken into account. We shall adopt the vacuum (free-space) parameters chosen in [Machleidt (1993)] without adjustments. This is a phenomenologically powerful approach, fit to experimental data in free space as well as in medium to the momentum/energy scale defined by the cutoff Λ_m, which is equivalent to the in-medium cutoff $\Lambda_{\rm FS}$.[15]

Given the $V_{lowk}(n)$, one can then do full nuclear dynamics that correspond to Fermi-liquid theory including higher-order correlations on top of the Fermi sea. There are various different approaches to do this in many-body theory. Our approach is based on the ring-diagram summation pioneered by Kuo and his colleagues [Song, Yang and Kuo (1987)]. This approach has been proven to work very well in finite-nuclear system. What's new is that this scheme can be incorporated in the second decimation in applying V_{lowk}RG to compact-star matter as described in [Dong, Kuo, Lee, Machledit and Rho (2013); Paeng, Kuo, Lee and Rho (2015)].

8.3.2.1 *Ring diagrams*

We will be interested in the EoS, in particular, the nuclear symmetry energy, $E_{sym}(n)$ up to $n \sim 6n_0$, so the appropriate cutoff scale is $\Lambda_m \sim 3$ fm^{-1}. The V_{lowk} so obtained preserves the low-energy phase shifts in the vacuum (up to energy Λ_m^2) and the deuteron binding energy of V_{NN}. (For example, the deuteron binding energy given by V_{lowk} of $\Lambda_m = 2.0$ and 3.0 fm^{-1} are

[15]We will comment later that this approach can capture the essential features of high-order χPT such as the role of three-body force etc. Hence this is as respectful or powerful an approach as χPT implementing three-body forces that figure at higher chiral order.

both -2.226 MeV.) By construction, V_{lowk} is a smooth 'tamed' potential which is suitable for being used directly in many-body calculations.

The first step in the procedure is to verify if in R-I the above V_{lowk} — with IDD suitably incorporated — interaction can satisfactorily reproduce the empirical nuclear matter saturation properties (saturation density $n_0 \simeq 0.16\ \mathrm{fm}^{-3}$ and average energy per nucleon $E_0/A \simeq -16$ MeV at saturation). To do this, one is to calculate n_0 and E_0/A using a low-momentum ring-diagram approach [Siu, Holt, Kuo and Brown (2009)], where the *pphh* ring diagrams are summed to all orders within a model space of the cutoff Λ_m.

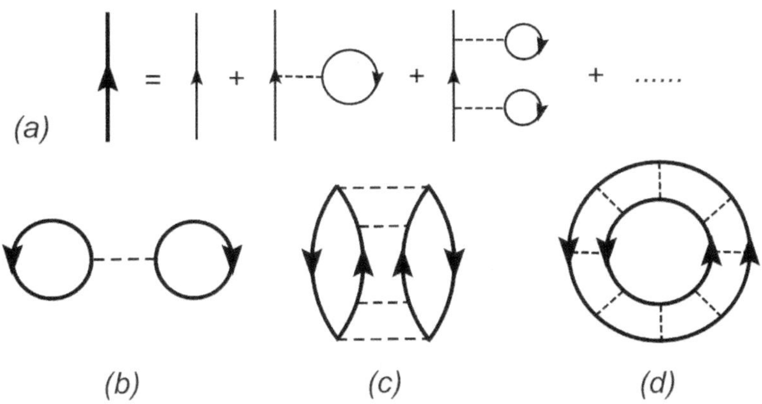

Fig. 8.7 Diagrams included in the *pphh* ring-diagram summation for the ground state energy of nuclear matter. Included are self-energy insertions on the single-particle propagator as indicated by (a), and *pphh* ring diagrams by (b,c,d).

Briefly the ring diagram technique is as follows [Siu, Holt, Kuo and Brown (2009)]. The ground state energy shift is defined as $\Delta E_0 = E_0 - E_0^{\mathrm{free}}$ where E_0 is the true ground-state energy and the corresponding quantity for the non-interacting system E_0^{free}. One considers ΔE_0 as given by the all-order sum of the *pphh* ring diagrams as shown in (b), (c) and (d) of Fig. 8.7, they being respectively a (1st-, fourth- and eighth-order) ring diagram. In the ring-diagram calculations, one also includes HF single-paticle insertions to all orders as illustrated by (a) of the figure. Each vertex of the diagrams is a V_{lowk} interaction obtained from an IDD-implemented V_{NN} potential. In general there are three types of ring diagrams, the proton-proton, neutron-neutron and proton-neutron ones. The proton and neutron Fermi momenta are, respectively, $k_{Fp} = (3\pi^2 n_p)^{1/3}$ and $k_{Fn} = (3\pi^2 n_n)^{1/3}$, where n_p and n_n denote respectively the proton- and neutron-density. With such ring diagrams summed to all orders, we have

$$\Delta E(n, \alpha) = \int_0^1 d\lambda_m \sum_m \sum_{ijkl<\Lambda} Y_m(ij, \lambda) \times Y_m^*(kl, \lambda) \langle ij | V_{\rm lowk} | kl \rangle, \quad (8.46)$$

where the transition amplitudes Y are obtained from a *pphh* RPA equation [Siu, Holt, Kuo and Brown (2009); Dong, Kuo and Machleidt (2009)]. Here λ is a strength parameter, integrated from 0 to 1. The above ring-diagram method reduces to the usual HF method if only the first-order ring diagram is included. In this case, the above energy shift becomes $\Delta E(n, \alpha)_{HF} = \frac{1}{2} \sum n_i n_j \langle ij | V_{\rm lowk} | ij \rangle$ where n_k=(1,0) if $k(\leq, >) k_{Fp}$ for proton and n_k=(1,0) if $k(\leq, >) k_{Fn}$ for neutron.

To be confident that the V_{lowk}RG is a reliable tool for compact star physics, it is helpful to recognize that the approach that incorporates the V_{lowk} ring-diagram technique has met with a great deal of success in wide-ranging phenomena in nuclear physics. It has been applied to symmetric and asymmetric nuclear matter [Siu, Holt, Kuo and Brown (2009); Dong, Kuo and Machleidt (2009)] and to the nuclear symmetry energy [Dong, Kuo and Machleidt (2011)]. This framework has also been tested by applying it to dilute cold neutron matter in the limit that the $^1S_0$ scattering length of the underlying interaction approaches infinity [Siu, Kuo and Machleidt (2008); Dong, Siu, Kuo and Machleidt (2010)]. This limit — which is a conformal fixed point — is usually referred to as the unitary limit, and the corresponding potentials are the unitarity potentials. For many-body systems at this limit, the ratio $\xi \equiv E_0/E_0^{free}$ is expected to be a universal constant of value ~ 0.44. The above ring-diagram method has been used to calculate neutron matter using several very different unitarity potentials (a unitarity CDBonn potential obtained by tuning its meson parameters and several square-well unitarity potentials) [Siu, Kuo and Machleidt (2008); Dong, Siu, Kuo and Machleidt (2010)]. The ξ ratios given by the approach for all these different unitarity potentials are all close to 0.44, in good agreement with the Quantum-Monte-Carlo result (see [Dong, Siu, Kuo and Machleidt (2010)] and references quoted therein). In fact the ring-diagram results for ξ are significantly better than those given by HF and BHF (Brueckner HF) [Siu, Kuo and Machleidt (2008)].

To summarize. The above many-body approach is essentially equivalent to doing Landau Fermi-liquid fixed-point theory with quasiparticle correlations on top of the Fermi sea treated with V_{lowk} with IDDs implemented, as discussed in [Holt, Brown, Holt and Kuo (2007)]. This procedure is a microscopic improvement on the relativistic mean-field treatment involving single decimation of *bs*HLS Lagrangian. In the application to denser regime going into R-II, the above procedure will be simply extrapolated. It is a valid procedure if the Fermi-liquid structure holds in R-II, which constitutes one of the main assumptions we make for application to compact-star matter in Chapter 9.

An important point to note here is that the same V_{lowk} RG formalism that addresses successfully the unitary limit in dilute cold neutron matter as discussed above is found to predict a pseudo-conformal structure at high density in compact-star matter [Ma, Lee, Paeng and Rho (2018)], which will be discussed in the next chapter. This implies, intriguingly, that a sort of pseudo-conformality persists in all range of density in neutron matter.

8.4 Applications to baryonic matter

In this section we apply the V_{lowk} formalism with the parameters of *bs*HLS determined above to nuclear matter and dense matter, covering both R-I and R-II. Both symmetric matter and asymmetric matter will be treated. Confronting compact stars will be given in Chapter 9.

8.4.1 *"Bare" parameters of bsHLS*

We first determine the specific values of the "bare" parameters applicable in both R-I and R-II. We will work with the simplified *bs*HLS Lagrangian (8.7) valid at the leading order in scale symmetry, i.e., "LOSS."

8.4.1.1 *R-I*

As given above, there is only one scaling factor in this density regime Φ_σ, Eq. (8.8), which follows the change of the dilaton condensate as a function of density. The coupling constants do not scale in R-I. *The "bare" parameters of bsHLS are completely controlled by this quantity once the parameters in the matter-free vacuum are adopted.* This scaling factor can be determined either by experiments or by χPT up to n_0. How it extrapolates beyond is not precisely known. In practice, it is reasonable in R-I to take the form

$$\Phi = \Phi_\sigma(n) = \frac{1}{1 + c_I \frac{n}{n_0}} \tag{8.47}$$

with c_I a constant. The range of c_I that gives a good fit to nuclear matter properties is found to be

$$c_I \approx 0.13 - 0.20 \tag{8.48}$$

with the upper value giving the measured pion decay constant (8.15). Given that the effective cutoff used for the decimation in the V_{lowk} RG procedure is $\sim (2-3)$ fm^{-1}, much lower than the matching scale Λ_M, what enters in (8.48) is IDD* that includes the induced density dependence $\mathrm{DD}_{\text{induced}}$. It is expected that there will be small fine-tuning within the range (8.48). All other parameters with their IDDs are fixed in (8.7).

8.4.1.2 *R-II*

Due to the topology change at $n_{1/2} \approx 2.0n_0$, some parameters undergo drastic modifications. The most important quantity is the hidden local gauge coupling g_ρ and hence the ρ mass related to g_ρ by the low-energy theorem, i.e., KSRF relation

$$\frac{m_\rho^*}{m_\rho} \approx \frac{g_\rho^*}{g_\rho} \equiv \Phi_\rho \to \left(1 - \frac{n}{n_{\rm VM}}\right) \quad \text{for} \quad n > n_{1/2}, \tag{8.49}$$

where $n_{\rm VM}$ is the putative VM fixed-point density. Here we have assumed a linear density dependence for simplicity.[16] Where $n_{\rm VM}$ is located is not known in QCD. In Chapter 9, it will come out to be $\gtrsim 20n_0$ from the consideration of the sound velocity of $\sim$ 2-solar mass star. We recall that this changeover in g_ρ^* is of crucial importance for the drastic behavior of the nuclear tensor forces observed above.

As for other parameters, apart from the properties of the ω meson, the scaling is very simple because in R-II, as in the 1/2-skyrmion phase, parity doubling emerges giving rise to the chiral-invariant mass m_0. It is associated with the dilaton condensate and the dilaton condensate becomes density independent in R-II. Therefore we have the effective nucleon mass becoming independent of density for $n > n_{1/2}$

$$\frac{m_N^*}{m_N} \approx \frac{f_\sigma^*}{f_\sigma} \sim \kappa \sim (0.6 - 0.9). \tag{8.50}$$

The dilaton mass also goes proportional to the dilaton condensate. This follows from the partially conserved dilatation current (PCDC) [Crewther and Tunstall (2015)] (see Chapter 6), so

$$\frac{m_\sigma^*}{m_\sigma} \approx \kappa. \tag{8.51}$$

[16]The precise density dependence of Φ_ρ is of course unknown. What matters is that g_ρ drops toward the VM fixed point.

The dilaton coupling to nucleon and other fields is unscaling to the leading order in scale-chiral symmetry, so it is a constant in R-II as in R-I.

It has already been mentioned that unlike other hadrons, the role of ω is not clear from the point of view of scale-chiral symmetry. While the flavor $U(2)$ symmetry seems good in the vacuum and possibly in R-I, it should be broken strongly in R-II. First of all it does not follow the ρ meson toward the VM fixed point. This is seen in mean-field theory [Paeng, Lee, Rho and Sasaki (2013)] and V_{lowk}RG [Paeng, Kuo, Lee and Rho (2015)]. On the other hand, it plays a highly important role in dense matter in providing repulsion. Some sort of fine-tuning is therefore needed in the d-scaling of mass and the coupling constant and we take it as

$$\frac{m_\omega^*}{m_\omega} \approx \kappa \frac{g_\omega^*}{g_\omega} \tag{8.52}$$

where g_ω is the $U(1)$ gauge coupling. What we can say from numerical analysis of Chapter 9 is that g_ω^* scales weakly in density. It is taken in the form

$$\Phi_\omega \equiv \frac{g_\omega^*}{g_\omega} \approx 1 - d\frac{n - n_{1/2}}{n_0} \tag{8.53}$$

with $d \approx 0.05$.

8.4.2 *bsHLS in the mean field: Parity doubling and DLFP*

Before we subject the bsHLS Lagrangian in the double-decimation approach to the EoS for both nuclear matter and compact star matter — in Chapter 9, we discuss what can be obtained in the mean-field approximation to bsHLS Lagrangian with IDDs implemented. We assume that the nuclear EFT in the mean field is equivalent to Landau Fermi-liquid theory in the vicinity of nuclear matter density n_0. Indeed this assumption fares well with a simplified bsHLS Lagrangian as discussed above and in Chapter 4. Here we return to the same problem with an objective to go to higher density, and study certain properties associated with the parity-doubling and the dilaton-limit fixed point discussed in Subsection 8.2.2.2 that are closely related to what is found in the V_{lowk}RG formalism given in Chapter 9.

We start with a bsHLS Lagrangian with the bare parameters endowed will the IDD* determined above. Since parity doubling is hidden in QCD, the chiral singlet nucleon mass term $\sim m_0$ should be put in *ab initio* in the bsHLS Lagrangian if one wants to apply the mean-field approximation (corresponding to a single-decimation RG) as in [Detar and Kunihiro (1989)]. We are interested in what one can learn as density exceeds $n \gtrsim n_{1/2} \sim 2n_0$.

8.4.2.1 *Parity-doublet* ***bs****HLS*

In writing down the form of bsHLS Lagrangian that will be analyzed, we take into account the indication from the V_{lowk}RG analysis (in Chapter 9) that the flavor $U(2)$ symmetry for the vector mesons ρ and ω breaks down in R-II. We will limit our considerations to two flavors ($N_f = 2$). The symmetry of the Lagrangian is $G_{\rm global} \times H_{\rm local}$, where $G_{\rm global} = [SU(2)_L \times SU(2)_R]_{\rm global}$ is the global chiral symmetry and $H_{\rm local} = [SU(2)_V \times U(1)_V]_{\rm local}$ is the hidden local symmetry. We take ρ as the gauge bosons of $[SU(2)_V]_{\rm local}$ and ω as of $[U(1)_V]_{\rm local}$. We use the notations defined in Chapter 2 with small differences. Since we are treating ρ and ω in two different gauge groups, the notations are somewhat more complicated than what was given before, say, in Chapters 2 and 4. Some are redundant but we will write them down here for completeness.

The basic quantities are the HLS gauge bosons, V_ρ^μ and V_ω^μ,

$$V_\rho^\mu = g_\rho \vec{\rho}^\mu \cdot \frac{\vec{\tau}}{2}, \tag{8.54}$$

$$V_\omega^\mu = g_\omega \frac{\omega^\mu}{2} \tag{8.55}$$

where g_ρ and g_ω are the HLS gauge couplings that will be taken unequal for the local $SU(2) \times U(1)$ symmetry concerned. The vectors transform

$$V_\rho^\mu \to h\, V_\rho^\mu\, h^\dagger - i\partial^\mu h \cdot h^\dagger\,, \tag{8.56}$$

$$V_\omega^\mu \to u\, V_\omega^\mu\, u^\dagger - i\partial^\mu u \cdot u^\dagger\,, \tag{8.57}$$

with $h \in [SU(2)_V]_{\rm local}$ and $u \in [U(1)_V]_{\rm local}$ in terms of $g_{L,R} \in [SU(2)_{L,R}]_{\rm global}$ and the two matrix valued variables ξ_L and ξ_R, combined in a 2×2 special-unitary matrix representing the pion field

$$U = \xi_L^\dagger \xi_R = e^{2i\pi/f_\pi} \tag{8.58}$$

transforming

$$U \to g_L U g_R^\dagger\,. \tag{8.59}$$

The variables ξs transform as

$$\xi_{L,R} \to uh\, \xi_{L,R}\, g_{L,R}^\dagger. \tag{8.60}$$

They may be parameterized as

$$\xi_{L,R} = e^{\frac{i}{2}\sigma_\omega/f_{\sigma\omega}} e^{i\sigma_\rho/f_{\sigma\rho}} e^{\mp i\pi/f_\pi}. \tag{8.61}$$

Here $\pi = \pi^a \frac{\tau_a}{2}$ denote the pseudoscalar NG bosons associated with the spontaneous breaking of $G_{\rm global}$ chiral symmetry, and σ_ω and $\sigma_\rho (= \sigma_\rho^a \frac{\tau_a}{2})$

are the NG bosons associated with the spontaneous symmetry breaking of $U(1)_{\rm local}$ and $SU(2)_{\rm local}$ respectively.[17] The σ_Vs are absorbed into the HLS gauge bosons through the Higgs mechanism, giving rise to their HLS boson masses. f_π, $f_{\sigma\omega}$ and $f_{\sigma\rho}$ are decay constants of the associated particles.

To construct hidden local symmetric Lagrangian, as explained in Chapter 2, we use the Maurer–Cartan 1-forms

$$\begin{aligned}\hat{\alpha}^\mu_\perp &= \frac{1}{2i}\left[D^\mu\xi_R\cdot\xi_R^\dagger - D^\mu\xi_L\cdot\xi_L^\dagger\right],\\ \hat{\alpha}^\mu_\parallel &= \frac{1}{2i}\left[D^\mu\xi_R\cdot\xi_R^\dagger + D^\mu\xi_L\cdot\xi_L^\dagger\right],\end{aligned} \tag{8.62}$$

with the covariant derivatives of $\xi_{L,R}$

$$D^\mu\xi_L = \partial^\mu\xi_L - iV^\mu_\rho\xi_L - iV^\mu_\omega\xi_L\,, \tag{8.63}$$

$$D^\mu\xi_R = \partial^\mu\xi_R - iV^\mu_\rho\xi_R - iV^\mu_\omega\xi_R \tag{8.64}$$

transforming homogeneously,

$$\hat{\alpha}^\mu_{\perp,\parallel} \to u\,h\,\hat{\alpha}^\mu_{\perp,\parallel}\,h^\dagger\,u^\dagger\,. \tag{8.65}$$

To the leading order in derivative expansion (i.e., to $\mathcal{O}(p^2)$), the HLS Lagrangian for the system is

$$\begin{aligned}\mathcal{L}_M = {} & f_\pi^2 \mathrm{Tr}\left[\hat{\alpha}_{\perp\mu}\hat{\alpha}^\mu_\perp\right] + f_{\sigma\rho}^2\mathrm{Tr}\left[\hat{\alpha}_{\parallel\mu}\hat{\alpha}^\mu_\parallel\right] + f_0^2\mathrm{Tr}\left[\hat{\alpha}_{\parallel\mu}\right]\mathrm{tr}\left[\hat{\alpha}^\mu_\parallel\right]\\ & - \frac{1}{2g_\rho^2}\mathrm{Tr}\left[V_{\mu\nu}V^{\mu\nu}\right] - \frac{1}{2g_0^2}\mathrm{Tr}\left[V_{\mu\nu}\right]\mathrm{Tr}\left[V^{\mu\nu}\right],\end{aligned} \tag{8.66}$$

with

$$\begin{aligned}f_0^2 &= \frac{f_{\sigma\omega}^2 - f_{\sigma\rho}^2}{2},\\ \frac{1}{g_0^2} &= \frac{1}{2}\left(\frac{1}{g_\omega^2} - \frac{1}{g_\rho^2}\right)\end{aligned} \tag{8.67}$$

and the field strength is given by

$$V^{\mu\nu} = \partial^\mu V^\nu - \partial^\nu V^\mu - i\left[V^\mu, V^\nu\right], \tag{8.68}$$

$$V^\mu = V^\mu_\rho + V^\mu_\omega\,. \tag{8.69}$$

We recover local $U(2)$ symmetry if we set $f_{\sigma\omega}^2 = f_{\sigma\rho}^2$ and $g_\omega = g_\rho = g$ in (8.66).

[17] A word of warning on notations. Note that the σ_V for $V = (\omega, \rho)$ here are the would-be NG bosons in hidden local symmetric theory, not to be confused with the dilaton σ figuring in scale-chiral symmetry.

The next step is to couple the nucleon parity-doublet field N to the HLS Lagrangian (8.66).

$$\mathcal{L}_N = \bar{Q}i\gamma^\mu D_\mu Q - g_1 f_\pi \bar{Q}Q + g_2 f_\pi \bar{Q}\rho_3 Q - im_0\bar{Q}\rho_2\gamma_5 Q \\ + g_{V\rho}\bar{Q}\gamma^\mu \hat{\alpha}_{\parallel\mu} Q + g_{V0}\bar{Q}\gamma^\mu \mathrm{Tr}\left[\hat{\alpha}_{\parallel\mu}\right] Q + g_A \bar{Q}\rho_3\gamma^\mu\hat{\alpha}_{\perp\mu}\gamma_5 Q\,, \tag{8.70}$$

with the nucleon doublet $Q = (Q_1, Q_2)^T$ and the covariant derivative of Q

$$D_\mu Q = (\partial_\mu - iV_{\rho\mu} - iV_{\omega\mu})\,Q, \tag{8.71}$$

and the ρ_i are the Pauli matrices acting on the parity-doublet. g_A, $g_{V\rho}$ and $g_{V0} \equiv \frac{1}{2}(g_{V\omega} - g_{V\rho})$ are dimensionless parameters. Diagonalizing the mass matrix and denoting the parity eigenstate fields as

$$N = \begin{pmatrix} N_+ \\ N_- \end{pmatrix} \tag{8.72}$$

that transform $N \to u\,h\,N$, the resulting parity-doublet baryonic HLS Lagrangian can be straightforwardly written down [Paeng, Lee, Rho and Sasaki (2013)]

$$\mathcal{L}_N = \bar{N}i\not{D}N - \bar{N}\hat{\mathcal{M}}N + g_A\bar{N}\gamma^\mu\hat{G}\hat{\alpha}_{\perp\mu}\gamma_5 N \\ + g_{V\rho}\bar{N}\gamma^\mu\hat{\alpha}_{\parallel\mu}N + g_{V0}\bar{N}\gamma^\mu\mathrm{Tr}\left[\hat{\alpha}_{\parallel\mu}\right]N\,, \tag{8.73}$$

$$\hat{\mathcal{M}} = \begin{pmatrix} m_{N_+} & 0 \\ 0 & m_{N_-} \end{pmatrix}, \quad \hat{G} = \begin{pmatrix} \tanh\delta & \gamma_5/\cosh\delta \\ \gamma_5/\cosh\delta & -\tanh\delta \end{pmatrix} \tag{8.74}$$

with

$$m_{N_\pm} = \mp g_2 f_\pi + \sqrt{(g_1 f_\pi)^2 + m_0^2}\,, \tag{8.75}$$

$$\cosh\delta = \frac{m_{N_+} + m_{N_-}}{2m_0}\,. \tag{8.76}$$

Note that the ω meson couples to nucleon and NG σ_ω, but there is no ω coupling to other mesons, i.e., π and ρ, at tree order. There can be tree-order $\omega\pi^3$ and ω-π-ρ couplings in the homogeneous Wess–Zumino term in the anomalous part of the HLS Lagrangian that could give rise to a one-loop correction to ω-nucleon coupling, but does not contribute at the order we are working with.

One can read off the vector meson mass and the $\rho\pi\pi$ coupling constant at tree level as

$$m_\rho^2 = g_\rho^2 f_{\sigma\rho}^2\,, \quad m_\omega^2 = g_\omega^2 f_{\sigma\omega}^2\,, \tag{8.77}$$

$$g_{\rho\pi\pi} = \frac{1}{2}a_\rho g_\rho\,, \tag{8.78}$$

$$a_\rho = \frac{f_{\sigma\rho}^2}{f_\pi^2}\,. \tag{8.79}$$

The vector mesons couple to nucleons with the coupling constants

$$g_{\rho N_+N_+} = g_{\rho N_-N_-} = (g_{V\rho} - 1)\, g_\rho\,, \tag{8.80}$$

$$g_{\omega N_+N_+} = g_{\omega N_-N_-} = (g_{V\omega} - 1)\, g_\omega \tag{8.81}$$

and the axial vectors coupling as

$$g_{AN_+N_+} = -g_{AN_-N_-} = g_A \tanh\delta \tag{8.82}$$

where the subscripts stand for the parity of the nucleon doublet.

8.4.2.2 *Going towards the DLFP*

To access the high density regime R-II, we need to incorporate scale symmetry into the parity-doublet HLS Lagrangian written above. We do this in the same way as explained in Chapter 6 by means of the conformal compensator field χ. There is one subtlety here with the term involving the chiral scalar mass m_0, namely the term $-im_0\bar{Q}\rho_2\gamma_5 Q$ in (8.70). The scale dimension for Q is 3/2, so as it stands, it is not scale invariant, which seems to require multiplication by χ by the procedure adopted. However because of the locking of f_π^* and f_χ^* for density $n \gtrsim n_{1/2}$, there appears an apparent inconsistency with the assumption that m_0 is a chiral invariant mass. How this locking occurs is totally ununderstood and constitutes an unsolved problem in the field. To avoid this inconsistency, we will simply ignore multiplying with χ factor on that form.[18]

The resulting parity-doubling bsHLS Lagrangian is

$$\mathcal{L} = \mathcal{L}_N + \mathcal{L}_M + \mathcal{L}_\chi\,, \tag{8.83}$$

$$\begin{aligned}\mathcal{L}_N &= \bar{Q}i\gamma^\mu D_\mu Q - g_1 f_\pi \frac{\chi}{f_\sigma}\bar{Q}Q + g_2 f_\pi \frac{\chi}{f_\sigma}\bar{Q}\rho_3 Q - im_0\bar{Q}\rho_2\gamma_5 Q\\ &\quad + g_{V\rho}\bar{Q}\gamma^\mu\hat{\alpha}_{\parallel\mu}Q + g_{V0}\bar{Q}\gamma^\mu \mathrm{Tr}\left[\hat{\alpha}_{\parallel\mu}\right]Q + g_A\bar{Q}\rho_3\gamma^\mu\hat{\alpha}_{\perp\mu}\gamma_5 Q\,,\end{aligned} \tag{8.84}$$

$$\begin{aligned}\mathcal{L}_M &= \frac{f_\pi^2}{f_\sigma^2}\chi^2\mathrm{Tr}\left[\hat{\alpha}_{\perp\mu}\hat{\alpha}_\perp^\mu\right] + \frac{f_{\sigma\rho}^2}{f_\sigma^2}\chi^2\mathrm{Tr}\left[\hat{\alpha}_{\parallel\mu}\hat{\alpha}_\parallel^\mu\right]\\ &\quad + \frac{f_{\sigma\omega}^2 - f_{\sigma\rho}^2}{2f_\sigma^2}\chi^2\mathrm{Tr}\left[\hat{\alpha}_{\parallel\mu}\right]\mathrm{Tr}\left[\hat{\alpha}_\parallel^\mu\right] - \frac{1}{2}\mathrm{Tr}\left[\rho_{\mu\nu}\rho^{\mu\nu}\right]\\ &\quad - \frac{1}{2}\mathrm{Tr}\left[\omega_{\mu\nu}\omega^{\mu\nu}\right]\,,\end{aligned} \tag{8.85}$$

$$\mathcal{L}_\chi = \frac{1}{2}\partial_\mu\chi\cdot\partial^\mu\chi - V(\chi) \tag{8.86}$$

[18]One way of handling this problem, in the absence of a rigorous justification, is to replace m_0 by a spurion operator ζ that transforms with scale dimension 1 and set $\zeta \to m_0$ at the end of the calculation.

where $V(\chi)$ is the dilaton potential that breaks scale symmetry both explicitly and spontaneously. Again we do not need its specific form for our purpose. Here, f_σ is the vacuum expectation value of χ at zero temperature and density. We are intentionally using here notations that are different from those in the *bs*HLS used elsewhere.

Now it is assumed that as density increases beyond $n_{1/2}$ — where in the skyrmion description, half-skyrmion phase sets in — the scale symmetry hidden in the nonlinear sigma model emerges in a scale-invariant linear sigma model as discussed in Chapter 5. This emergence is controlled by a constant which can go from one condition to another depending on complex scale symmetry properties. In our model, this "dialing" is assumed to be done by the IDDs encoded in the parameters of the Lagrangian.

To move towards a chiral symmetric Gell-Mann–Lévy-type linear sigma model, we do the field re-parametrization $\Sigma = U\chi\frac{f_\pi}{f_\sigma} = s + i\vec{\tau}\cdot\vec{\pi}$ — that also defines scalar s — and

$$\psi_{1,2} = \frac{1}{2}\left[\left(\xi_R^\dagger + \xi_L^\dagger\right) \pm \gamma_5\left(\xi_R^\dagger - \xi_L^\dagger\right)\right] Q_{1,2}, \tag{8.87}$$

or equivalently

$$\psi = \frac{1}{2}\left[\left(\xi_R^\dagger + \xi_L^\dagger\right) + \rho_3\gamma_5\left(\xi_R^\dagger - \xi_L^\dagger\right)\right] Q, \tag{8.88}$$

$$Q = \frac{1}{2}\left[\left(\xi_R + \xi_L\right) + \rho_3\gamma_5\left(\xi_R - \xi_L\right)\right] \psi. \tag{8.89}$$

With these reparameterized fields and going to parity eigenstates, one finds a complicated expression for (8.83) composed of a part that is regular, $\mathcal{L}_{\rm reg}$, and a part that is singular, $\mathcal{L}_{\rm sing}$, as $\mathrm{Tr}(\Sigma\Sigma^\dagger) \equiv \kappa^2 = 2\left(s^2 + \pi^{a\,2}\right) \to 0$, where a is iso-spin index. The singular part that arises solely from the scale invariant part of the original Lagrangian (8.83) has the form

$$\mathcal{L}_{\rm sing} = \left(g_{V\rho} - g_A\right)\mathcal{A}\left(1/\mathrm{Tr}\left[\Sigma\Sigma^\dagger\right]\right) + (\alpha - 1)\,\mathcal{B}\left(1/\mathrm{Tr}\left[\Sigma\Sigma^\dagger\right]\right), \tag{8.90}$$

where $\alpha \equiv \frac{f_\pi^2}{f_\sigma^2}$ and

$$\begin{aligned}
\mathcal{A} = & -\frac{i}{4}\mathrm{Tr}\left(\Sigma\Sigma^\dagger\right)^{-2}\bar{\psi}\mathrm{Tr}\left(\not{\partial}\left(\Sigma\Sigma^\dagger\right)\right)\left\{\Sigma, \Sigma^\dagger\right\}\psi \\
& + \frac{i}{2}\mathrm{Tr}\left(\Sigma\Sigma^\dagger\right)^{-2}\bar{\psi}\mathrm{Tr}\left(\Sigma\Sigma^\dagger\right)\left(\Sigma\not{\partial}\Sigma^\dagger + \Sigma^\dagger\not{\partial}\Sigma\right)\psi \\
& - \frac{-i}{2}\mathrm{Tr}\left(\Sigma\Sigma^\dagger\right)^{-1}\bar{\psi}\rho_3\gamma_5\left(\Sigma\not{\partial}\Sigma^\dagger - \Sigma^\dagger\not{\partial}\Sigma\right)\psi
\end{aligned} \tag{8.91}$$

$$\mathcal{B} = \frac{-1}{16\alpha}\mathrm{Tr}\left(\Sigma\Sigma^\dagger\right)^{-1}\mathrm{Tr}\left[\partial_\mu\left(\Sigma\Sigma^\dagger\right)\right]\mathrm{Tr}\left[\partial^\mu\left(\Sigma\Sigma^\dagger\right)\right]. \tag{8.92}$$

That $\mathcal{L}_{\rm sing}$ be absent leads to the conditions that

$$g_{V\rho} - g_A \to 0\,, \quad \alpha - 1 \to 0\,. \tag{8.93}$$

Using large N_c sum-rule arguments [Beane and van Kolck (1994)] and the RGE given in the next section, we infer[19]

$$g_A - 1 \to 0\,. \tag{8.94}$$

In the density regime where GML-type linear sigma model is valid, the nucleon mass can be given as

$$m_{N_\pm} = \mp g_2\langle s\rangle + \sqrt{(g_1\langle s\rangle)^2 + m_0^2}\,, \tag{8.95}$$

where $\langle s\rangle$ is the vacuum expectation value of s. As the chiral symmetry restoration point is approached, $\langle s\rangle \to 0$, so in the limit $\mathrm{Tr}(\Sigma\Sigma^\dagger) \to 0$, we expect

$$m_{N_\pm} \to m_0\,. \tag{8.96}$$

These are the constraints that lead to the dilaton limit found in [Sasaki, Lee, Paeng and Rho (2011)] and announced above. It follows then that

$$g_{\rho NN} = g_\rho(g_{V\rho} - 1) \to 0. \tag{8.97}$$

We thus find that in the dilaton limit, the ρ meson decouples from the nucleon. In contrast, the limiting $\mathrm{tr}(\Sigma\Sigma^\dagger) \to 0$ *does not* give any constraint on $(g_{V\omega} - 1)$. The ω-nucleon coupling remains non-vanishing in the Lagrangian which in unitary gauge (with $\sigma_\omega = \sigma_\rho = 0$) and in terms of fluctuations $\tilde{s}$ and $\tilde{\pi}$ around their expectation values, takes the form

$$\begin{aligned}\mathcal{L}_N &= \bar{\mathcal{N}} i\partial\!\!\!/ \mathcal{N} - \bar{\mathcal{N}}\hat{M}\mathcal{N} - g_1\bar{\mathcal{N}}\left(\hat{G}\tilde{s} + \rho_3\gamma_5 i\vec{\tau}\cdot\vec{\tilde{\pi}}\right)\mathcal{N} \\ &\quad + g_2\bar{\mathcal{N}}\left(\rho_3\tilde{s} + \hat{G}\gamma_5 i\vec{\tau}\cdot\vec{\tilde{\pi}}\right)\mathcal{N} + (1 - g_{V\omega})\, g_\omega \mathcal{N}\frac{\omega\!\!\!/}{2}\mathcal{N}\,,\end{aligned} \tag{8.98}$$

which is the same as the Lagrangian given in [Detar and Kunihiro (1989)] except for the ω-nucleon interaction. This is just the nucleon part of the linear sigma model in which the ω is minimally coupled to the nucleon, applicable infinitesimally below the critical density n_c with the mass $\hat{M}$ replacing m_0. What is significant with this result is that it shows that the suppression of the ω-repulsion predicted with $U(2)$ [Sasaki, Lee, Paeng and Rho (2011)] be absent due to potentially significant $U(2)$ symmetry

[19] It is most plausible that $g_A - 1 = 0$ is reached only after $g_A - g_{V\rho} = 0$ is reached. This is because $\mathrm{Tr}(\Sigma\Sigma^\dagger) \to 0$ gives the constraint $g_A - g_{V\rho} = 0$, but not $g_A - 1 = 0$. However the RGE has the infrared fixed point $g_{V\rho} - 1 = 0$, so the point at which $g_A - g_{V\rho} = 0$ could be very near the chiral restoration point at which g_A is close to 1.

breaking as is indicated in the phenomenology of EoS for dense baryonic matter.

In summary, as the DLFP is approached

$$m_{N_\pm} \to m_0, \quad f_\chi \to f_\pi, \quad g_A \to 1, \; g_{\rho NN} \to 0. \tag{8.99}$$

A RG analysis confirmed this IR fixed point [Paeng, Lee, Rho and Sasaki (2013)]. Note also that in the half-skyrmion phase, the soliton mass becomes a constant proportional to m_0, hence a sort of parity doubling symmetry "emerges" from the many-body correlations. This also resembles what happens in the V_{lowk}RG given below and also in Chapter 9. Finally due to the breakdown of $U(2)$ symmetry for the vector mesons, the property of ω-nucleon interaction in dense matter remains complicated at high density.

8.4.3 *Equation of state in V_{lowk}RG*

One can make a detailed prediction by the V_{lowk}RG formalism with the bsHLS determined above of the EoS of dense matter ranging up to $n \sim 6n_0$. Here we limit to R-I, $0 \lesssim (n/n_0) \lesssim 2$ and check how the calculation fares in particular in normal nuclear matter. Going beyond $n_{1/2}$ will be discussed in Chapter 9 where, it should be remarked, results quite similar to what's found here in the mean field with parity-doubling bsHLS model will be found in the V_{lowk}RG with the bsHLS Lagrangian *without* parity-doubling put in *ab initio*.

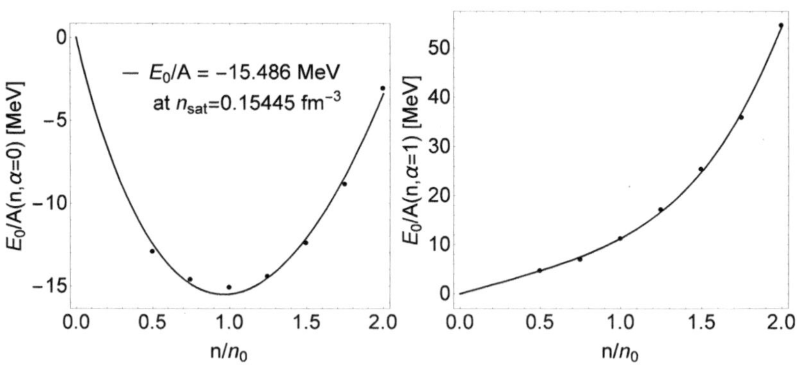

Fig. 8.8 E_0/A for $\alpha = 0$(left) and $\alpha = 1$(right) matter, where $\alpha = (N_n - N_p)/A$ and $A = N_n + N_p$ with the proton/neutron number $N_{p/n}$. The dots are from the V_{lowk} RG. The saturation density is estimated as $n_{\rm sat} = 0.154{\rm fm}^{-3}$.

In R-I, the scaling is governed entirely by the d-scaling of $f_\sigma^* = \langle\chi\rangle^*$. Although the explicit forms of the d-scaling masses involved are not known in terms of the condensates, i.e., $\langle\bar{q}q\rangle$ and $\langle G^2\rangle$, we learned [Crewther and Tunstall (2015)] that the masses of nucleon and dilaton are dominated by $\langle G^2\rangle$ (this will be contrasted in R-II by that the masses of the vector mesons are more crucially controlled by $\langle\bar{q}q\rangle$). Using the master scaling formula (8.41), the V_{lowk}RG makes an extremely simple prediction for the EoS in the vicinity of the normal nuclear matter n_0. With the hadron masses listed in the particle data booklet taken as the "bare" parameters in the vacuum of the *bs*HLS Lagrangian those given at the matching scale Λ_M, the EFT calculation gives the energy per particle for the symmetric and asymmetric nuclear matter as shown in Fig. 8.8.

Just to give an idea as to how this one-parameter prediction fares, we list those quantities that can be compared with experiments at $n = n_0$. The predicted values are: Equilibrium density $n_0 = 0.154$ fm^{-3}, binding energy BE = 15.5 MeV, compression modulus $K = 215.2$ MeV. All these are consistent with empirical values.

It deserves to be stressed that this one-parameter prediction can be vastly improved by recognizing that there are varieties of freedom in the parameter choice. In fact the calculation given in Figs. 8.8 requires a fine-tuning, as pointed out in [Paeng, Kuo, Lee, Ma and Rho (2017)], with the constant c_I in the range (8.48). One can think of this fine-tuning as arising from possibly different scales involved with the hadrons involved, i.e., mesons and baryons. It also underlines the *unnaturalness* in fine-tuning of the parameters involved with the properties of the normal nuclear matter. What is qualitatively more significant is that without the effect of IDD, the V_{lowk} procedure does not give nuclear matter at the correct equilibrium density and with the correct binding energy [Siu, Holt, Kuo and Brown (2009); Dong, Kuo, Lee, Machleidt and Rho (2013)]. That three-body forces, via $\mathrm{DD}_{\mathrm{induced}}$ in Subsection 8.2.3.4, could remedy this problem is discussed in the literature [Holt, Rho and Weise (2016)].

Chapter 9

Compact Stars

In this chapter, we confront the effective field theory (EFT) formalism formulated in the previous chapters with compact-star phenomena. The formalism has been checked for the properties of nuclear matter, and is expected to be valid up to the density $n_{1/2} \sim 2n_0$ at which topology change, crucial in the present formalism, takes place. How to extrapolate the formalism beyond $n_{1/2}$ has been discussed in the preceding chapters. We will therefore skip the details and go to the crucial point.

We will not enter here either into detailed discussions on other approaches, such as various energy density functionals, hybrids combining standard nuclear many-body models with simplified quark models (e.g., MIT bag, strong coupling NJL model, color superconductivity etc.,) other than succinct comparisons whenever feasible. There are many reviews[1] on these approaches, but given the poor knowledge of the density regime involved in massive compact stars and the absence of model-independent constraints, none can be considered as giving trustful hints to what's actually going on in Nature. We will adhere strictly to the formalism based on one single effective Lagrangian, *bs*HLS, with the parameters fixed with IDDs inferred directly — via correlators — or assumed from QCD. The approach has only a small number of parameters, highly constrained from what is considered to be consistent with QCD. Although far from the solid ground available up to nuclear matter density n_0 for formulating EFT à la

[1]An exhaustive up-to-date summary of the situation as of the date of this volume can be glimpsed in [Tolos, Centelleles and Ramos (2017)]. A fairly complete list of reviews on the matter related to what is involved in this chapter is found in this article. There are some 200 formulations in several different approaches, such as, mostly, the Skyrme energy density functional and relativistic mean-field (RMF) approaches and some chiral perturbation approaches. For getting an overview of the development, see [Nazarewicz, Reinhard, Satula and Vretenar (2014)].

Folk Theorem (FT), the goal here is to adher as faithfully as possible to the FT strategy in accessing the density range $\sim (5-7)n_0$ relevant to compact stars. Given the limited number of free parameters involved, less than 5 or 6 at most, the principal aim is to have a qualitatively correct understanding of an overall structure of highly dense systems, instead of going for high-precision fits to available experimental data. It is to zero-in on unambiguous predictions that would allow to make as unique a statement as feasible on what takes place in a totally uncharted regime of wildness exemplified by the symmetry energy shown in Fig. 8.1 in Chapter 8. We quote Fig. 9.1, similar to Fig. 8.1, to remind the readers how widely uncontrollable the symmetry energy can be within a single (energy density functional) model when parameters are varied [Chen (2015)]. This "wilderness" illustrates the difficulty to pin down viable theoretical models among many available in the literature that experimental efforts, including the tidal deformability in gravitational waves from coalescing neutron stars discussed below, will face.

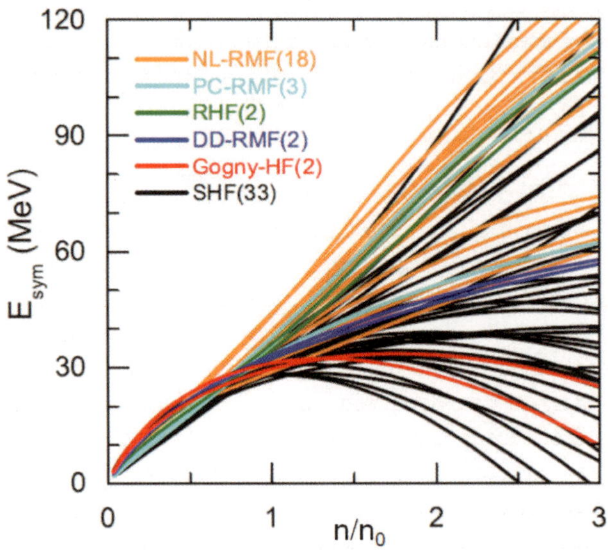

Fig. 9.1 A sample of predicted $S = E_{sym}$ by models that are fit to the matter properties up to n_0 and extrapolated to higher densities, showing the wilderness at high densities.

We should underline that our aim is basically different in spirit from the currently dominant theme in the nuclear astrophysics circle. The aim is not to construct a model with (numerous) free parameters to explain accurately a selected — and incomplete — set of what's observed but to stick to one simple unified scheme based on unambiguously defined premises with as few a number as possible of free parameters,[2] and make unambiguous predictions on compact stars within a highly constrained and possibly consistent framework. In what is described below, there are essentially no free parameters to play with.

9.1 Basic issues

We address, with the V_{lowk}RG approach precisely formulated in Chapter 8 with all the parameters of the theory determined once and for all, a variety of compact-star properties, including tidal deformability of neutron star as probed by gravitational waves. We compute the EoS for densities up to $\sim 6n_0$ with the assumption that the formalism is applicable *without explicit QCD variables* for the density regime involved in the massive stars. There are only a limited number of unknown parameters in the approach once the "bare parameters" in the vacuum of *bs*HLS Lagrangian are fixed at the matching scale in the matter-free space, taken from the PDG's Particle Physics Booklet.[3]

The standard approach in nuclear and astrophysical communities imposes certain constraints that the candidate theory for compact-star matter at density $\sim 6n_0$ must accurately satisfy the properties of the normal nuclear matter at n_0 and dismisses models that cannot achieve the accuracy demanded by the normal-matter constraints. In the literature special emphasis is put on various quantities associated with the E_{sym} at normal nuclear matter density n_0, such as $E_{sym}(n_0)$, $L(n_0)$, K_{sym} etc. — defined below. To illustrate the state of art involved in this matter, we quote in Fig. 9.2 the analyses made by [Fortin *et al.* (2016)] for a large variety of currently popular models, both Skyrme energy density functional and RMF models. The question posed here is: Are these constraints *indispensable* for the physics of compact stars? After all, the symmetry energy E_{sym},

[2]Whether or not what we presume to be in the spirit of the FT in R-II makes sense should of course be re-examined as the contact with Nature becomes mature. At present, the whole issue is totally open.

[3]Just for comparison, a reliable energy functional formalism — such as the Skyrme energy functional or RMF models — requires typically $\gtrsim 10$ parameters for a reasonable fit to measured quantities at normal density, some giving what's considered to be acceptable properties of neutron stars when extrapolated to high densities.

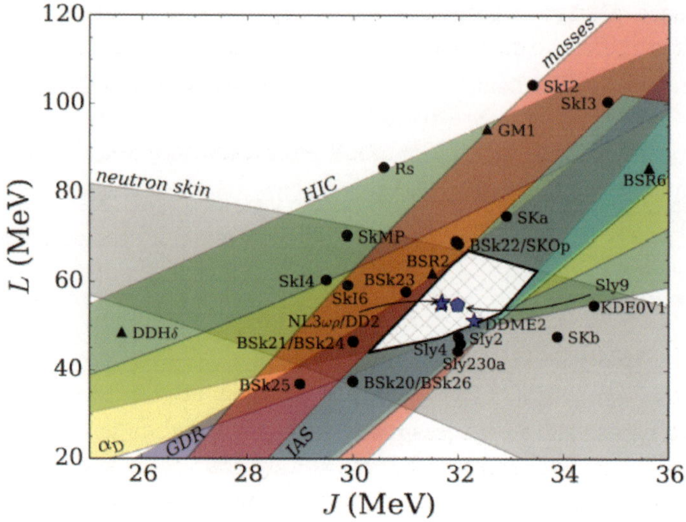

Fig. 9.2 The symmetry energy $J \equiv E_{sym}$ vs. its derivative $L = 3n_0(dE_{sym}/dn)$ at $n = n_0$ given by a large number of Skyrme energy-density models and RMF models. The "constraints" give the white region. Those models that satisfy the constraints are indicated by the blue diamond (3 RMF models) and the pentagon (1 Skyrme energy functional). Taken from [Fortin *et al.* (2016)].

Fig. 9.1 [Chen (2015)], is indisputably a total mess in R-II. So weeding out this wilderness seems to be a formidable, if not impossible, task. Given a ~ 2 solar-mass star can have the central density of order $\sim (5-7)n_0$, there seems to be no reason why the strict adherence to the known data at n_0 is *absolutely* called for.

We will find indeed that the present (4-parameter) theory discussed in this volume (in Chapter 8) may not post-dict certain *normal-matter parameters* as "precisely" as demanded within theoretical error bars, but can fully satisfactorily describe the maximum mass, M vs R relation etc. Given the paucity of our understanding of the density regime involved — and since what is needed is a qualitative overview — such a requirement for the "accurate fit" seems to make no sense. In our approach, we eschew the strict adherence to the constraints made at the normal nuclear matter density n_0. We instead stick to simplicity, without sacrificing the quality of the fit, and focus on predictions that are novel or basically different from the predictions of the conventional models available in the literature. *One of the most prominent predictions made, which is qualitatively different from what's found in the literature, is the sound velocity that signals a precursor to emergent symmetries hidden in QCD.*

9.2 Prediction of star properties

Given the energy density $\epsilon(n) = E(n)/V$ and the pressure $P(n)$ as obtained in the preceding chapter, one has to solve the general relativistic hydrodynamic equation. The most commonly used equation is the Tolman–Oppenheimer–Volkov (TOV) equation. The procedure is well established and amply explained in the books and review articles [Shapiro and Teukolski (1983)]. We will be very brief on it.

9.2.1 *New results*

Before going into the main discussion, we briefly summarize the results obtained that are qualitatively different from what's found in the literature. The input properties that do not figure explicitly in other approaches are the existence of the vector manifestation (VM) fixed point for the iso-vector vector meson ρ and axial-vector meson a_1, the hidden scale and local symmetries and the topology change in the possible guise of hadron-quark continuity à la Cheshire Cat (CC) at $n \sim 2n_0$. With the few unknown parameters in the approach that are fixed with reasonable assumptions based on the symmetries of QCD, the resulting EFT Lagrangian is found to satisfactorily reproduce *all* normal nuclear matter properties, and extrapolated to densities beyond $n_{1/2} \sim 2n_0$, predict the properties of $\sim$ 2-solar mass stars consistent with observations. The formalism has, as described in the preceding chapter, the simple mechanism for hardening of the EoS at $n \sim n_{1/2} \approx 2n_0$ associated with the topology change present in the skyrmion description of dense matter. What is also noteworthy is the tendency of the tensor force softening toward $n_{1/2}$ — manifested in the C14 dating Gamow–Teller matrix element — making the symmetry energy softer before hardening at $n > n_{1/2}$. As discussed below, this can have an impact on the toroidal deformability measured from gravitational waves emitted from coalescing neutron stars. There is arguably a more important consequence, namely, the apparent emergence of hidden symmetries and parity doubling for densities $n \gtrsim n_{1/2}$ and the approach of the sound velocity v_s to the "conformal velocity" $\sqrt{1/3}$ — in the unit $c = 1$ — for $n \gtrsim 3n_0$. This last result is a strikingly different prediction from all other models available in the literature that manage to arrive at the properties of $\sim$ 2-solar mass stars. To quote the ensemble of results based on the Skyrme energy-density models, the typical value of v_s listed in the literature is $0.5 \lesssim v_s^2 \lesssim 0.8$ for $M = 2M_\odot$.

9.2.2 *Tolman–Oppenheimer–Volkoff (TOV) equations*

With given pressure and energy density, the structure of a spherically symmetric neutron star is determined by solving the TOV equations

$$\begin{aligned} \frac{dM(r)}{dr} &= 4\pi r^2 \epsilon(r)\,, \\ \frac{dP(r)}{dr} &= -\frac{GM(r)\epsilon(r)}{r^2}\left[1 + \frac{P(r)}{\epsilon(r)}\right]\left[1 + \frac{4\pi r^3 P(r)}{M(r)}\right]\left[1 - \frac{R_s(r)}{r}\right]^{-1} \end{aligned} \tag{9.1}$$

where $M(r)$ is the enclosed gravitational mass, $\epsilon(r)$ is the energy density inclusive of the rest mass density, $P(r)$ is the pressure, and $R_s(r) = 2GM(r)$ is the Schwarzschild radius with a given mass $M(r)$. The hydrostatic equilibrium is assumed in the above equations.

Various equations of state are available in the literature. Once the central density $n(r = 0)$ is given, solving Eq. (9.1) numerically leads to the mass and size of neutron stars. By varying $n(r = 0)$, the maximum mass of neutron stars can be obtained. It is generally accepted that what truly matters is how soft or stiff the EoS is, independently of what mechanism is at work. A stiff equation of state gives higher maximum mass, and a soft equation of state, such as kaon condensation naively treated (see Chapter 10 for subtlety), could give a lower maximum mass. Empirical constraints on the EoS are available only in the low-density ($n \lesssim n_0$) region, so the extrapolation to higher densities is required. In our approach, the extrapolation is provided by the formalism of V_{lowk} with bsHLS.

9.2.3 *Equation of state*

What enters into the TOV equation (9.1) is the energy density $\epsilon(n)$ with which the pressure P is given by $P(n) = n^2 \frac{d(\epsilon(n)/n)}{dn}$. This energy density is to be calculated in the V_{lowk}RG formalism in bsHLS, the parameters of which are unambiguously given in terms of IDDs (supplemented by $\mathrm{DD_{induced}}$) in both R-I and R-II determined in Chapter 8. The EoS for symmetric nuclear matter and neutron matter were calculated in that chapter. A particularly important quantity needed is the symmetry energy E_{sym} that we recall from Chapter 8

$$E(n) = E_0(n, \alpha = 0) + \alpha^2 E_{sym}(n) + O(\alpha^4) \tag{9.2}$$

with $\alpha = (N - Z)/(N + Z)$. α is determined by the charge neutrality and beta stability involving leptons, which make the system deviate from pure neutron matter with $\alpha = 1$. In Fig. 9.3 are given the predicted results for the ground-state energy for symmetric nuclear matter and pure neutron matter and in Fig. 9.4 the symmetry energy E_{sym}. They are compared with the presently available data. One can safely say that they are consistent with the data.

It is to be noted that as observed before, the symmetry energy, which appears as a cusp in the skyrmion crystal picture and in the mean field at $n_{1/2} \sim 2n_0$, now manifests in V_{lowk}RG as a changeover from softness to hardness, providing stiffening in the EoS. This is the crucial ingredient in the framework of the formulation made here for supporting massive stars. Note that the key mechanism here is the topology change described in Chapter 7 from skyrmions to half-skyrmions in the solitonic matter description of dense matter which affects drastically the nuclear tensor forces. There a pivotal role is played by the assumed property of VM fixed point with the ρ mass going massless as an emerging symmetry. As mentioned at various places, there is an analogous changeover from soft to hard in description where strong-coupled quark interactions are appended to the standard nuclear EoS at a density in the vicinity of $2n_0$ [Masuda, Hatsuda and Takatsuka (2016); Fukushima and Kojo (2016)]. In Chapter 3, a conjecture was made about the potential Cheshire Cat phenomenon operative here. More on this point below.

9.2.4 *"Constraints" at n_0*

Although we do not attach much importance to the significance of the so-called "constraints" at normal nuclear matter density n_0 for the EoS of compact stars, we give what our formulation predicts for the quantities involved: $E_0(n_0)/A - m_N \simeq -15.5$ MeV, $n_0 \simeq 0.154$ fm^{-3}, $K(n_0) \simeq 215$ MeV, $J = E_{sym}(n_0) \simeq 26$ MeV, $L(n_0) \simeq 49$ MeV. They are more or less consistent — although not at all fine-tuned — with the presently quoted empirical values $E_0/A - m_N = -15.9 \pm 0.4$ MeV, $n_0 = 0.164 \pm 0.007$ fm^{-3}, $K = 240 \pm 20$ MeV or 230 ± 40 MeV, $30 \lesssim E_{sym} \lesssim 35$ MeV, $20 \lesssim L \lesssim 66$ MeV.

Now the rationale for those "constraints" is that they are available experimentally, so they would give information of the EoS around the nuclear matter density. Our point of view is that the quantities at near n_0 would give constraints in the density regime R-I, namely $n \lesssim n_{1/2}$ but not

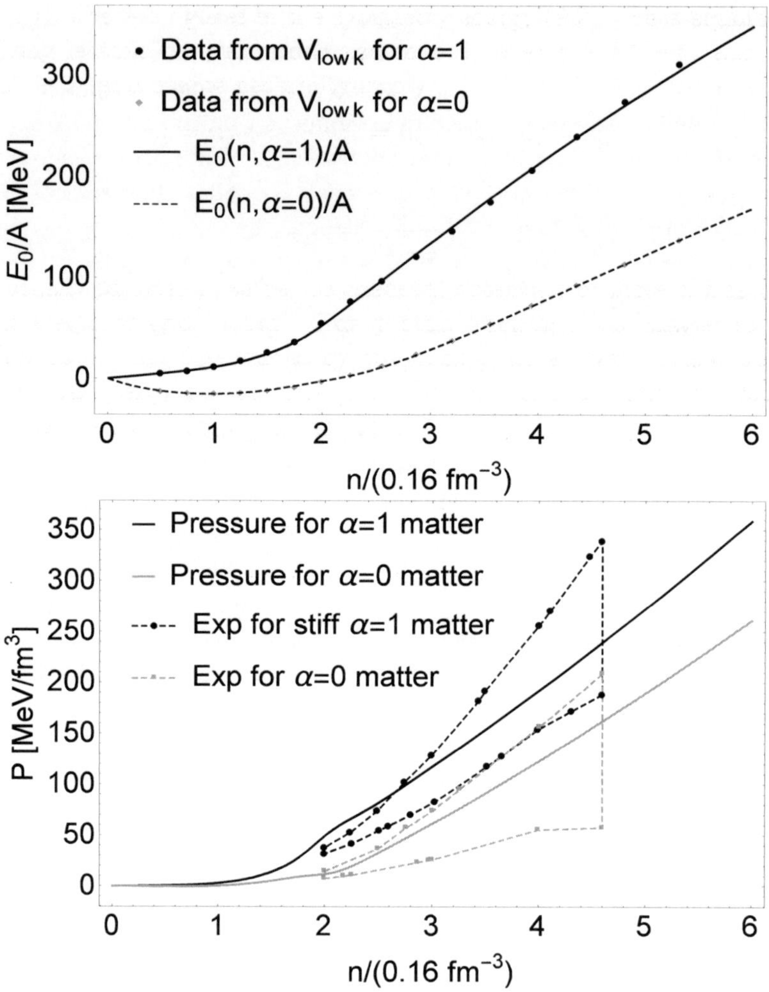

Fig. 9.3 The predicted E_0/A (upperpanel) and P (pressure) (lower panel) for $\alpha = 1$ and $\alpha = 0$ matter. The empirical values taken from Danielewicz [Danielewicz, Lacey and Lynch (2002)], Li and Chen [Li and Chen (2005)] and Tsang *et al.* [Tsang *et al.* (2009)] are also shown.

necessarily in R-II which can involve different "phase(s)." It is the topology change in our formulation and the merging of hadronic degrees of freedom and quark-gluon degrees of freedom in hybrid baryon-quark descriptions that delineate the two phases that are not directly connected in the

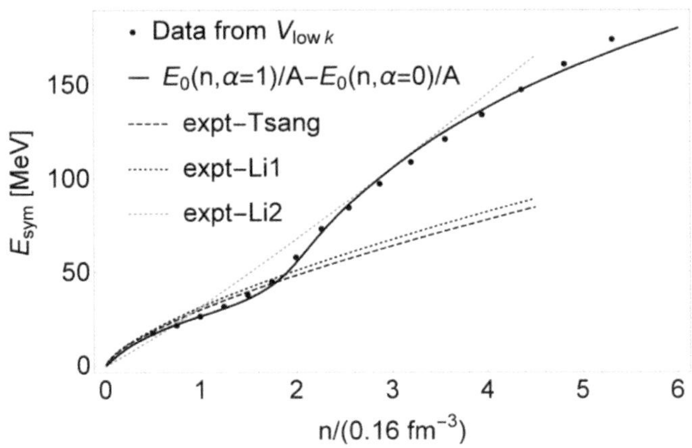

Fig. 9.4 The symmetry energy E_{sym}. Same as Fig. 9.3 for the empirical values.

theoretical framework, which is effective theory. In these descriptions, there is nothing to suggest that the phase R-II is constrained by what's in R-I. Hence the numerically precise "constraints" at $n \lesssim n_{1/2}$ cannot "dictate" what takes place in R-II. This point of view is at odds with what's currently espoused by many in the field, e.g., [Kolomeitsev, Lattimer, Ohnishi and Tews (2016)]. There is an analogy to this in nuclear physics: Approaching dense matter (1) in chiral perturbation theory on the one hand and (2) in Wilsonian renormalization group built on the Fermi surface on the other. The former is constrained the low-energy theorems valid for $k_F/\Lambda \ll 1$ and the latter is constrained by Fermi-liquid fixed point valid for $k_F/\tilde{\Lambda} \gg 1$. There are no direct relations in the parameters figuring in EFTs with scale changes, although QCD proper must necessarily have the connection.

9.2.5 *Massive stars*

Given $\epsilon(n)$ and $P(n)$, the solution of the TOV equations gives the mass $M(R)$ vs. the radius R of the star. Low-mass stars are subject to the crust structure which should be implemented in supplement to the EoS effective at densities $n \gtrsim n_0$. Our formulation cannot address the low-density structure — such as "pastas" etc., so that part is incorporated with EoSs given elsewhere, such as, e.g., [Baym, Pethick and Sutherland (1971)].

The leptons that participate in beta stability need to be taken into account as they determine the proton fraction $x \equiv n_p/n = Z/(N+Z) =$

$(1-\alpha)/2$ where α figures in the symmetry energy (9.2) . Beta equilibrium involving leptons, i.e., electrons and muons, $n \leftrightarrow p + e^- + \bar{\nu}_e$ and $n \leftrightarrow p + \mu^- + \bar{\nu}_\mu$ and charge neutrality imply

$$n_p = n_e + n_\mu, \quad \mu_n = \mu_p + \mu_e, \quad \mu_e = \mu_\mu \tag{9.3}$$

where

$$\mu_{n,p} = \left(\frac{\partial \epsilon}{\partial n_{n,p}}\right)_V. \tag{9.4}$$

One usually assumes μ_l to be the chemical potentials of those of free Fermi gas of electrons and muons. The proton fraction x for matter in beta equilibrium is then determined by minimizing, for a given nucleon density n, the total energy per particle (9.2) plus the contributions from leptons and from the rest mass of the nucleons.

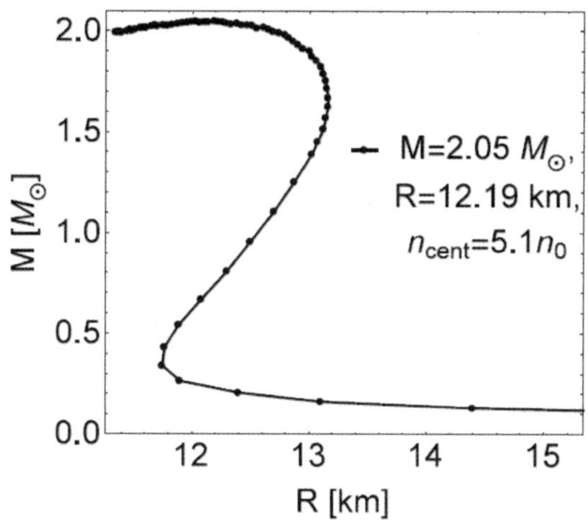

Fig. 9.5 The mass vs. radius of the neutron star in beta equilibrium.

9.2.5.1 *M vs. R*

Given in Fig. 9.5 is M vs. R calculated in V_{lowk}RG with beta equilibrium properly taken into account. In Fig. 9.6 is M vs. $n_{\rm cent}$, central density predicted. The predicted values are: $M_{max} \simeq 2.05 M_\odot$ and $R \simeq 12.2$ km with the central density $n_{cent} \simeq 5.1 n_0$. The maximum mass predicted is consistent with the observed value $M = 2.01 \pm 0.04\, M_\odot$[Antoniadis (2013)].

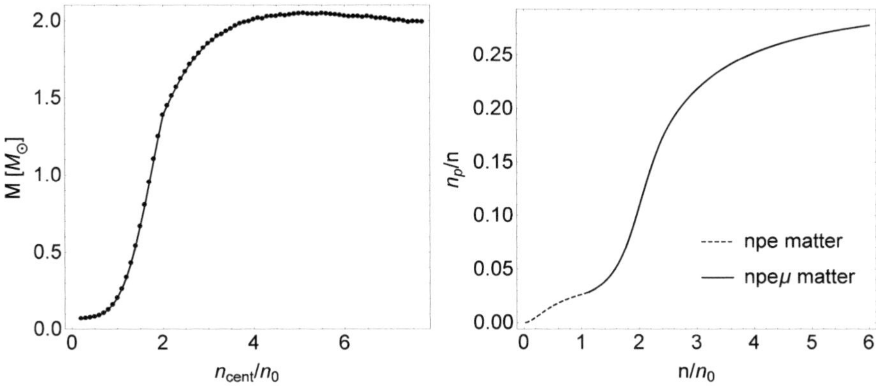

Fig. 9.6 Massive neutron star vs. the central density of the neutron star(left) is shown and the proton fraction $x = \frac{N_P}{N_P+N_N}$ (right) in neutron star matter with beta equilibrium. "npe matter" and "npeμ matter" are composed of neutron, proton, electron and neutron, proton, electron, muon respectively.

As for radius, it is not well fixed for various reasons that we shall not go into. For massive stars of ~ 2 solar mass, it ranges $\sim 9-14$ km. The prediction is compatible with the range. One should mention, however, that the radius is sensitive to densities near n_0 and our approach with little fine-tuning does not address this regime of densities for which our description is in any case of the same quality with other field theoretic or phenomenological models.

We now give a few predictions of the theory: sound velocity and tidal deformability from gravitational wave.

9.2.5.2 *Sound velocity*

To address the sound velocity of the star, the quantity that is crucial is the VeV of the trace of energy-momentum tensor (TEMT), $\langle\theta_\mu^\mu\rangle$. It is related to the sound velocity v_s by

$$\frac{\partial}{\partial n}\langle\theta_\mu^\mu\rangle = \frac{\partial\epsilon(n)}{\partial n}(1-3v_s^2) \tag{9.5}$$

with

$$v_s^2 = \frac{\partial P(n)}{\partial n}/\frac{\partial\epsilon(n)}{\partial n}. \tag{9.6}$$

To arrive at the sound velocity, we look at the TEMT. This can be done in two ways: (1) The mean-field approximation and (2) the V_{lowk}RG, both with bsHLS.

• Mean-field approximation

We first look at θ^μ_μ in RMF.

The RMF approach taken to be equivalent to Landau Fermi-liquid fixed-point theory should become more accurate as density increases. This is because in Wilsonian RG approach, the Fermi-liquid fixed point is approached as $\tilde{\Lambda}/k_F \to 0$ where $\tilde{\Lambda} = \Lambda - k_F$ (Λ being the cutoff scale on top of the Fermi surface from which the decimation is done), provided of course, as we assume, the Fermi-liquid structure continues to hold.

We apply the mean-field approximation with the Lagrangian (8.1) which is valid at the leading order scale symmetry "LOSS."[4] For readers' ease, we write it out explicitly, here for the case of $SU(2) \times U(1)$ symmetry for the vector mesons.

$$\mathcal{L} = \mathcal{L}_{\rm inv} + \mathcal{L}_{\rm SB} \tag{9.7}$$

where

$$\mathcal{L}_{\rm inv} = \mathcal{L}_N + \mathcal{L}_M \tag{9.8}$$

$$\begin{aligned}
\mathcal{L}_N = {} & \bar{N} i\gamma^\mu \left(\partial_\mu - i g_\rho \vec{\rho}_\mu \cdot \frac{\vec{\tau}}{2} - i g_\omega \frac{\omega_\mu}{2}\right) N \\
& - \frac{m_N}{f_\sigma}\chi \bar{N} N + g_A \bar{N}\gamma^\mu \alpha_{\perp\mu}\gamma_5 N \\
& + g_{V\rho}\bar{N}\gamma^\mu \left(\alpha_{\parallel\mu} - g_\rho \vec{\rho}_\mu \cdot \frac{\vec{\tau}}{2}\right) N \\
& + g_{V\omega}\bar{N}\gamma^\mu \left(\frac{\partial_\mu \sigma_\omega}{2 f_{\sigma\omega}} - g_\omega \frac{\omega_\mu}{2}\right) N\,,
\end{aligned} \tag{9.9}$$

$$\begin{aligned}
\mathcal{L}_M = {} & \frac{f_\pi^2}{f_\sigma^2}\chi^2 {\rm Tr}\left[\alpha_{\perp\mu}\alpha_\perp^\mu\right] + \frac{f_{\sigma\rho}^2}{f_\sigma^2}\chi^2 {\rm Tr}\left[\left(\alpha_{\parallel\mu} - g_\rho \vec{\rho}_\mu \cdot \frac{\vec{\tau}}{2}\right)^2\right] \\
& + \frac{f_{\sigma\omega}^2}{2 f_\sigma^2}\chi^2 \left(\frac{\partial_\mu \sigma_\omega}{f_{\sigma\omega}} - g_\omega \omega_\mu\right)^2 \\
& - \frac{1}{2}{\rm Tr}\left[\rho_{\mu\nu}\rho^{\mu\nu}\right] - \frac{1}{4}\left[\omega_{\mu\nu}\omega^{\mu\nu}\right] + \frac{1}{2}\partial_\mu \chi \cdot \partial^\mu \chi
\end{aligned} \tag{9.10}$$

$$\mathcal{L}_{\rm SB} = -V(\chi) + \frac{f_\pi^2}{4}{\rm Tr}\left(\mathcal{M} U^\dagger + h.c.\right)\left(\frac{\chi}{f_\sigma}\right)^3 \tag{9.11}$$

where $V(\chi)$ is the scale symmetry breaking potential, $\mathcal{M}$ is the (quark) mass matrix and

$$\rho^{\mu\nu} = \partial^\mu \vec{\rho}^\nu \cdot \frac{\vec{\tau}}{2} - \partial^\nu \vec{\rho}^\mu \cdot \frac{\vec{\tau}}{2} - i g_\rho \left[\vec{\rho}^\mu \cdot \frac{\vec{\tau}}{2}, \vec{\rho}^\nu \cdot \frac{\vec{\tau}}{2}\right], \tag{9.12}$$

[4]We use the same notation as in (8.1) which is different from the rest of the volume.

$$\omega^{\mu\nu} = \partial^\mu \frac{\omega^\nu}{2} - \partial^\nu \frac{\omega^\mu}{2} \tag{9.13}$$

and

$$\alpha_\perp^\mu = \tfrac{1}{2i}\left[\partial^\mu \tilde{\xi}_R \cdot \tilde{\xi}_R^\dagger - \partial^\mu \tilde{\xi}_L \cdot \tilde{\xi}_L^\dagger\right], \tag{9.14}$$

$$\alpha_\parallel^\mu = \tfrac{1}{2i}\left[\partial^\mu \tilde{\xi}_R \cdot \tilde{\xi}_R^\dagger + \partial^\mu \tilde{\xi}_L \cdot \tilde{\xi}_L^\dagger\right] \tag{9.15}$$

with

$$\tilde{\xi}_{L,R} = e^{i\sigma_\rho/f_{\sigma\rho}} e^{\mp i\pi/f_\pi}. \tag{9.16}$$

For the purpose of the mean-field calculation, this Lagrangian can be greatly simplified by linearization.

Consider the thermodynamic potential $\Omega = E - TS - \mu N$ at zero temperature taken in mean-field with (9.7) for symmetric nuclear matter[5]

$$\begin{aligned}\left.\frac{\Omega(T=0)}{V}\right|_{\substack{\omega_0=\langle\omega_0\rangle \\ \chi=\langle\chi\rangle}} &= \frac{1}{4\pi^2}\left[2E_F^3 k_F - m_N^{*2} E_F k_F - m_N^{*4}\ln\left(\frac{E_F+k_F}{m_N^*}\right)\right] \\ &\quad + \left[g_\omega\left(g_{V\omega}-1\right)\langle\omega_0\rangle - \mu\right]\frac{2}{3\pi^2}k_F^3 \\ &\quad - \frac{1}{2} f_{\sigma\omega}^2 g_\omega^2 \frac{\langle\chi\rangle^2}{f_\sigma^2}\langle\omega_0\rangle^2 + V(\langle\chi\rangle)\end{aligned} \tag{9.17}$$

where $\langle\omega_0\rangle$ and $\langle\chi\rangle$ are the "vacuum" (medium) expectation value of $\omega_{\mu=0}$ and χ, $m_N^* = \frac{\langle\chi\rangle}{f_\sigma} m_N$ and $E_F = \sqrt{k_F^2 + m_N^{*2}}$. The nucleon number density is

$$n \equiv N/V = -\frac{\partial(\Omega/V)}{\partial\mu} = \frac{2}{3\pi^2}k_F^3 \tag{9.18}$$

and the chemical potential μ given by the condition $\frac{\partial(\Omega/V)}{\partial n} = 0$ is

$$\mu = E_F + g_\omega\left(g_{V\omega} - 1\right)\langle\omega_0\rangle. \tag{9.19}$$

The energy density ϵ and the pressure P at $T = 0$ are given by

$$\begin{aligned}\epsilon &= \frac{1}{4\pi^2}\left[2E_F^3 k_F - m_N^{*2} E_F k_F - m_N^{*4}\ln\left(\frac{E_F+k_F}{m_N^*}\right)\right] \\ &\quad + g_\omega\left(g_{V\omega}-1\right)\langle\omega_0\rangle n - \frac{1}{2} f_{\sigma\omega}^2 g_\omega^2 \frac{\langle\chi\rangle^2}{f_\sigma^2}\langle\omega_0\rangle^2 + V(\langle\chi\rangle)\end{aligned} \tag{9.20}$$

[5]For ease of writing in this part, we drop the $*$ indicating in-medium quantities — density-dependent due to the IDDs — apart from the renormalized nucleon mass that gets a tadpole contribution in medium. No confusion arises for the result we are getting at.

and

$$P = -\left.\frac{\Omega}{V}\right|_{\omega_0=\langle\omega_0\rangle,\,\chi=\langle\chi\rangle} \tag{9.21}$$

$$= \frac{1}{4\pi^2}\left[\frac{2}{3}E_F k_F^3 - m_N^{*2}E_F k_F + m_N^{*4}\ln\left(\frac{E_F+k_F}{m_N^*}\right)\right]$$
$$+\frac{1}{2}f_{\sigma\omega}^2 g_\omega^2\frac{\langle\chi\rangle^2}{f_\sigma^2}\langle\omega_0\rangle^2 - V(\langle\chi\rangle)\,. \tag{9.22}$$

Now we look at the stationarity conditions that give the gap equations for χ and ω

$$\left.\frac{\partial\Omega}{\partial\chi}\right|_{\omega_0=\langle\omega_0\rangle,\,\chi=\langle\chi\rangle} = 0\,, \quad \left.\frac{\partial\Omega}{\partial\omega_0}\right|_{\omega_0=\langle\omega_0\rangle,\,\chi=\langle\chi\rangle} = 0\,. \tag{9.23}$$

They lead to

$$\frac{m_N^2\langle\chi\rangle}{\pi^2 f_\sigma^2}\left[k_F E_F - m_N^{*2}\ln\left(\frac{k_F+E_F}{m_N^*}\right)\right] - \frac{f_{\sigma\omega}^2}{f_\sigma^2}g_\omega^2\langle\omega_0\rangle^2\langle\chi\rangle + \left.\frac{\partial V(\chi)}{\partial\chi}\right|_{\chi=\langle\chi\rangle} = 0 \tag{9.24}$$

$$g_\omega\left(g_{V\omega}-1\right)n - f_{\sigma\omega}^2 g_\omega^2\frac{\langle\chi\rangle^2}{f_\sigma^2}\langle\omega_0\rangle = 0\,. \tag{9.25}$$

One obtains[6] from (9.24) and (9.25) the VeV of the TEMT θ_μ^μ

$$\langle\theta_\mu^\mu\rangle = \langle\theta^{00}\rangle - \sum_i\langle\theta^{ii}\rangle = \epsilon - 3P$$
$$= 4V(\langle\chi\rangle) - \langle\chi\rangle\left.\frac{\partial V(\chi)}{\partial\chi}\right|_{\chi=\langle\chi\rangle}. \tag{9.26}$$

This is just what one gets in the chiral limit by taking the mean-field value of the TEMT θ_μ^μ,

$$\theta_\mu^\mu = 4V(\chi) - \chi\frac{\partial V(\chi)}{\partial\chi} + \frac{f_\pi^2}{4}\mathrm{Tr}\left(\mathcal{M}U^\dagger + h.c.\right)\left(\frac{\chi}{f_\chi}\right)^3. \tag{9.27}$$

One notes that in the mean field of bsHLS, the TEMT is given solely by the dilaton condensate as required. This shows that the Fermi surface does not spoil scale symmetry. In fact we will arrive in V_{lowk}RG at the conclusion that strongly-correlated hadronic interactions do not modify the dilaton potential either. Thus if the dilaton condensate is a constant of, or depends weakly on, density as we found in Chapter 8, i.e., Eq. (8.50), and also below, then $\frac{\partial}{\partial n}\langle\theta_\mu^\mu\rangle$ will be zero or approximately zero. This feature will account for the emergent scale symmetry in compact-star matter.

[6] Unless otherwise stated, we will work in the chiral limit.

• **V_{lowk}RG double decimation**

In the sense of Fermi-liquid fixed-point approach, the RMF result would be accurate in the limit that $\tilde{N} = k_F/(\overline{\Lambda} - k_F) \to \infty$. In nuclear interactions, high-order correlations corresponding to $1/\tilde{N}$ corrections cannot be ignored. To capture these corrections, the renormalization-group approach with V_{lowk} (V_{lowk}-RG) proves to be powerful with the scheme developed as in [Dong, Kuo, Lee, Machledit and Rho (2013); Paeng, Kuo, Lee, Ma and Rho (2017)]. Arriving at the Fermi-liquid fixed point corresponds to the first decimation in [Brown and Rho (2004)] in terms of the V_{lowk}-RG which is shown in detail in [Holt, Brown, Holt and Kuo (2007)]. Going beyond the fixed-point structure, i.e., the second decimation, involves sophisticated high order correlation calculations. In what follows, we will simply follow the procedure used in [Paeng, Kuo, Lee, Ma and Rho (2017)] to do the double-decimation calculation. What's new is in the way the IDDs are implemented — as described in detail in Chapter 8 — in the effective (bsHLS) Lagrangian in the new development.

The basic premise, we reiterate, is that when embedded in the medium characterized by density n, formally interpreted in terms of baryon field operators as described above, the Lagrangian preserves the same symmetry structure — apart from the non-Lorentz covariance — as in the medium-free vacuum and the effect of changes in the vacuum structure caused by density, including the topology change, is entirely encoded in the way the parameters of the Lagrangian behave as density changes. Needless to say, should there be phase transitions along the way, this scheme will break down. We will assume that this does not happen in the density domain relevant in compact stars.

We adopt the IDDs defined in terms of the properties of scale symmetry involving the dilaton χ and chiral symmetry involving π and V as spelled out in detail in Chapter 8. The first thing to do is then to incorporate the IDD structure of the "bare parameters" of the Lagrangian.[7] The "bare" mass terms for the RG decimations are given by

$$\mathcal{L}_{\text{mass}} = -m_N^* \bar{N}N + \frac{m_\rho^{*\,2}}{2}\left(\rho_\mu^a\right)^2 + \frac{m_\omega^{*\,2}}{2}\left(\omega_\mu\right)^2 - \frac{m_\sigma^{*\,2}}{2}\sigma^2 - \frac{m_\pi^{*\,2}}{2}\pi^{a\,2}, \quad (9.28)$$

where σ and π are redefined as $\sigma \equiv \frac{\langle\chi\rangle}{f_\sigma}\sigma$ and $\pi^a \equiv \frac{\langle\chi\rangle}{f_\sigma}\pi^a$ to get the kinetic

[7] We recall that since the RG (first) decimation is done from $\Lambda < \Lambda_M$, certain many-body induced effects, such as $\text{DD}_{\text{induced}}$, will join the IDDs. This will make the effective density dependence most likely somewhat different from the IDDs given by the matching at Λ_M.

term in the form as

$$\mathcal{L}_{\rm scalar} = \frac{1}{2}\partial_\mu\sigma\partial^\mu\sigma + \frac{1}{2}\partial_\mu\pi^a\partial^\mu\pi^a\,. \tag{9.29}$$

Then, the masses are related to $\langle\chi\rangle$, g_ρ and g_ω — all of which are carrying IDDs — in the form of the master formula given in Chapter 8,

$$\frac{m_N^*}{m_N} \approx \frac{g_V}{g_V^*}\frac{m_V^*}{m_V} \approx \frac{m_\sigma^*}{m_\sigma} \approx \left(\frac{m_\pi^*}{m_\pi}\right)^2 \approx \frac{f_\pi^*}{f_\pi} \approx \frac{\langle\chi\rangle^*}{f_\sigma} \equiv \Phi^*\,, \tag{9.30}$$

where

$$m_\sigma^{*\,2} \equiv \left.\frac{\partial^2 V(\sigma)}{\partial\sigma^2}\right|_{\sigma=\langle\sigma\rangle^*}\,. \tag{9.31}$$

We note that the master formula (9.30) *completely* defines the sliding vacuum behavior of the Lagrangian in the whole range of densities — including the topology change — relevant for compact stars. They involve only two scaling quantities, Φ and $\mathcal{R} \equiv \Phi_V = g_V^*/g_V$. Now due to the topology change at $n = n_{1/2} \sim 2n_0$, they differ in the two regions R-I and R-II. Overall there are, therefore, only four vacuum-sliding quantities $\Phi_{\rm I,II}$ and $\mathcal{R}_{\rm I,II}$. In making the analysis yield a fit close to Nature, particularly to the precision data at n_0, there is some need for fine-tuning in R-I, but as discussed in [Paeng, Kuo, Lee, Ma and Rho (2017)], the fine-tuning is minimal, of the size to account for higher-order corrections to the V_{lowk}RG approach used in the analysis — such as functional-renormalization-group treatment that goes beyond what's included in the V_{lowk}RG. The most important constraint in R-II for a density-independent TEMT suggested by the mean-field calculation (which is equivalent to the single-decimation RG) as defined in Chapter 8 is that the VM fixed point be reached at higher density, say, $\gtrsim 10n_0$. This implies that the VM fixed point is reached at a higher density than what is often quoted in the literature, say, $\lesssim 5n_0$. What is significant for this critical density — which is unknown at present — is that as assumed, the phase transition, if any, from hadrons to quarks/gluon does not take place in the density regime relevant to compact stars.

The TEMT $\langle\theta_\mu^\mu\rangle$ (upper panel) and the sound velocity (lower panel) obtained in the V_{lowk}RG approach are given in Fig. 9.7. Note that $\langle\theta_\mu^\mu\rangle$ is a constant independent of density up to $n \sim 6n_0$ for both symmetric matter and pure neutron matter. This confirms the result obtained above in the mean-field (single decimation) approximation.

- **Sound velocity:** $v_s = \sqrt{1/3}$

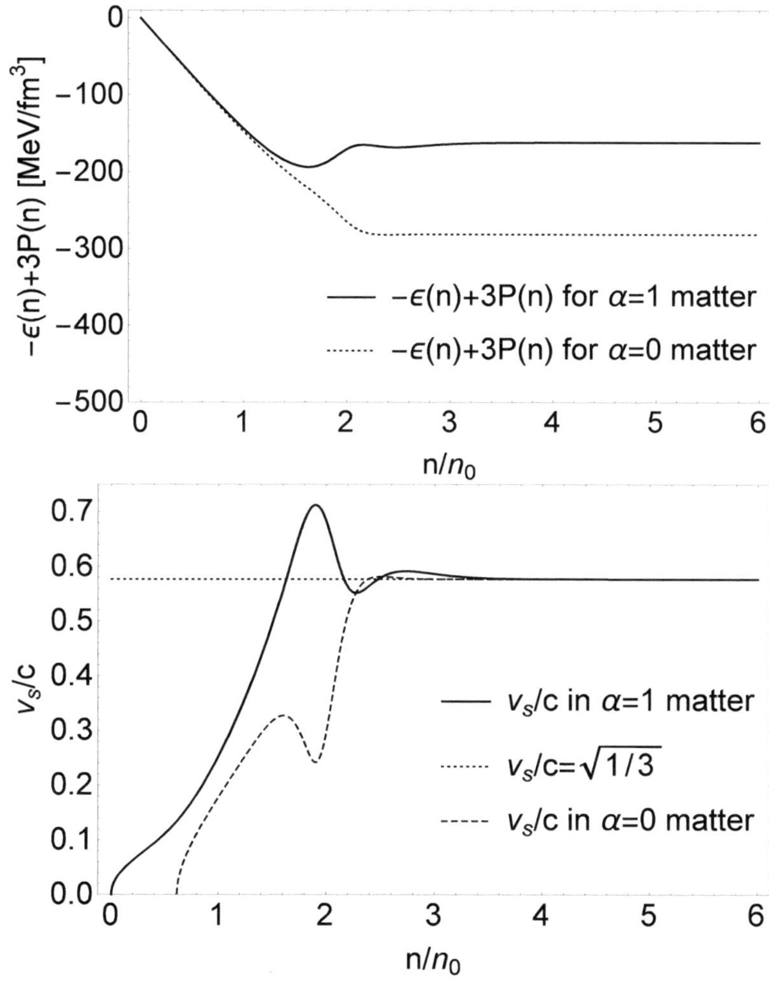

Fig. 9.7 $\langle\theta^\mu_\mu\rangle$ (upper panel) and v_s/c (lower panel) for both nuclear and pure neutron matter.

The TEMT and the sound velocity are related by

$$\frac{\partial}{\partial n}\langle\theta^\mu_\mu\rangle = \frac{\partial\epsilon(n)}{\partial n}(1 - 3v_s^2) \tag{9.32}$$

with $v_s^2 = \frac{\partial P(n)}{\partial n}/\frac{\partial\epsilon(n)}{\partial n}$. Since we have, from Fig. 9.7 (left panel), the TEMT independent of density, the left-hand side of (9.32) is zero. If we assume that there is no extremum in the energy density for density $n > n_{1/2}$ —

which we believe is justified,[8] then $\frac{\partial \epsilon(n)}{\partial n} \neq 0$. It then follows that

$$v_s^2 \approx 1/3. \tag{9.33}$$

One can see in Fig. 9.7 (right panel) that this indeed is verified in the calculation.

It is interesting to see how the "conformal velocity" 1/3 is approached. It starts with $v_s^2 < 1/3$ at low density $n < 2n_0$, goes up to $v_s^2 > 1/3$ at $n \sim 2n_0$, drops below 1/3 and then climbs to and asymptotes at 1/3 at higher density $n \gtrsim 3n_0$. This feature closely resembles the scenario arrived at by Bedaque and Steiner [Bedaque and Steiner (2015)] in the study of the sound velocity based on their analysis of neutron stars with mass around two solar masses with various phenomenological equations of state. In our theory, there is a rather abrupt changeover of the *bs*HLS Lagrangian parameters due to a topology change (i.e., the cusp in the skyrmion crystal description), so one might imagine that such a behavior could be an artifact of the sharp transition. One of the characteristic features arising from the topology change is the stiffening of the symmetry energy at higher density $n > 2n_0$. It is responsible for the relatively high proton fraction of nuclear matter in beta equilibrium. It might render the direct URCA process to set in precociously and trigger too rapid a star cooling which might be at odds with observation. If it turned out to be the case, then that could indicate within our formalism that we need to improve on the VM property of the ρ meson at high density. On the other hand, the fact that the dense baryonic matter in our description is in the precursor state to an emerging scale invariance and hence manifesting a conformal-type sound velocity as we are proposing is highly suggestive of the intricate mechanism of the Bedaque–Steiner scenario. What is surprising in our description is, however, that what attributes to "conformality" with $v_s^2 = 1/3$ sets in so "precociously" in density and for $\theta_\mu^\mu \neq 0$. We find it appealing to identify this phenomenon as an emergence of conformality. In [Bedaque and Steiner (2015)], in contrast, the matter with $v_s^2 > 1/3$ should prevail up to density $n \sim 5n_0$, the maximum central density of $\sim$ 2-solar mass objects, because of the strong hadronic interactions intervening in the phenomenological models they relied on. As noted before, all phenomenological models fit "accurately" to the properties of nuclear matter (at $n = n_0$) and consistent

[8] So far there is no indication for such abnormal states as those suggested by Lee and Wick [Lee and Wick (1974)].

with the symmetry "constraints" have $v_s^2 > 1/2$. The conformal velocity should of course appear at asymptotic density. Since the sound velocity becomes "conformal" even though the TEMT is not equal to zero, it would be more appropriate to call it "pseudo-conformal" sound velocity.

As argued in [Tews, Carlson, Gandolfi and Reddy (2018)], it is unlikely that the sound velocity converge at non-asymptotic density to, and remain at, $v_s^2 = 1/3$ unless there is a change of degrees of freedom from hadrons to other form, say, quarks and gluons. What we have in this theory is that the change of degrees of freedom is effectuated by the skyrmion-to-half-skyrmion transition in a form analogous to the Cheshire Cat phenomenon [Ma, Lee, Paeng and Rho (2018)].

9.2.5.3 *Gravitational wave and tidal deformability*

Although the density $n_{1/2}$ for the skyrmion-half-skyrmion cross-over played an important role in the formalism developed in the volume, it remained undeterminable theoretically. The question one can raise is whether it can have crucial impacts on certain aspects of dense matter that have not yet been explored. In this connection we discuss the possible link between the precocious onset of the pseudo-conformal sound velocity at $\sim n_{1/2}$ discussed above and the tidal deformability seen in gravitational waves coming from coalescing neutron stars as recently measured in [Abbott *et al.* (2017)]. Since the measurements — and analyses thereof — are on-going in this research, it is premature to go into detail in this volume. We will therefore make only a very brief discussion on the matter, with a focus on what the present formalism says about the issue raised above.

The tidal deformability measured in gravitational waves involves the range of star masses $1.3 - 1.5 M_\odot$, within which most of neutron star masses so far discovered are populated. The central density of this range of stars is expected to be around $\sim 2n_0$. This is not high enough density to probe the central density of massive stars. But what makes this of particular relevance to the issue is that it coincides with the range of density involved in the topology change (or hadron-quark transition) going from soft to hard in the equation of state.

The question then is: Can the tidal deformability give a hint at the density at which there is a changeover of degrees of freedom?

To address this question, we recall the cusp structure in the symmetry energy E_{sym} given in the topology change. The symmetry energy is "soft" in the region R-I ($n < n_{1/2}$) and "hard" in R-II ($n \gtrsim n_{1/2}$). Since the tidal deformability is highly sensitive to the symmetry energy,[9] it will reflect the density regime involved, i.e., R-I or R-II. What distinguishes E_{sym} in R-I from that in R-II is that going toward $n_{1/2}$ from below, the symmetry energy is to decrease whereas in R-II it should increase. So depending on whether the density regime involved is in R-I or R-II, the behavior in the tidal deformability will be qualitatively different.

In [Paeng, Kuo, Lee, Ma and Rho (2017)] discussed above with $n_{1/2} = 2n_0$, the dimensionless tidal deformability,[10] $\bar{\Lambda}$, for the star mass $M \approx 1.4M_\odot$ was found to be ~ 790, close to the upper bound set by [Abbott *et al.* (2017)], $\bar{\Lambda} < 800$. It is most likely this bound will be tightened to a lower value in the future. Given that the EoS in R-I in the V_{lowk}RG formulation [Paeng, Kuo, Lee, Ma and Rho (2017)] correctly reproduces overall nuclear matter properties at n_0 and E_{sym} is supposed to be "soft" in R-I, the large $\bar{\Lambda}$ implies that the density probed is in R-II. This suggests that $n_{1/2}$ must be higher than $2n_0$.

$$n_{1/2} > 2n_0. \tag{9.34}$$

As shown in Fig. 9.7, the pseudo-conformal sound velocity sets in slightly above $n_{1/2}$. This implies that the precise determination of the bound for $\bar{\Lambda}$ could also indicate where hidden scale symmetry can get "unhidden" in compact-star matter. We turn to this matter in the next subsection.

9.2.6 *Pseudo-conformal model*

To explore the correlation between the tidal deformability, pseudo-conformal sound velocity and topology change, we consider a simple model that we will call "pseudo-conformal model (PCM)." The model is constructed in the following way.

Assuming that one can extrapolate what's well reproduced at the normal matter density n_0 to $n_{1/2}$, we simply take the EoS given by the V_{lowk}RG in R-I and then replace the EoS of the R-II sector by a simple EoS that gives $v_s^2 = 1/3$ but has a non-vanishing θ_μ^μ and is independent of density. In [Paeng, Kuo, Lee, Ma and Rho (2017)], it was shown that this procedure precisely reproduces the V_{lowk}RG result. Although there is a cusp in the symmetry energy as seen in the skyrmion crystal simulation — and also in the closure sum approximation with the tensor force, since there is no

[9]There are a large number of discussions on this matter in the literature, a recent of which is [Krastev and Li (2018)].

[10]$\bar{\Lambda}$ is related to the dimensionful tidal deformability λ as $\bar{\Lambda} = \lambda G \left(\frac{c^2}{GM_\odot}\right)^5 \left(\frac{M_\odot}{M}\right)^5$ with the gravitational constant G.

phase transition belonging to the Ginzburg–Landau–Wilsonian paradigm, going from R-I to R-II should have no discontinuity of the sort known in condensed physics as "boojum" [Mermin (1977)].

To do this, we consider E/A of the form for R-II

$$E/A = -m_N + ax^b + cx^d \quad \text{with } x \equiv n/n_0 \tag{9.35}$$

where a, b, c and d are parameters to be fixed. The sound velocity takes the form

$$v_s^2 = \frac{dP/dx}{d\epsilon/dx} = \frac{ab(b+1)x^b + cd(d+1)x^d}{a(b+1) + c(d+1)x^d} \tag{9.36}$$

where P is the pressure and ϵ is the energy density. If we choose $d = -1$ and $b = 1/3$, then the E/A given by

$$E/A = -m_N + ax^{1/3} + cx^{-1} \quad \text{with } x \equiv n/n_0 \tag{9.37}$$

has the sound velocity

$$v^2 = 1/3 \tag{9.38}$$

independently of a and c. Now what we call pseudo-conformal model (PCM) is the sum of V_{lowk}RG in R-I and (9.37) in R-II with the coefficients a and c for $\alpha = 0$ and $\alpha = 1$ completely determined by the continuity conditions at $n_{1/2}$

$$\mu_{\text{R-I}} = \mu_{\text{R-II}}, \; P_{\text{R-I}} = P_{\text{R-II}} \tag{9.39}$$

with μ being the chemical potential. The trace of energy-momentum tensor from the PCM is, by construction, non-zero and independent of density in R-II. Thus there is no manifest scale invariance at any density. Yet it develops the sound velocity $v_s^2 = 1/3$, manifesting possible emergence of scale symmetry.

Now the question is: Does this EoS constructed in what might appear to be too simplistic represent correctly the compact-star structure? More precisely, is the precocious onset of the pseudo-conformal sound velocity inconsistent with *any* observables available up to date of massive compact stars?

In answering this question, we pick the transition density $n_{1/2}$ at a higher value than $2n_0$ which seems, as it stands, give too high a tidal deformability, so as to explore the effect on the tidal deformability. It has been observed in the past studies that putting $n_{1.2}$ anywhere between $\sim 1.5n_0$ and $\sim 2n_0$ affect very little overall star properties. We expect this to be true also between $2n_0$ and $3n_0$. We take, just for illustration,

$n_{1/2} = 2.6n_0$. We expect, based on what's discussed above, that this will lead to a $\bar{\Lambda}$ lower than what was found for $n_{1/2} = 2n_0$ and also a central density higher than $\sim 2n_0$ found.

In Fig. 9.8 are given the trace of energy-momentum tensor (TEMT) (left panel) and the sound velocity given by the PCM, a simplified model of the V_{lowk} RG with $n_{1/2}$ set at $2.6n_0$. They represent the inputs of the pseudo-conformal model.

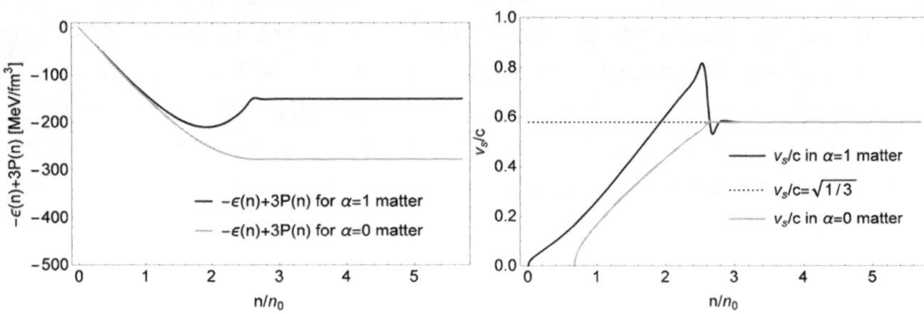

Fig. 9.8 The trace of energy mementum tensor and the sound velocity.

Now the star properties of the model.

In Fig. 9.9 are plotted E/A and the symmetry energy E_{sym} vs. density. The results are uncannily similar to those obtained in V_{lowk} RG given above. We note in particular the same property in E_{sym}, i.e., the continuous changeover from soft-to-hard at $\sim (2-3)n_0$ hinting at a smoothed cusp structure. The mass vs. R and the central density of the star with beta equilibrium taken into account are given in Fig. 9.10. The maximum mass comes out to be $M_{max} = 2.02M_{\cdot}$, with radius $R = 11.86$ km and the central density $n_c = 5.6n_0$. Given the approximations made, one could perhaps conclude that the results are the same as what was obtained in the V_{lowk} RG calculation with $n_{1/2} = 2n_0$, viz, $M_{max} = 2.05M_{\cdot}$, $R = 12.19$ km and $n_c = 5.1n_0$.

What's found here is highly nontrivial.

First of all the pseudo-conformal model does reproduce closely the V_{lowk} RG, insensitive to the location of $n_{1/2}$ in the vicinity of $n_{1/2} \gtrsim 2n_0$, capturing the intricate interplay of the topology change and symmetries emerging from nuclear dynamics in R-II, and second it is consistent with the observations despite the drastic simplification made.

Finally we look at what the location of $n_{1/2}$ does to the tidal deformability $\bar{\Lambda}$. As expected intuitively, it comes out that the higher the topology

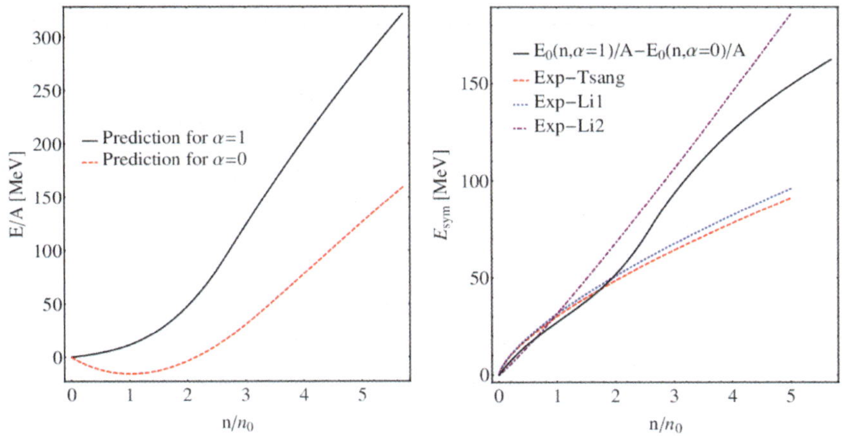

Fig. 9.9 E/A (left) and E_{sym} (right) vs density n.

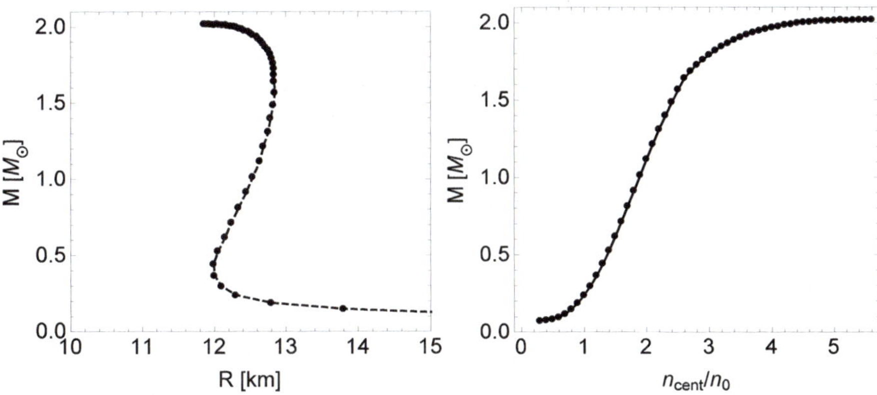

Fig. 9.10 The mass vs. radius of the neutron star(left) in beta equilibrium and the mass vs. the central density of the neutron star(right).

change density the lower the $\bar{\Lambda}$. This suggests that the precise value of $\bar{\Lambda}$, once measured, could pin down the density $n_{1/2}$ at which the topology change (or more generally degrees-of-freedom change) takes place. The location of $n_{1/2}$ has so far been accessible neither in theory — with QCD properties not accessible — nor in experiment, so it is appealing that the gravitational wave could ultimately provide the means to determine $n_{1/2}$. The model predictions are summarized in Fig. 9.11 and in Table 9.1. The model predicts $\bar{\Lambda} \simeq 644$ for an $1.4M_\odot$ star with $n_{\rm cnetral} \simeq 2.3n_0$. It thus confirms the expectation. While one expects little change from $n_{1/2} = 2n_0$ in other star properties, it does make a big difference in the tidal deformability. Indeed the sensitivity of $\bar{\Lambda}$ to $n_{1/2}$ reflects the sensitivity to the cusp structure intrinsic to the topology change.

For the future effort in correlating $\bar{\Lambda}$ with — and in determining — $n_{1/2}$, it is critically important to recognize that $\bar{\Lambda}$ is highly sensitive to the star mass concerned. In the discussion made above for a given value $M = 1.4M.$, an increase of M by less than 4% in the PCM decreases $\bar{\Lambda}$ by $\sim$ 24%. The $\bar{\Lambda}$ bound given in the measurement is for the mass which is a linear approximation around a "canonical" reference mass and it would require an extreme accuracy in pinning down the masses involved to arrive at a precise tidal deformability needed for the purpose.

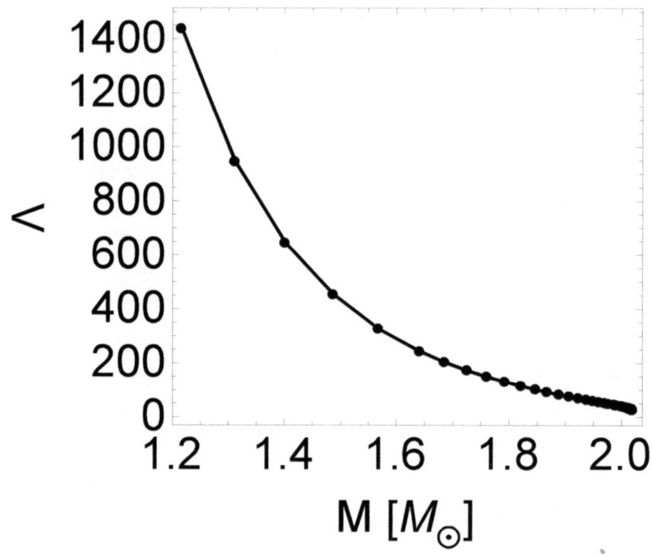

Fig. 9.11 The dimensionless tidal deformability Λ

9.3 Hadron-quark continuity in stars: Cheshire Cat strikes again

There is a highly pertinent question as to where and how the degrees of freedom present in QCD could enter in the description of compact stars.

Table 9.1 dimensionless tidal deformability

$M/M_\odot$	n_c/n_0	$\lambda/(10^{36}\text{gcm}^2\text{s}^2)$	$\bar{\Lambda}/100$	R/km
1.12	2.0	4.16	22.5	12.64
1.22	2.1	4.02	14.4	12.69
1.31	2.2	3.86	9.45	12.76
1.40	2.3	3.67	6.44	12.79
1.49	2.4	3.46	4.52	12.83
1.57	2.5	3.26	3.26	12.85

The density reached in the center of the $\sim$ 2-solar mass star is $\sim(5-7)$ times the normal nuclear matter density n_0. When nucleons are squeezed by gravity in the interior of massive stars, if one thinks of the nucleon in the matter as having a size of $\sim$ 1 fm in radius as in the MIT bag model, then pictorially, the nucleons must overlap and quarks will percolate between them [Baym and Chin (1976)]. At high density, say, $n \gtrsim 10n_0$, one then would expect the matter to be in the form of all quarks and gluons. This suggests a picture of massive stars in terms of at least three states, purely hadronic, hybrid hadron-quarkish or cross-over, purely quarkish. We encountered a similar picture when we considered the nucleon structure, namely, the Cheshire Cat structure of the proton: purely skyrmionic, half-skyrmrion-half bag in the chiral bag at magic angle, purely MIT bag. Here, they are connected to each other by a sort of "gauge choice."

Could this Cheshire Cat phenomenon manifest also in the EoS of compact stars? In this closing section, we conjecture that a Cheshire Cat phenomenon does indeed apply here.

One illustrative argument for our — admittedly speculative — thesis of a Cheshire Cat mechanism for EoS can be made with the scenario of 3 density regimes proposed semi-phenomenologically in [Kojo, Powell, Song and Baym (2016)]. To be specific, let us take Fig. 9.12 from this reference. The three-window model is roughly characterized by three density regimes in the M-vs-R curve for massive stars. The low density regime $n \lesssim 2n_0$ is to be mediated by hadrons, the intermediate density regime $2n_0 \lesssim n \lesssim 4n_0$ is populated by a hybrid of hadrons and quarks with strong coupling quark interactions and the third region $n \gtrsim 4n_0$ in which the $\sim$ 2 solar-mass stars are found is to involve quark percolation leading to a hadron-quark dual state, with the perturbative QCD regime setting in at much larger density, presumably at $\gg 10n_0$. What characterizes — physically — the EoS is that there is in the EoS a transition from soft to hard at $n \sim (2-4)n_0$, with the softness at low density to account for the star radius and the hardness at high density to support the massive star.

9.3.1 *Duality: Half-skyrmion phase and quarkyonic phase*

To develop the idea of the possible parallel, let us recall the essential features of the topology-impacted EoS discussed in this volume that gave the star structure for massive stars. Up to the density at which the half-skyrmions start figuring — "low-density" regime, the matter is hadronic. This can be described equally well in the sense of Weinberg's FT described in Chapter 2 by standard nuclear theory, i.e., sophisticated phenomenological approach implemented with accurate numerical procedures (such as e.g., [Akmal, Pandharipande and Ravenhall (1998)])) and refined energy density functionals, or high-order chiral EFT (with pions only) [Holt, Kaiser and Weise (2013); Holt, Rho and Weise (2016)] or topology-implemented V_{lowk}RG [Paeng, Kuo, Lee, Ma and Rho (2017)]. In terms of topology this region corresponds to the skyrmion phase. At density $n_{1/2} \sim 2n_0$, the skyrmion-half-skyrmion changeover takes place. At this density, the bilinear quark condensate $\langle\bar{q}q\rangle$, non-zero locally — hence supporting chiral density wave — vanishes on the average in the lattice simulation, $\hat{\Sigma} \equiv \overline{\langle\bar{q}q\rangle} = 0$. In this regime, the nucleon has a chiral invariant mass $m_0 \sim (0.6-0.9)m_N$

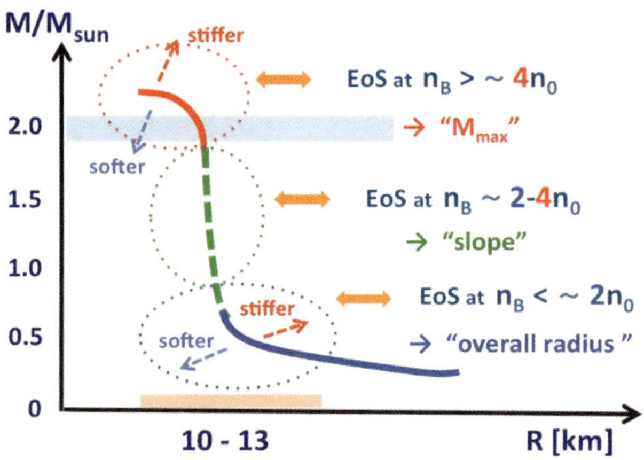

Fig. 9.12 Schematic M vs. R for a massive star: 3 density regimes from [Kojo, Powell, Song and Baym (2016)].

given mainly by the chiral-invariant dilaton condensate. Now while the bilinear chiral condensate vanishes, there are pions excited in this region with a complicated order parameter representing a non-vanishing order parameter for chiral symmetry, e.g., $\langle \bar{q}\bar{q}qq \rangle$. There emerges, *albeit* partial, parity-doubling that accounts not only for the chiral invariant nucleon mass, but also the degeneracy of π and σ at the dilaton limit fixed point at a density at which the VM takes place, most likely at $n \gtrsim 10n_0$. The topology-based scenario as formulated in this volume ends at the point where the VM fixed point and DLFP are reached — together or separately. However it can be continued into the color-flavor-locked quark phase by a superqualiton picture [Hong, Rho and Zahed (1999)] to be mentioned below.

In Chapter 3, the pure skyrmion and the pure MIT bag were connected by a "gauge transformation" which would then imply the same physics at the two extremes. The pure skyrmion description would be most effectively probed at low energy-momentum scale, while the MIT bag structure will be probed at high scale. Going from low to high scale is mediated by the amount of physics attributed to the partition into the two sectors which depends on how to access the relevant regimes. One can think of the hadron-quark description of [Kojo, Powell, Song and Baym (2016)] in a similar scenario, with the regime $n \gtrsim 2n_0$ simulated by going into the quarkyronic phase [Fukushima and Kojo (2016)] with hadrons and quarks co-existing in a chiral symmetric and confined phase. Indeed the chiral bag model described in Chapter 3 can capture the transition to a quarkyonic structure with the bag boundary conditions providing a close connection between the discrete half-skyrmion symmetry and the continuous chiral symmetry as shown in [Mantovani Sarti and Vento (2014); Vento (2017)]. The half-skyrmion phase is achieved by delocalization of the baryonic charge, i.e., by a modification of the baryon density profile, as one moves to high densities. The analysis presented in [Mantovani Sarti and Vento (2014); Vento (2017)] is built upon a series of approximations, but the delocalization mechanism, equivalent to the percolation invoked in the quarkyonic structure [Fukushima and Kojo (2016)], will survive more sophisticated analyses, because it is physically very intuitive. It should be feasible to translate rigorously — as illustrated in the case of the flavor-singlet axial charge discussed in Chapter 3 — by adroit boundary conditions, including both chiral and scale symmetries, what leads to the percolation phenomenon in the quarkyonic picture into the skyrmion-half-skyrmion picture by the Cheshire Cat Principle developed in Chapter 3.

9.3.2 *Duality: Superqualiton and color-flavor locking*

The color-flavor-locked (CFL) phase of QCD, somewhat academic for compact stars, amusingly has a Cheshire Cat description. This raises the question as to where the QCD degrees of freedom become indispensable in hadronic matter under extreme density conditions.

The three-window phenomenological description of [Kojo, Powell, Song and Baym (2016)] involves color-superconducting quark matter, including even the CFL phase expected in the perturbative QCD (pQCD) regime at high density $n \gg 10n_0$. The strength of the CFL phase to which the 2-flavor superconductivity transits at high density would — in the NJL framework — be required to be much stronger in massive compact stars [Fukushima and Kojo (2016)] than what is expected for pQCD applicable to CFL if the CFL phase were to be brought down, in the spirit of hadron-quark continuity, to the hadronic regime. An intriguing question is how far up in density one can go with the skyrmion description. This question was raised a long time ago [Hong, Rho and Zahed (1999)]. Let us just mention in what sense it is relevant to this volume.

As already discussed in Chapter 3, the Skyrme model can be formulated in the CFL phase, with the baryons arising as skyrmions, dubbed as "superqualitons." The spectrum was found to be identical to the color-flavor-locked model of QCD [Alford, Rajagopal and Wilczek (1999); Schäfer and Wilczek (1999)]. One can even hidden-local-symmetrize the Skyrme Lagrangian obtaining the corresponding HLS. It may be an interesting exercise to formulate the structure of hadrons arising from this Cheshire Cat Principle all the way to the CFL density — which seems eminently feasible. Among many fascinating — perhaps academic for compact stars — issues is what happens at densities exceeding the VM/DL fixed-point density $n \gtrsim 10n_0$.

It should be noted that the three-window model of [Kojo, Powell, Song and Baym (2016); Fukushima and Kojo (2016)] requires the vector coupling in the NJL model to be strong at $n \sim 10n_0$ and even in the CFL regime. It is simply inaccessible by pQCD. This is in agreement with what's found in the V_{lowk}RG analysis and goes toward the Cheshire Cat picture. In this connection, one can ask how the pion in the half-skyrmion phase is related to the "pion" in the quarkyonic phase? This question could also be answered in the NJL model mimicking the quarkyonic structure as in [Song and Baym (2017)]. In the half-skyrmion picture, the bilinear quark condensate globally vanishes though locally non-zero. Therefore one expects the condensate $\langle \bar{q}\bar{q}qq \rangle$ to contribute importantly to a generalized GMOR relation for the pion. Thus in the simple two-component picture of [Song and Baym (2017)], the pion in the half-skyrmion phase is more $\tilde{\pi}$-like. In reality it could be a much more complicated mixture. This picture should apply to all other degrees of freedom, such as the dilaton, vector mesons etc., in the Cheshire Cat mechanism.

Chapter 10

Strangeness in Stars

10.1 "Strangeness problem"

In accessing massive compact stars found in nature employing the effective field theory based on scale-invariant hidden local symmetry, bsHLS, fully defined in the preceding chapters, the strangeness degree of freedom has been simply left out. The basic premise in the volume was the notion of Cheshire Cat operative in a sort of smooth transition from hadron degrees of freedom in quantum effective field theory to quark-gluon degrees of freedom of QCD as suggested in Chapter 9. The formalism encountered no difficulty in explaining satisfactorily both low-density regime in the neighborhood of the normal nuclear matter as well as in high density regime relevant to the interior of massive compact stars, $\sim (5-7)n_0$. There was no compelling reason to resort to other degrees of freedom than what's included in bsHLS — nucleons, pions, vector mesons (or infinite tower of vector mesons) and scalar dilaton. However as explained in Chapter 6, in the way that the dilaton degree of freedom, which is very important as illustrated above, is implemented via the Crewther–Tunstall (CT) approach, all three flavors, u, d and s must figure in bsHLS. The strangeness flavor comes in the setup of scale-chiral symmetry, χPT_σ, and is essential for placing the dilaton in the appropriate symmetry scheme. It however does not play a direct role. Apart from the scalar mass comparable to the kaon mass, what entered *directly* in the compact-star EoS involves only the $SU(2)$ flavors. On the other hand, kaons and hyperons could intervene directly under certain conditions and could not in principle be ignored in the high density regime involved in massive compact stars. Furthermore in the presence of temperature — which has not been considered in the volume, strangeness flavor could in principle figure importantly as for instance in the star cooling. As argued below, it is possible to think that *explicit* strangeness degrees of freedom are banished beyond the range of densities involved in the massive stars concerned.

In standard chiral perturbation approaches in flavor $SU(3)$ as first considered by Kaplan and Nelson [Kaplan and Nelson (1988)] and studied extensively as recounted in numerous reviews, e.g., [Brown, Lee and Rho (2008)], the $SU(3)$ chiral Lagrangian would predict, at the mean field level, s-wave kaons condensing at a low density, typically at $\sim 3n_0$. Such low-density kaon condensation would have a large impact on a variety of astrophysical phenomena including possible low-mass black holes [Brown, Lee and Rho (2008)]. The recent discovery of ~ 2 solar-mass compact stars [Demorest *et al.* (2010); Antoniadis (2013)] raised the poignant issue as to whether such kaon condensation and equally hyperon presence, triggering softness in the EoS as naively suspected, are not unequivocally ruled out by the observed massive compact stars. Indeed, the verdict pronounced in [Demorest *et al.* (2010)] was that one of the manifestation of strangeness, namely, kaon condensation, is indeed ruled out by the observation.

This raises the question: Can the discovery of $\sim$2-solar-mass stars be taken as the smoking-gun signal for the death of kaon condensation or equivalently of hyperons in compact star matter?

There are a variety of scenarios, either invoking or not invoking (up-to-date unknown) repulsive interactions in strange-flavor channels, the so-called strangeness problem, both hyperon problem and kaon condensation problem, could somehow be avoided. It is fair however to say that up to date, given the paucity of experimental information and consistent theoretical tools, there are no generally accepted scenarios. Thus this issue remains an open problem to be resolved.

We believe the verdict pronounced in [Demorest *et al.* (2010)], though it may turn out to be valid, is however too premature at present to be taken definitive, given the large uncertainty in the theoretical treatment of strangeness in dense matter. The theoretical results the conclusion made in [Demorest *et al.* (2010)] is drawn from naively attaching tree-order chiral Lagrangian treatments of kaon-nucleon interactions to highly-correlated nuclear dynamics, both treated separately at a vastly different level of sophistication and with no feedbacks between the two taken into account, which could play a crucial role. Given the preeminent role of the dilaton and vector-meson degrees of freedom in the EoS for matter at density $n > n_{1/2}$ as described above, which is not taken into account in the conventional treatment of kaon condensation, such a verdict is much too hasty.

As we will see below, the situation can be a lot subtler than one would summarily think. In fact, there is an indication known since a long time that a unified treatment of *both meson and baryon sectors on the same footing* is necessary for a reliable description. For example, when the strongly repulsive short-range correlations among the nucleons are suitably taken into account, the naive chiral perturbative estimate can be qualitatively wrong [Pandharipande, Pethick and Thorsson (1995)]. Indeed the repulsion operative in the nuclear sector, well recognized in nuclear dynamics, is of the same origin as the attraction between kaons and nucleons responsible for boson condensation and is found to push the condensation threshold density beyond the density relevant for compact stars. Therefore both the baryon sector and meson sector and their interconnections, i.e., back-reactions, must be treated on the same footing.

Within the formalism adopted in this volume, it should be in principle feasible to approach the problem with the same V_{lowk}RG approach employed above for two flavors generalized to three flavors, say, $V_{lowk}^{SU(3)}$RG. It would rely on V_{lowk}s in coupled channels involving hyperons and kaons. The procedure, although clearly involved, should be straightforward. The proper treatment will require V_{lowk}s in the octet baryon sector, octet pseudo-scalar sector and their couplings, suitably taking into account the IDDs. Working this out to confront nature is however beyond the present capability, mainly due to the paucity of experimental information that does not allow the large number of "bare" parameters (with IDDs endowed properly) of three-flavor *bs*HLS to be adequately controlled. In what follows, we will work with a drastically simplified version of the approach and try to arrive at a qualitative confirmation of the possibility that strangeness — via hyperons and kaons — does not figure in compact-star equation of state. For this objective we first argue using a large N_c reasoning that kaon condensation and hyperon presence must appear more or less at the same density in compact-star matter.

10.2 Kaon condensation vs. hyperons

So far in the literature, the appearance of hyperons and the onset of kaon condensation are treated separately. This cannot be trusted. With our Lagrangian *bs*HLS, it is possible to treat them simultaneously.

Let us first see whether kaons condense before or after the appearance of hyperons. At present, the only way this problem can be addressed in a tractable approximation in consistency with the basic premise of QCD —

such as large N_c — is the skyrmion description in which both kaons and hyperons can be treated on the same footing with the same Lagrangian. This matter was first discussed in [Lee and Rho (2013)] employing the successful Callan–Klebanov bound-state model [Callan and Klebanov (1985)]. In this model, anti-kaons K^- are bound to the $SU(2)$ skyrmion to yield hyperons. A rather surprising observation is that this model can interpolate kaons between the chiral limit ($m_K \to 0$) and the Isgur–Wise heavy-quark limit ($m_K \to \infty$) [Klebanov (1989)]. The model can be applied to nuclear matter by putting the Callan–Klebanov skyrmions on crystal lattice [Klebanov (1985)]. It was shown [Lee and Rho (2013)] that put on a crystal lattice, the energy difference between the lowest-lying hyperon Λ and the nucleon N in medium comes out as

$$E_\Lambda^* - E_N^* = \omega_K^* + \mathcal{O}(N_c^{-1}) \tag{10.1}$$

where the asterisk represents medium-dependence. Note that the in-medium kaon mass is of $O(N_c^0)$ in the N_c counting. It is fortunate that the leading $\mathcal{O}(N_c)$ term and the flavor singlet $\mathcal{O}(N_c^0)$ Casimir energy term — which is extremely difficult to calculate — cancel out in the difference.

Now, the Λs will appear in the compact star matter when

$$\mu_e \geq E_\Lambda^* - E_N^* \tag{10.2}$$

where μ_e is the electron chemical potential which is equal to $\mu_n - \mu_p$ in weak equilibrium. On the other hand, kaons will appear by the weak process $e^- \to K^- + \nu_e$ when

$$\mu_e \geq \omega_K^*. \tag{10.3}$$

Therefore to the leading order in N_c in QCD, hyperons and condensed kaons populate compact stars simultaneously. Which one appears first in the single Lagrangian description then depends on $\mathcal{O}(1/N_c)$ hyperfine corrections, namely, when the skyrmion-kaon system is rotationally quantized. A simple quasi-particle approximation leads to

$$E_\Lambda^* - E_N^* = \omega_K^* + \frac{3}{8\mathcal{I}^*}(c^{*2} - 1) \tag{10.4}$$

where $\mathcal{I}^* > 0$ is the in-medium moment of inertia of skyrmion rotator and c^* is the in-medium hyperfine coefficient multiplying the effective spin operator of strangeness -1. The coefficient c is highly model-dependent even in the matter-free space [Callan and Klebanov (1985)], so it is unknown in dense matter except in the large N_c limit and also in the chiral limit. In either or both of these limits, $c^* \to 1$. In the matter-free space, it is found to be $c^2 \sim 0.5$. Although presently there is no firm proof, it seems likely that ${c^*}^2 < 1$ in medium, approaching 1 from below near chiral restoration. If this is the case, it may be that hyperons could appear before kaons condense and they ultimately join in the vicinity of chiral restoration. It is however difficult to be more precise on this point since the effect is at $\mathcal{O}(N_c^{-1})$ and at that order there are many other corrections, such as higher-order nuclear correlations, that go beyond the mean-field order, something that could be done in the three-flavor V_{lowk} approach. In the absence of a realistic calculation with kaon mass and $1/N_c$ corrections, it seems reasonable to assume that kaon condensation and hyperons appear at about the same density.

10.3 Hyperon problem

It will be shown below that kaon condensation will *inevitably* take place at some high density above n_0. The precise value for the density at which kaons start condensing cannot be pinned down theoretically unless a full V_{lowk}RG analysis is performed. Here we develop an argument based on a mean-field treatment — more or less equivalent to single-decimation Fermi-liquid approximation — with *bs*HLS that hyperons do not destabilize massive stars. With kaons condensing at about the same density strangeness will not affect the EoS calculated in the preceding chapter.

Simple energetic considerations do suggest that hyperons could be present at high density in compact-star matter. Specifically, the lowest-lying hyperon Λ, with its attractive interaction, is estimated to appear at matter density $\sim 2n_0$ with the others possibly appearing at higher density. What this suggests is that the hyperons could appear at about the same density as the one at which the half-skyrmion phase appears in the skyrmion matter as discussed in the preceding chapters. If this were the case, then the prediction made in the previous chapter would make no sense without the hyperonic degree of freedom taken into account.

In an admittedly phenomenological study using a Monte Carlo simulation over parameters that enter in the EoS for symmetric and asymmetric nuclear matter Bedaque and Steiner obtain the range of density — denoted Δ — constrained by hydrodynamic stability of the system that ensures that stars with $M > 2M_\odot$ could be supported [Bedaque and Steiner (2015)]. The EoS properties taken into account were the compression moduli K and L, symmetry energies $E_{sym} = S$ and $E^\Lambda_{sym} = S_\Lambda$. The Δ so deduced is the range of density beyond which the in-medium Λ mass becomes greater than the vacuum value. One expects — and it is confirmed experimentally — that the Λ-nucleon interaction is attractive at normal nuclear matter density, so Λs can be bound in nuclear matter. In compact star matter, as density increases, the chemical potential difference between neutron and proton increases and it can become energetically favored to have spontaneous creating of hyperons in the system. This can happen when density reaches roughly twice the normal nuclear matter density unless the attractive interaction is overpowered by repulsive mechanisms. The instability generated by the presence of hyperons at such low density is the "hyperon problem." The Δ then stands for the range of density at which the Λ-interactions must be repulsive enough to make the in-medium Λ mass be greater than the vacuum mass. The analysis by Bedaque and Steiner establishes that the range of Δ required is $1 < \Delta/n_0 \lesssim 3$. We refer to this as "Bedaque–Steiner constraint."

We can address this problem using the single-decimation RG procedure as an approximation to $V_{lowk}^{SU(3)}$RG. The line of reasoning developed in Section 4.4.1 is applied to three flavors. We take the mean-field approximation with the three-flavor *bs*HLS endowed with the IDDs — assumed to be $SU(3)$ symmetric for a qualitative treatment — as equivalent to Fermi-liquid fixed-point theory as in the flavor $SU(2)$ case considered in the preceding chapters.

In the relativistic mean field approximation with the Lagrangian (8.7) extended to flavor $SU(3)$, the chemical potential for the Λ in medium gets contributions from two sources, one from the intrinsic density dependence (IDD) in the "bare" mass parameter $m_\Lambda^* \approx \Phi m_\Lambda$ and the other from the potential terms coming from Λ-nuclear coupling via σ and ω exchanges as depicted in Fig. 10.1,

$$\mu_\Lambda = m_\Lambda^* - \frac{g_{\sigma\Lambda}^* g_{\sigma N}^*}{m_\sigma^{*\,2}} n_s + \frac{g_{\omega\Lambda}^* g_{\omega N}^*}{m_\omega^{*\,2}} n \tag{10.5}$$

where n_s and n are, respectively, nucleon scalar density and nucleon number density as defined in Fig. 10.1. The notations for (σ, ω) coupling to Λ and N are self-evident. The asterisk stands for IDD-scaling parameters.

Apart from the Λ coupling to the mesons, the large cancelation between the dilaton (σ) attraction and the vector (ω) repulsion responsible for small binding energy for nuclear matter must take place also in this case. In fact, using the standard constituent quark (equivalent to quasi-quark) counting, we may take $g_{\sigma\Lambda} \approx \frac{2}{3} g_{\sigma N}$ and $g_{\omega\Lambda} \approx \frac{2}{3} g_{\omega N}$, the 2/3 factor accounting for

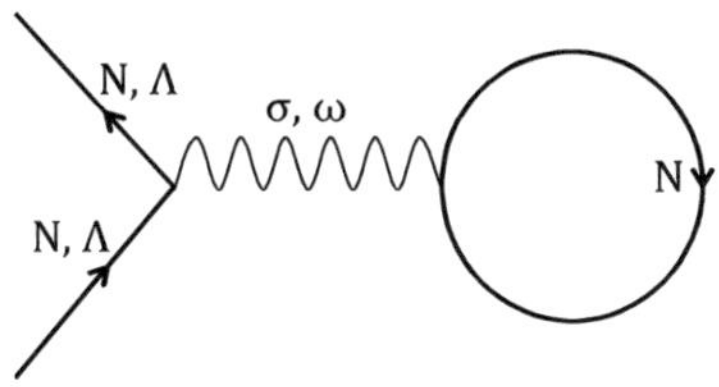

Fig. 10.1 Tadpole diagram for self-energies for the nucleon N and the hyperon Λ in medium. The loop corresponds to the nucleon scalar density $n_s = \langle \bar{N}N \rangle$ for coupling to σ and the nucleon number density $n = \langle N^\dagger N \rangle$ for coupling to ω.

the two non-strange quarks in Λ vs. 3 in nucleon. Then

$$\mu_\Lambda = m_\Lambda^* + \frac{2}{3}\left(-\frac{g_{\sigma N}^{*\,2}}{m_\sigma^{*\,2}} n_s + \frac{g_{\omega N}^{*\,2}}{m_\omega^{*\,2}} n\right). \tag{10.6}$$

This indicates that the Λ effective mass shift $\mu_\Lambda - m_\Lambda$ will become positive near $\sim 2n_0$ as in the symmetric nuclear matter.

To make a numerical estimate of the Λ mass shift in dense medium, we take into account the IDD-scaling — assuming flavor $SU(3)$ symmetry — of the parameters in the bsHLS Lagrangian in the mean-field calculation which corresponds to the "single-decimation procedure" of [Brown and Rho (2004)]. In doing this, one should recognize that the scaling parameter c_I in this procedure could well be different, i.e., renormalized, from the IDD coefficient entering into the double-decimation procedure with V_{lowk} employed in the previous chapter addressed to compact stars. This is because in the single-decimation procedure, the scaling function Φ_I is related to the Fermi-liquid fixed-point parameters as shown in [Friman and Rho (1996)] and encodes certain nonperturbative quasi-particle interactions on top of the IDD effects, termed "induced density dependence" or $\text{DD}_{\text{induced}}$ in Chapter 8 — manifesting scale-chiral symmetry. We ignore this subtlety that we expect to be of higher-order fluctuation corrections. Thus we take c_I as used in the V_{lowk} calculation, $c_I \approx 0.13$. The scaling and the constants for the dilaton and ω used in the calculation are the same as those determined in Chapter 8.

The result is plotted in Fig. 10.2. We see that $\mu_\Lambda - m_\Lambda$ crosses zero at density $n \sim 1.7n_0$. The result is insensitive to the demarcation density for the topology change. In fact what comes out in the mean field in bsHLS[1]

[1] We stress that here "mean field" endowed with IDDs goes beyond what is used in the standard mean-field theory.

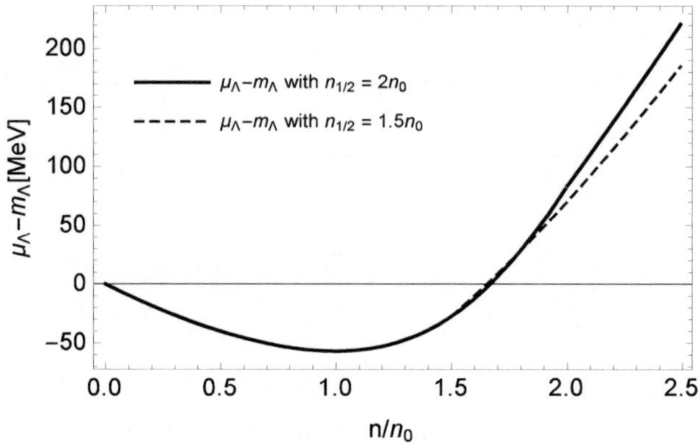

Fig. 10.2 $\mu_\Lambda - m_\Lambda$ vs. n/n_0 calculated with the scaling parameters determined in the theory (Chapter 9). The demarcation density was chosen for $n_{1/2} = (1.5, 2.0)n_0$. The density at which the mass shift crosses zero corresponds to Δ of [Bedaque and Steiner (2015)].

is quite easy to understand. Since m^*_Λ stops dropping with f^*_σ stabilizing at $2n_0$, what matters is the interplay of the ratio $(g^*/m^*)^2$ for the scalar and vector mesons — with an opposite sign — multiplied, respectively, by the scalar density n_s and by the baryon number density n. The vector repulsion wins over the scalar attraction as density increases in the same way as it does in nuclear matter. Although the estimate is admittedly approximate, it is most likely robust qualitatively. We conclude from this that within the formalism developed in [Lee, Paeng and Rho (2015)], the BS constraint [Bedaque and Steiner (2015)] is met for avoiding the hyperon problem.

10.4 Kaons in skyrmion crystal

We return to the skyrmion crystal description of dense matter of Section 10.2 and examine what the dilaton does to kaon condensation. In considering where hyperons and condensed kaons appear in compact stars, we have treated dense asymmetric matter in beta equilibrium. In the Callan–Klebanov model on which the reasoning is anchored, the topological Wess-Zumino term plays the dominant role. Viewed in terms of the bsHLS Lagrangian, it is, albeit roughly, equivalent to the ω exchange term which localized, corresponds to the Weinberg–Tomozawa term in standard chiral perturbation theory. This is in fact the term used in the literature for describing anti-kaon (K^-) binding to nuclei. For nuclear systems at low density, this is indeed the principal mechanism for bound kaon-nuclear systems seen in experiments.

But this interaction is not the relevant interaction at high density. The impact of the role of dilaton missing in the normal density kaonic-nuclear systems needs to be incorporated. In the next subsection, it will be shown that the scalar (dialton) degree of freedom brings a totally new picture to dense kaonic matter, in particular to the putative s-wave kaon condensation.

Here as a warm-up we illustrate what can happen using the skyrmion crystal picture. Since we are primarily interested in qualitative rather than numerically accurate features, we stick to the Skyrme model (skyrmion$_\pi$ with Nambu–Goldstone bosons only). Strangeness is involved, so we will deal with three flavors with the topological Wess–Zumino term included. The dilaton will be introduced in the way described in Chapter 7. Other heavy-meson fields, while quantitatively relevant, are not expected to modify the general features, so they will be ignored. This means that the density at which the skyrmion-half-skrymion transition, which here again will figure importantly, takes place will not be given reliably but this does not affect the main argument.

10.4.1 *The Lagrangian*

We take the scale-chiral Lagrangian valid in the leading order in scale symmetry ("LOSS") written down in Chapter 6

$$\begin{aligned}\mathcal{L}_{sk} = {} & \frac{f_\pi^2}{4}\left(\frac{\chi}{f_\sigma}\right)^2 \mathrm{Tr}(L_\mu L^\mu) + \frac{1}{32e^2}\mathrm{Tr}[L_\mu, L_\nu]^2 \\ & + \frac{f_\sigma^2}{4}\left(\frac{\chi}{f_\sigma}\right)^3 \mathrm{Tr}\mathcal{M}(U + U^\dagger - 2) \\ & + \frac{1}{2}\partial_\mu\chi\partial^\mu\chi + V(\chi) \end{aligned} \tag{10.7}$$

where $V(\chi)$ is the ilaton potential, $L_\mu = U^\dagger\partial_\mu U$, with U the chiral field taking values in $SU(3)$. We shall ignore the pion mass for simplicity, so the explicit chiral symmetry-breaking mass term is given by the mass matrix $\mathcal{M} = \mathrm{diag}\left(0, 0, 2m_K^2\right)$. In flavor $SU(3)$, the anomaly term, i.e., the Wess–Zumino term, $\mathcal{S}_{WZ} = -\frac{iN_C}{240\pi^2}\int d^5x\varepsilon^{\mu\nu\lambda\rho\sigma}\mathrm{Tr}\left(L_\mu L_\nu L_\lambda L_\rho L_\sigma\right)$, turns out to play a crucial role in our approach:

10.4.2 Fluctuating kaons in the skyrmion matter

We consider the fluctuation of kaons in the background of the skyrmion matter u_0

$$U(\vec{x},t) = \sqrt{U_K(\vec{x},t)}U_0(\vec{x})\sqrt{U_K(\vec{x},t)}, \tag{10.8}$$

$$U_K(\vec{x},t) = e^{\frac{i}{\sqrt{2}f_\pi}\begin{pmatrix} 0 & K \\ K^\dagger & 0 \end{pmatrix}}, \quad U_0(\vec{x}) = \begin{pmatrix} u_0(\vec{x}) & 0 \\ 0 & 1 \end{pmatrix}. \tag{10.9}$$

Substituting (10.8) into (10.7) and the Wess–Zumino term, we get the kaon Lagrangian in the background matter field $u_0(x)$

$$\begin{aligned}
\mathcal{L}_K = & \left(\frac{\chi_0}{f_\sigma}\right)^2 \dot{K}^\dagger G \dot{K} - \left(\frac{\chi_0}{f_\sigma}\right)^2 \partial_i K^\dagger G \partial_i K - \left(\frac{\chi_0}{f_\sigma}\right)^3 m_K^2 K^\dagger K \\
& + \frac{1}{4}\left(\frac{\chi_0}{f_\sigma}\right)^2 \left(\partial_\mu K^\dagger V^\mu(\vec{x}) K - K^\dagger V_\mu(\vec{x}) \partial^\mu K\right), \\
& + \frac{iN_c}{4f_\pi^2} B^0 \left(K^\dagger G \dot{K} - \dot{K}^\dagger G K\right)
\end{aligned} \tag{10.10}$$

where $\chi_0(\vec{x})$ is the classical dilaton field and

$$V_\mu(\vec{x}) = \frac{i}{2}[(\partial_\mu u_0^\dagger)u_0 - (\partial_\mu u_0)u_0^\dagger], \tag{10.11}$$

$$G(\vec{x}) = \frac{1}{4}(u_0 + u_0^\dagger + 2), \tag{10.12}$$

$$B^\mu(\vec{x}) = \frac{1}{24\pi^2}\varepsilon^{\mu\nu\lambda\sigma}\mathrm{Tr}\left(u_0^\dagger \partial_\nu u_0 u_0^\dagger \partial_\lambda u_0 u_0^\dagger \partial_\sigma u_0\right). \tag{10.13}$$

In the spirit of mean field approximation, we will take the space average on the background matter fields u_0 and χ_0 and obtain

$$\mathcal{L}_K = \alpha(\partial_\mu K^\dagger \partial^\mu K) + i\beta(K^\dagger \dot{K} - \dot{K}^\dagger K) - \gamma K^\dagger K \tag{10.14}$$

where

$$\alpha = \langle \kappa^2 G \rangle, \beta = \frac{N_c}{4f_\pi^2}\langle B^0 G \rangle, \gamma = \langle \kappa^3 G \rangle m_K^2 \tag{10.15}$$

with $\kappa = \chi_0/f_\chi$. Lagrangian (10.14) yields a dispersion relation for the kaon in the skyrmion matter as

$$\alpha(\omega_K^2 - p_K^2) + 2\beta\omega_K + \gamma = 0. \tag{10.16}$$

Solving this for ω_K and taking the limit of $p_K \to 0$, we have

$$m_K^* \equiv \lim_{p_K \to 0} \omega_K = \frac{-\beta + \sqrt{\beta^2 + \alpha\gamma}}{\alpha}. \tag{10.17}$$

This equation will be used for evaluating the in-medium effective kaon mass.

10.4.3 *Skyrmion crystal*

The key element in Eq. (10.14) for the kaonic fluctuation is the background u_0 which reflects the "vacuum" modified by the dense skyrmion matter. The classical dilaton field tracks the quark condensate affected by this skyrmion background u_0 that carries information on chiral symmetry of dense medium. Our approach here is to describe this background u_0 in terms of crystal configuration. The pertinent u_0 has been worked out in detail in [Lee *et al.* (2003)], from which we shall simply import the results for this work. As shown there, skyrmions put on an FCC — which is the favored crystal configuration — make a transition at a density $n_{1/2}$ to a matter consisting of half-skrymions. With the parameters of the Lagrangian picked for the model, we find $n_{1/2} \sim 1.3n_0$ but this value should not be taken seriously. It depends on the hidden gauge fields not taken into account. As noted before, the reasonable range of density is $n_{1/2} \sim (2-3)n_0$.

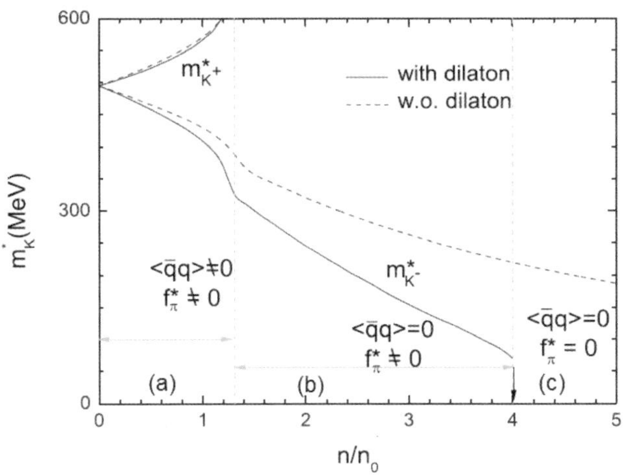

Fig. 10.3 $m^*_{K\pm}$ vs. n/n_0 (where $n_0 \simeq 0.16$ fm^{-3} is the normal nuclear matter density) in dense skyrmion matter which consists of three phases: (a) $\langle\bar{q}q\rangle \neq 0$ and $f^*_\pi \neq 0$, (b) $\langle\bar{q}q\rangle = 0$ and $f^*_\pi \neq 0$ and (c) $\langle\bar{q}q\rangle = 0$ and $f^*_\pi = 0$. The parameters are fixed at $\sqrt{2}ef_\pi = m_\rho = 780$ MeV and dilaton mass $m_\sigma = 720$ MeV.

10.4.4 *Effective kaon mass*

We now discuss the results given in Fig. 10.3 in some detail.

The results are given for the dilaton mass $m_\sigma \approx 720$ MeV. The dilaton mass is currently known neither experimentally nor theoretically. We take $m_\sigma \approx 720$ so as to give ~ 600 MeV in nuclear matter consistent with the Fermi-liquid description of nuclear matter implementing IDDs [Song, Brown, Min and Rho (1997)]. The density $n_{1/2}$ at which the half-skyrmion matter appears turns out to be more or less independent of the dilaton mass [Lee *et al.* (2003)]. In fact, the $n_{1/2}$ is entirely dictated by the parameters that give the background u_0 (for which the hidden local symmetric mesons cannot be ignored). For the skyrmion background, we pick $\sqrt{2}ef_\pi$ — which is the vacuum ρ mass at tree order– to be ~ 780 MeV which fixes e for the given $f_\pi \approx 93$ MeV. What is controlled by the dilaton mass are the density at which the dilaton condensate vanishes together with the pion decay constant, i.e., the dilaton limit fixed point (DLFP) and more significantly, the rate at which the kaon mass drops as density exceeds $n_{1/2}$.

What is novel in this model is that the behavior of the kaon is characterized by three different phases, not two as in conventional approaches. In the low density regime up to $\sim n_{1/2}$, the kaon interaction can be described by standard chiral symmetry treatments. For instance, the binding energy of the kaon at nuclear matter density comes out to be $\sim (80 - 100)$ MeV for $m_\sigma = 720$ MeV. Going above $n_{1/2}$, however, as the matter enters the half-skyrmion phase with vanishing quark condensate (on average) and non-zero pion decay constant, the kaon mass starts dropping more steeply. This form of matter — which is undoubtedly highly nonperturbative missing in SχPT — is most likely not captured by the Weinberg-Tomozawa term. Finally at n_c at which both the quark condensate and the pion decay constant vanish, the kaon mass vanishes. This happens in this model at density $\approx 4.0n_0$ for the dilaton mass $m_\sigma = 720$ MeV. The reason for this is that the dilaton condensate, $\langle\chi\rangle$, vanishes at the DLFP. This means that "kaon condensation" in symmetric nuclear matter seems to take place at the point at which the scale symmetry associated with the soft dilaton is restored, that is, the DLFP discussed in the preceding chapters. In the realistic treatment given in the previous chapter in conjunction with the EoS of compact stars, it was found that the vector manifestation (VM) fixed point and the DLFP are to arrive at $n >\sim 10n_0$. Therefore this anomalous behavior of the kaon mass at the skyrmion-1/2-skyrmion transition must take place at a much higher density than what's estimated in the simple model given above. In fact this observation will be supported in the renormalization-group (RG) treatment given below.

10.4.5 *Compact-star matter and kaon condensation*

Thus far, the kaon is treated at the semiclassical level as a quasiparticle bound to the skyrmion matter. To understand what this represents in nature, we should note that when one quantizes the system where a kaon is bound to a single skyrmion, the bound system gives rise to the hyperons, Λ and Σ, as shown by Callan and Klebanov [Callan and Klebanov (1985)]. This would suggest that kaons bound to a skyrmion matter, when collectively quantized, would correspond to hyperonic matter. On the other hand, one can interpret the system in Ginzburg–Landau mean-field theory focusing on the kaon fluctuation and identify the vanishing of the effective kaon mass in medium, ${m_K}^*$, as the signal for kaon condensation. Thus in this picture, kaon condensation and hyperon condensation are physically equivalent giving a support to the argument based on the Callan–Klebanov model given above.

Now a physically interesting question is what does the half-skyrmion matter do to compact stars? What we have done above is to subject the kaon to the skyrmion background which is given in the large N_c limit. In the large N_c limit, there is no distinction between symmetric matter and asymmetric matter. Compact stars have neutron excess which typically engenders repulsion at densities above n_0, and hence the asymmetry effect needs to be taken into account. This effect will arise when the system is collective quantized which gives rise to the leading $1/N_c$ correction to the energy of the bound system. Most of this correction could be translated into a correction to the effective mass of the kaon. We have no estimate of this correction but it is unlikely to be big. What is more important is the energy that arises from neutron excess, called symmetry energy in nuclear physics — discussed in a different context in Chapter 8 — of the neutron-rich system appropriate for compact stars,

$$E(n,x) = E(n,x=0) + \alpha^2 E_{sym}(n) + \cdots \tag{10.18}$$

with $\alpha = (P-N)/(N+P)$ with $N(P)$ standing for the neutron(proton) number. The ellipsis stands for higher order terms in $|\alpha| \leq 1$.

In compact-star matter, kaons will condense at a density below the critical density n_c because of the electron chemical potential which tends to increase as the matter density increases [Brown, Thorsson, Kubodera and

Rho (1992)]. Now the electron chemical potential is known to be entirely controlled by the symmetry energy in charge-neutral beta equilibrium systems. Thus collective quantization is required to address at what density kaons will condense. Quantizing the moduli space of multi-skyrmion systems (or multi-instanton systems in hQCD) has not yet been worked out fully. As mentioned above, the effect of hyperons on electron chemical potential will be automatically taken into account once the moduli quantization is made. To take realistically into account this "strange" effect of the half-skyrmion phase into the EoS, it would be necessary to "translate" this phenomenon, as was done with the cusp in the symmetry energy, into the parameters of *bs*HLS Lagrangian in V_{lowk}RG approach. It would be absolutely necessary to take into account the repulsive interactions that resolve the hyperon problem discussed above. Otherwise one would wind up with too low a critical density as in the naive application of SχPT mentioned above.

10.5 Kaons in Fermi liquid

The crucial role of the dilaton in the strangeness problem in dense compact stars, discussed above, should be unravelled in a $V_{lowk}^{SU(3)}$RG treatment. In the absence of such a treatment, we look at a simplified model in which what is considered essential in such a treatment is captured.

The model we shall investigate is formulated as follows. Noting that the V_{lowk}RG in two flavors employed in Chapter 9 is equivalent to doing Landau-Fermi liquid theory with *bs*HLS Lagrangian endowed with IDDs [Holt, Brown, Holt and Kuo (2007)], we extend it to three flavors in the baryon sector and take effective four-Fermi interactions built on the Fermi surface. The key requirement is to account for both nuclear interactions in the Fermi liquid and kaonic fluctuations on top in Wilsonian RG, with kaon properties warped by the baryonic background in dense medium as indicated by the result of Pandharipande *et al.* [Pandharipande, Pethick and Thorsson (1995)]. There are in principle many terms in the action. For illustrative purpose we will drastically simplify the action.

We assume that the half-skyrmion phase is in Fermi liquid as indicated in Chapter 8 and look at the kaon coupling to the Fermi liquid via exchange of a scalar that can be tuned to criticality (i.e., zero mass). We can think of the dilaton as that scalar and the criticality as the DLFP discussed in the volume.

We write the Euclidean action for kaon-Fermi-liquid nuclear systems in momentum space [Paeng and Rho (2015)]

$$Z = \int [d\phi][d\phi^*][d\psi][d\bar{\psi}]e^{-\tilde{S}^E} \tag{10.19}$$

with the actions

$$\tilde{S}^E = \tilde{S}^E_\psi + \tilde{S}^E_\phi + \tilde{S}^E_{\psi\phi} \tag{10.20}$$

$$\tilde{S}^E_\psi = \int \frac{d\epsilon d^3\vec{k}}{(2\pi)^4}\,\bar{\psi}_\sigma \left\{-i\epsilon + (e(\vec{k}) - e_F)\right\}\psi_\sigma - \int \left(\frac{d\epsilon d^3\vec{k}}{(2\pi)^4}\right)^4 \lambda\bar{\psi}_\sigma\bar{\psi}_{\sigma'}\psi_\sigma\psi_{\sigma'}\delta^4(\epsilon, \vec{k}) \tag{10.21}$$

$$\tilde{S}^E_\phi = \int \frac{d\omega d^3\vec{q}}{(2\pi)^4}\{\phi^*(\omega^2 + q^2)\phi + m^2_\phi\phi^*\phi + \cdots\} \tag{10.22}$$

$$\tilde{S}^E_{\psi\phi} = -\int \left(\frac{d\epsilon d^3\vec{k}}{(2\pi)^4}\right)^2 \left(\frac{d\omega d^3\vec{q}}{(2\pi)^4}\right)^2 h\,\phi^*\phi\bar{\psi}_\sigma\psi_\sigma\delta^4(\omega,\,\epsilon,\,\vec{q},\,\vec{k}) \tag{10.23}$$

where ψ and $\bar{\psi}$ are the fermion coherent states. Although our notations for the fields are general as they can apply to other systems like pions/nucleons and electrons/phonons, we should keep in mind that we are specializing to the K^- field for the boson ϕ and the proton and neutron for the fermion ψ. We note that the action (10.20) is essentially the same model studied in [Lee, Rho and Sin (1995)] except for the bosonic action. The bosonic action used there is different from (10.22) in that the system considered there was assumed to be a compact-star matter in weak and chemical equilibrium (with the boson approximated to satisfy a quadratic dispersion relation is most likely incorrect for kaons) whereas here we are dealing with relativistic bosons which could be tuned to criticality by attractive KN interactions. It is interesting that a same-type action figures in condensed matter physics. In fact, (10.20) is quite similar to the action studied for quantum critical metals in [Fitzpatrick, Kachru, Kaplan and Raghu (2013)] from which we shall borrow various scaling properties for the renormalization group.

We have taken here a local KN interaction which in terms of our *bs*HLS Lagrangian can be interpreted as arising from the exchanges of the dilaton and ω meson with the mesons integrated out. In the $V^{SU(3)}_{lowk}$ RG formulation they could be treated in terms of KN potentials coming from the exchanges of those mesons. Similarly we have taken only one generic four-Fermi interaction. More realistically it would have various channels, including the channels corresponding to the dilaton exchange and the ω exchange. The intricate interplay between the various channels, in particular the scalar and vector channels, must play a crucial role in giving the mechanisms the resolution of the hyperon problem and kaon condensation. One notable observation is that in the nucleon-nucleon interaction, the attractive scalar exchange and repulsive vector exchange compete whereas in the KN interaction (where N is a neutron) both are attractive and add. We will not go into these intricacies here but just give a generic feature of what a consistent treatment might mean in a simple situation. It is clear that this simple treatment cannot possibly be truly realistic.

It is instructive to compare the constant h in (10.23) with what is given in the standard chiral Lagrangian. In terms of the chiral counting, the leading term ($O(p)$) is "Weinberg–Tomozawa-type" term, coming from the exchange of the ω meson and gives, for the s-wave kaon,

$$h_{WT} \propto q_0/f_\pi^2 \tag{10.24}$$

where q_0 is the fourth component of the kaon 4-momentum and the next chiral order ($O(p^2)$) term is the "KN sigma-type" term

$$h_\Sigma \propto \Sigma_{KN}/f_\pi^2 \tag{10.25}$$

which can be thought of arising from a scalar meson exchange. In our approach, both terms can be considered as resulting from integrating out the "heavy" mesons from the bsHLS Lagrangian.

We should stress that the fermion ψ is a quasiparticle field in Fermi-liquid near its fixed point, so the coupling h encodes more than just the "bare" couplings in the bsHLS Lagrangian. This point was made in [Rho (1991)]. For convenience, we shall use these chiral Lagrangian terms for identifying the two terms contributing to h. It should be remembered that they are not the same.

As written, the action for the fermion system (10.21) is for a Fermi liquid with marginal four-Fermi interactions [Shankar (1994)]. So the energy-momentum of the quasiparticle (fermion) is measured with respect to the Fermi energy e_F and Fermi momentum k_F. The bosonic action is for massive Klein–Gordon field, which later will be associated with a pseudo-Goldstone field, with higher-field terms ignored. In contrast to that of the quasi-particle, the boson energy-momentum will be measured from zero. We would like to look at meson-field fluctuations on the background of a fermionic matter given by the Fermi-liquid theory. The Fermi-liquid structure is expected to be valid as long as $\tilde{N} \equiv k_F/\bar{\Lambda} \gg 1$ where $\bar{\Lambda} = \Lambda - k_F$ where Λ is the (momentum) cutoff scale from which the mode decimation — in the sense of Wilsonian — is made. For nucleon systems, the action (10.21) gives the nuclear matter stabilized at the (Landau Fermi-liquid) fixed point, with corrections suppressed by $1/\tilde{N}$ [Shankar (1994)]. The four-Fermi interactions and the effective mass of the quasiparticle are the fixed-point parameters. We expect that as long as $\tilde{N} \gg 1$, the Fermi-liquid action can be trusted even at higher densities than n_0. We should recall that the skyrmion-half-skyrmion "transition" which figures importantly is not a phase transition in the sense of Ginzburg–Landau–Wilson paradigm. This is because the half-skyrmions are confined [Nitta (2015); Zhang, Kimm, Zou and Cho (2017)] although their presence makes a dramatic impact on the EoS.

For simplicity, we shall assume a spherically symmetric Fermi surface in which case we can set for the quasiparticle momentum

$$\vec{k} = \vec{k}_F + \vec{l} \approx \hat{\Omega}(k_F + l). \tag{10.26}$$

Then for $\bar{\Lambda} \ll k_F$, we have

$$e(k) - e_F \approx \vec{v}_F \cdot \vec{l} + O(l^2) \tag{10.27}$$

where v_F is the Fermi velocity $v_F = k_F/m_L^*$ with m_L^* the effective quasiparticle (Landau) mass.

10.5.1 *Scaling*

To do the Wilsonian decimation, we need to set the cutoff scale Λ at which the classical action has bare parameters. Quantum effects are calculated by doing the mode elimination by lowering the cutoff from Λ to $s\Lambda$ with $s < 1$. We shall do this closely following the procedure given by [Fitzpatrick, Kachru, Kaplan and Raghu (2013)]. The important point to note is that the boson and fermion fields have different kinematics. While the boson momentum is measured from the origin — and hence the momentum cutoff is Λ, the fermion momentum is measured from the Fermi momentum k_F. Therefore the mode elimination for the fermion involves lowering the fermion momentum from $\bar{\Lambda}$ to $s\bar{\Lambda}$. As noted above, the strategy in the fermion sector is to take the large $\tilde{N}$ limit.[2]

We define the scaling laws of fields and other quantities like the delta functions by requiring that the kinetic energy terms of the fermion and the

[2]We remind the readers that the larger the k_F or density, the more $\bar{\Lambda}$ shrinks, which would mean that the large $\tilde{N}$ argument would hold better in the fermion sector as the density increases. This suggests that the mean field approximation with effective Lagrangians — chiral Lagrangian or hidden local symmetry Lagrangian — would get better the higher the density. The possibility was suggested in Chapter 7 for the "quasiparticles" in the half-skyrmion phase.

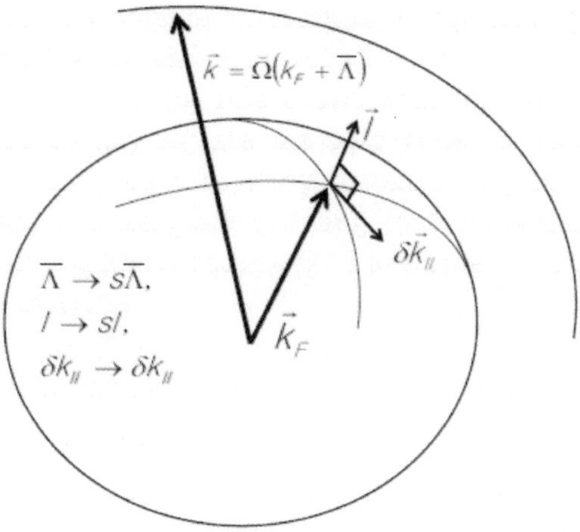

Fig. 10.4 Only the orthogonal component of the momentum to the fermi surface scales as $l \to s\,l$, where $\delta\vec{k}_{\|}$ is the parallel component of the momentum to the fermi surface.

boson be invariant under the scaling

$$\epsilon \to s\epsilon,\ \omega \to s\omega,\ l \to sl,\ \vec{q} \to s\vec{q}. \tag{10.28}$$

As is seen from Fig. 10.4, only the fermion momentum orthogonal to the Fermi surface scales in the way the fermion energy and all components of the boson momentum do. Since the quasiparticle mass m^* is a fixed-point quantity, the Fermi velocity does not scale. Therefore we have the fields scaling as

$$\begin{aligned} \phi &\to s^{-3}\phi, \\ \psi &\to s^{-3/2}\psi, \end{aligned} \tag{10.29}$$

for which we denote the scaling dimensions as $[\phi] = -3$ and $[\psi] = -3/2$[3] and the meson mass term scales

$$\int d\omega d^3q\, m_\phi^2 \phi^*\phi \to \int d\omega d^3q\, m_\phi^2 s^{-2}\phi^*\phi = \int d\omega d^3q\, m_\phi'^2\phi^*\phi, \tag{10.30}$$

with $m'_\phi \equiv s^{-1}m_\phi$. This shows the well-known fact that the meson mass term is "relevant." Using the procedure of scaling toward the Fermi surface, we have

$$[d\epsilon d^3k] = 2,\ [\delta(\epsilon, k)] = -2\,. \tag{10.31}$$

This confirms that the four-Fermi interaction term in (10.21) is marginal.[4]

[3]This notation will be used in what follows.

[4]For two or three spatial dimensions, this applies to only forward scattering and BCS-type scattering. Others are trivially marginal.

The scaling of the coupling term (10.23) is more subtle. Using the scaling dimensions

$$[\phi] = -3,\ [\psi] = -3/2,\ [d\epsilon d^3\vec{k}] = 2,\ [d\omega d^3\vec{q}] = 4, \tag{10.32}$$

we find the scaling of the integrand $I_{\psi\phi}$ of the action (10.23) written as $\int I_{\psi\phi}$ is

$$[I_{\psi\phi}] = 3 + [h] + [\delta(\omega, \epsilon, \vec{q}, \vec{k})], \tag{10.33}$$

where the bare coupling constant h will have the scaling dimension $[h_{WT}] = [q_0] = 1$ and $[h_\Sigma] = 0$. The $\psi\phi$ coupling will be "relevant" if $[I_{\psi\phi}] < 0$, marginal if $= 0$ and "irrelevant" if > 0. It is thus the scaling of the delta function that determines the scaling of the coupling h.

It will follow from arguments based on *bs*HLS developed so far that the δ function scales as

$$[\delta(\omega, \epsilon, \vec{q}, \vec{k})] = -4. \tag{10.34}$$

This then leads to

$$[I_{\psi\phi}]_{h_{WT}} = 0,\ [I_{\psi\phi}]_{h_\Sigma} = -1. \tag{10.35}$$

Thus the Weinberg–Tomozawa (WT)-type coupling is marginal,[5] while the sigma-term-type coupling is relevant. The former does not lead to anything interesting in the RG analysis (as shown in [Paeng and Rho (2015)]), so we focus on the latter.

Briefly the scaling (10.34) is arrived at as follows. (For details, we refer to [Paeng and Rho (2015)].)

First note that the KN coupling giving rise to (10.23) differs from that leading to the four-Fermi interaction (10.21) by that one of the two $\psi\psi\phi$ vertices in the former is replaced by a $KK\phi$ vertex. Within the scaling rule we adopt, the $\psi\psi\phi$ vertex is marginal but the $KK\phi$ vertex is relevant. All other quantities are the same. Since in our approach, we want the four-Fermi interaction to remain marginal so that the baryonic matter remains in Fermi liquid, we need to enforce that the presence of the ϕ field leave unaffected the marginal four-Fermi interaction. This is achieved by assigning a definite scaling property to the ϕ propagator. By imposing this scaling condition on the ϕ propagator that figures in the KN-KN interaction, one sees immediately that the Σ-term type KN coupling in (10.23) must be relevant. This is simply because the $KK\phi$ vertex is relevant. One

[5]The ω exchange contributes dominantly to the WT term and since as seen in Chapter 8, the behavior of ω is not affected significantly by density. Thus it is not expected to significantly affect the RG properties of the coupling (10.23).

can verify this result by an explicit counting of the scalings involved in the Feynman diagrams [Paeng and Rho (2015)].

The same simple counting rule shows that the WT term should be marginal. For this, simply replace the $\psi\psi\phi$ vertex of the scalar-exchange term by a $\psi\psi\omega$ coupling and the $KK\phi$ vertex by a $q_0 KK\omega$ vertex, which amounts to replacing a marginal coupling by a marginal coupling and a relevant coupling by a marginal coupling. The net result is marginal because $q_0 KK\omega$ vertex is marginal.

10.5.2 *Renormalization group*

We are interested in the flows of the kaon mass and the kaon-nucleon coupling and the effect of kaon condensation on the Fermi liquid structure of the baryonic system. For this we need the scaling properties of the diagrams. At tree order, we have

$$h' = s^{-1} h\,, \tag{10.36}$$

i.e., relevant as found above. The kaon mass scaling is of course relevant,

$$m_\phi'^2 = s^{-2} m_\phi^2. \tag{10.37}$$

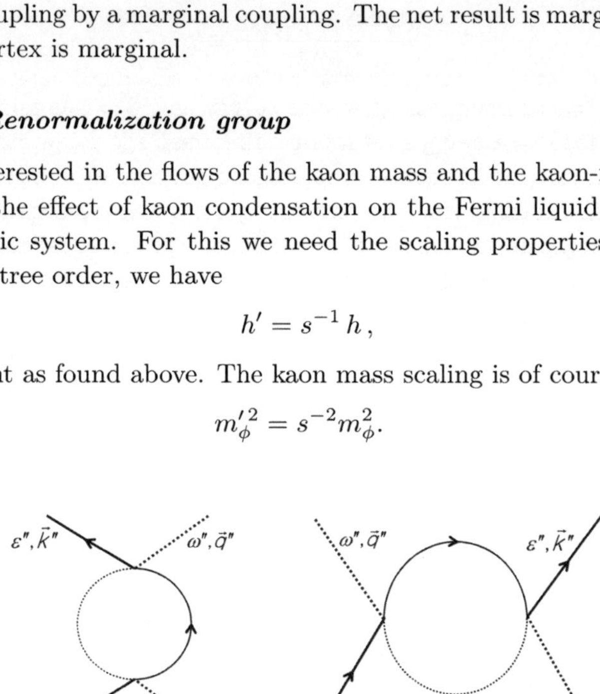

Fig. 10.5 One-loop graphs contributing to h. The process corresponds to f(ermion) $(\epsilon, \vec{k})$+ m(eson) $(\omega, \vec{q})$→ f(ermion) $(\epsilon'', \vec{k}'')$+ m(eson) $(\omega'', \vec{q}'')$.

At one-loop order, there are two diagrams, Figs. 10.5, contributing to h and one diagram, Fig. 10.6, to m_ϕ. The one-loop graph Fig. 10.6 contributing to the ϕ mass is easy to evaluate. By decimating from $\bar{\Lambda}$ to $s\bar{\Lambda}$ in the nucleon loop, we get

$$\int \frac{d\omega d^3\vec{q}}{(2\pi)^4} \phi^* \phi s^{-2} \left[-m_\phi^2 + \frac{\gamma h k_F^2}{2\pi^2} \int_{s\bar{\Lambda}<|l'|<\bar{\Lambda}} dl' \; |\mathrm{sgn}\,(l')| \right], \tag{10.38}$$

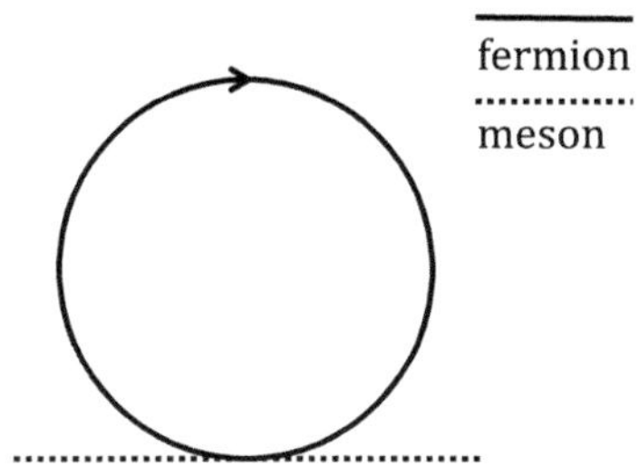

Fig. 10.6 One-loop graph contributing to m_ϕ.

where γ is the degeneracy factor for flavor (=2) and spin (=2): $\gamma = 4$ for nuclear matter and $\gamma = 2$ for neutron matter. It follows from above that

$$\delta\left[m_\phi^2\right] = s^{-2}\left[m_\phi^2 - \frac{\gamma h\bar{\Lambda}k_F^2}{\pi^2}(1-s)\right] - m_\phi^2\,. \tag{10.39}$$

As for the contribution to the $h\phi\phi\psi\psi$ coupling, it turns out that the two diagrams (a) and (b) of Fig. 10.5 are related to each other and they vanish [Paeng and Rho (2015)]. Two-loop corrections are further suppressed by $1/N$. Thus the (relevant) tree term dictates the RG flow.

10.5.3 *Flow analysis*

From what's discussed above, it is straightforward to write down the RG equations. Setting $t \equiv -\ln s$,

$$\frac{d\,m_\phi^2}{dt} = 2\,m_\phi^2 - A\,h\,, \tag{10.40}$$

$$\frac{d\,h}{dt} = h\,, \tag{10.41}$$

where

$$A = \frac{\gamma k_F^2\bar{\Lambda}}{\pi^2} = \frac{\gamma k_F^3}{N\pi^2}\,. \tag{10.42}$$

It is easy to get the analytic solutions for m_ϕ and h

$$m_\phi^2(t) = \left(m_\phi^2(0) - h(0)A\right)\mathrm{e}^{2t} + h(0)A\mathrm{e}^t\,, \tag{10.43}$$

$$h(t) = h(0)\mathrm{e}^t\,. \tag{10.44}$$

Given the analytic solution, we can work out how m_ϕ^2 and h flow as t increases (s decreases) for given values of A, $h(0)$ and $m_\phi^2(0)$.

From Eqs. (10.43) and (10.44), we have the relation,

$$\frac{m_\phi^2(t) - Ah(t)}{[h(t)]^2} = \frac{m_\phi^2(0) - A\,h(0)}{[h(0)]^2}, \tag{10.45}$$

which is satisfied for any value of t. It is convenient to introduce the new parameter c which is independent of t,

$$c \equiv \frac{m_\phi^2(0) - A\,h(0)}{[h(0)]^2}. \tag{10.46}$$

The quantities at $t = 0$, i.e., $m_\phi^2(0)$ and $h(0)$, are given at $s = 1$, that is, at the scale from which the decimation starts for a given k_F. These parameters in the bare action — and the parameter c — depend only on density. Since the flow depends on c, the RG properties of dense medium depend on the density dependence of the parameters of the bare chiral Lagrangian at the scale Λ. This is equivalent to the IDD that is obtained in relativistic mean field treatment of effective Lagrangians of the HLS type.

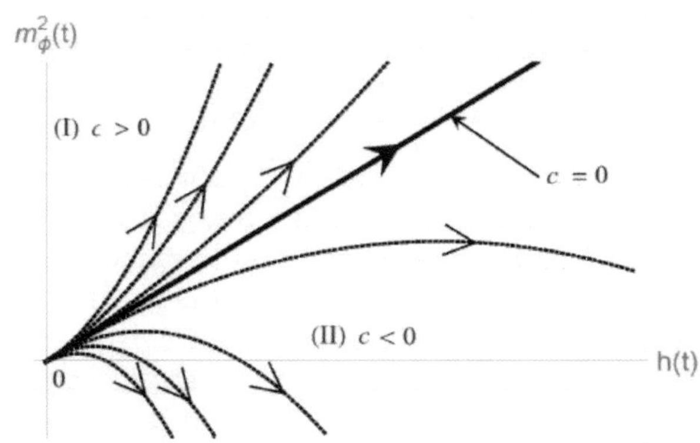

Fig. 10.7 RG flows of the parameters, $m_\phi^2(t)$ and $h(t)$, are shown in the parameter space. The lines are drawn with the fixed value of A and for different c's. The black solid line stands for the "critical line (or surface)" that delineates the parameter spaces.

Now using Eqs. (10.45) and (10.46), we readily obtain

$$m_\phi^2(t) = ch(t)^2 + A\,h(t). \tag{10.47}$$

The flows $m_\phi(t)$ vs. $h(t)$ are plotted in Fig. 10.7 for various values of c that depend on $\tilde{N} \equiv k_F/\bar{\Lambda}$. We note that $m_\phi^2(t)$ flows toward zero and becomes negative for $c < 0$. This signals the instability toward kaon condensation.

What happens after the condensation is a matter that goes beyond the model. It could be stabilized by terms higher order in the kaon field not taken into account in our analysis.

The conclusion is that only the $m_\phi(0)$ and $h(0)$ that satisfy

$$m_\phi^2(0) < Ah(0) \tag{10.48}$$

will be in the parameter space for the condensation to take place. The value $c = 0$ defines the critical line in the $h(t) - m_\phi^2(t)$ plane, with any point on which the parameters flow from the origin satisfying

$$m_\phi^2(t) = A\,h(t). \tag{10.49}$$

In the RG flows of $m_\phi(t)$ and $h(t)$, as t goes up, the attractive interaction gets stronger.

10.5.4 *Instability toward the condensation*

While one cannot determine the critical density at which kaons condense from the RG flow analysis, it is feasible to indicate whether and how the condensation can take place. This can be done by looking at the "parameter window" that signals the instability toward the condensation. In principle strangeness could be incorporated in the EoS in the formalism in which compressed matter is treated in the previous chapter and the properties of kaon condensation, i.e., the critical density and interconnection with hyperons, could be mapped out. In practice, it is not feasible to do reliable calculations due to too large a number of parameters to be controlled.

At some density, according to the values of c, the RG flows of m_ϕ^2 and $h(t)$ will follow certain of the flow lines in Fig. 10.7. If the value of c is changed by changing k_F to k'_F, the RG flow of m_ϕ^2 and $h(t)$ follows from the line (in k_F) to the line (in k'_F). By locating the sign change of

$$m_\phi^2(0) - A\,h(0)\,, \tag{10.50}$$

one can then locate the density at which the phase is moved to the region that signals the instability toward the kaon condensation. It gives the lower bound of the critical density. Let's use Eq. (10.48) to make this estimate. The sign change of c takes place when

$$n \equiv \gamma\frac{k_F^3}{6\pi^2} = \frac{\tilde{N}}{6}\frac{m_\phi^2(0)}{h(0)}. \tag{10.51}$$

What this says is that since the better the quasiparticle picture is in the baryon sector which would be the case the larger $\tilde{N}$ is, the higher the critical *density* n_K. This indicates that if the half-skyrmion phase, as argued in Chapters 8 and 9, is in Fermi liquid at its fixed point and if kaons do condense from the half-skyrmion phase which sets in at $n \gtrsim n_{1/2} \sim 2n_0$, then this implies n_K will naturally be very high given $1/\tilde{N} \to 0$. In compact stars in beta equilibrium, the electron chemical potential μ_e that increases with density, would help reduce n_K but the qualitative result that we obtain is not expected to be modified seriously.

Just to have an idea of what sort of density range that is involved, let us take $h(0) = \Sigma_{KN}/f^2$ from chiral Lagrangian. This is just to get a rough idea since the parameter h should in our formulation represent coupling of a kaon on top of the Fermi sea to a quasiparticle warped by the background, both IDD and $\text{DD}_{\text{induced}}$, which of course could be quite different from Σ_{KN}/f^2. From (10.51), replacing ϕ by K and f by f_π, we find the condensation density n_K bound by

$$n_K > \frac{\tilde{N}}{6} \frac{m_K^2 f_\pi^2}{\Sigma_{KN}}. \tag{10.52}$$

Let us take $m_K \approx 500$ MeV and $\Sigma_{KN} \approx 250\,\text{MeV}$ using a recent lattice measurement for Σ_{KN} [Alexandrou (2014)]. This gives

$$n_K > \tilde{N} m_\pi^3/3 \approx \tilde{N} n_0. \tag{10.53}$$

In order to have an idea what $\tilde{N}$ is, let's take the cutoff used in the V_{lowk}RG approach applied in Chapters 8 and 9. The optimal cutoff turns out to be $\Lambda \sim (2-3)$ fm^{-1} [Bogner, Kuo and Schwenk (2003)]. If one takes $\Lambda \sim 2$ fm^{-1}, then in the half-skyrmion phase, $n \gtrsim 2n_0$, $\tilde{N}$ will exceed $\sim 5n_0$, moving rapidly towards the Fermi-liquid fixed point. In the density range $(6-7)n_0$ relevant to the interior of massive compact stars, we will have $\tilde{N} \to \infty$. This implies that the condensation density n_K will be pushed way beyond the maximum interior density of massive stars.

10.6 Remarks

Combining the skyrmion description of dense matter with possible smooth hadron matter-quark matter transition traded in for topology change at a density $n \sim (2-3)n_0$ according to the Cheshire Cat Principle and an Wilsonian RG analysis of coupled baryon-meson systems in dense medium, we suggested that (1) hyperons and condensed kaons must appear simultaneously in the large N_c limit and (2) kaon condensation is banished to very high density beyond the density relevant to the interior of massive compact stars. The conclusion is that strangeness is unlikely to figure *explicitly* and cause the strangeness (hyperons and condensed kaons) problem in compact stars. But in arriving at this result, a crucial assumption is made on the state of the half-skyrmion matter from which kaons are considered to condense is a Fermi liquid near the Fermi-liquid fixed point. However the four-Fermi interaction corresponding to a scalar exchange in (10.21), though a fairly good approximation near normal nuclear matter density [Gelmini and Ritzi (1995)], cannot be valid when the dilaton limit with the vanishing dilaton mass is approached. In other words, the Fermi-liquid structure must break down as in condensed matter physics [Fitzpatrick, Kachru, Kaplan and Raghu (2013)]. To avoid this problem, one has to do a full-scale $V_{lowk}^{SU(3)}$RG calculation. What this implies is that the solution of the strangeness problem requires a basically different approach going beyond what has been done so far in the literature.

Epilogue

The principal objective of this volume is to formulate one and *only* one effective field theory (EFT) of quantum chromodynamics (QCD) for nuclear physics invoking *only* hadronic degrees of freedom that is to encompass from finite nuclei and (infinite) nuclear matter at densities in the vicinity of $n_0 \simeq 0.16\text{fm}^{-3}$ to compact-star matter at densities $\sim (6-7)n_0$. The strategy is to adhere as closely as feasible to the Folk Theorem (FT) on effective quantum field theory for QCD (Chapter 2). For this the QCD degrees of freedom, quarks and gluons, are traded in for hadron degrees of freedom with the help of topology in the spirit of hadron-quark continuity. This trading is formulated à la Cheshire Cat Principle (Chapter 3) involving pions, the (pseudo) Nambu–Goldstone bosons, and heavier mesons associated with hidden symmetries in QCD, i.e., scale symmetry for scalar meson(s) and hidden local symmetry for vector mesons (Chapter 5).

The FT is proven convincingly in nuclear physics in terms of chiral perturbation theory involving pions only, referred to as SχPT, in high-precision calculations — with uncertainty $\lesssim 2\%$ — in light nuclei (Chapter 2). It is also beginning to fare well in *ab initio* calculations in medium and heavy nuclei. This success is usually heralded as the beginning of a "fundamental" approach to nuclear physics. However going to denser systems as relevant to compact-star phenomena has defied long-standing efforts in the nuclear and astrophysics communities and remains as a challenge for an EFT à la FT.

This volume addresses this problem by means of three crucial and distinctively novel ways.

First, degrees of freedom more massive than the pion are incorporated within the spirit of the FT: The vector mesons of mass $\gtrsim 700$ MeV, $V = (\rho, \omega)$, and a dilaton scalar, σ, of mass ~ 700 MeV (Chapter 6). These massive degrees of freedom emerge from hidden local symmetry for the vector mesons and from hidden scale symmetry for the scalar meson. The former is supplemented by an infinite tower implied in holographic QCD with the gauge-gravity duality.

Second, the "bare" parameters of the effective Lagrangian, termed bsHLS, incorporating baryons ("b"), scalar meson ("s") and vector mesons ("HLS"), defined with a cutoff at $\Lambda_M \lesssim \Lambda_\chi \sim 4\pi f_\pi$, are matched to QCD in medium, thereby inheriting the properties of the vacuum change caused by the baryonic matter. The resulting parameters of the Lagrangian are then endowed with the "intrinsic density dependence (IDD)" of the QCD condensates (Chapter 8). In doing quantum theory calculations with bsHLS, performed along the line of Wilsonian renormalization group (RG) technique (Chapter 4), the cutoff taken in practice, say, $\overline{\Lambda}$, is at a scale lower than Λ_M, so certain "induced density dependence" ($\mathrm{DD}_{\mathrm{induced}}$ for short) reflecting spontaneous breaking of Lorentz symmetry in medium, needs to be added to the IDD, which is Lorentz-invariant. The sum of the IDD and $\mathrm{DD}_{\mathrm{induced}}$ gives what has been referred to in the literature as "Brown-Rho scaling."

Perhaps the most important and novel element in the approach is the third that has to do with the role of topology in nuclear interactions. While the Cheshire Cat Principle applies exactly to the baryon charge and fairly well, though not exactly, to long-wavelength processes, such as the flavor singlet axial charge g_A^0 (Chapter 3), the simplest Cheshire Cat with the pion field only does not work well for non-negligible momentum transfer processes. A notable example is the nucleon electromagnetic form factor for the nucleon. What is found to work well is the chiral bag with the magic chiral angle, with the baryon charge shared equally between quarks and topological soliton (Chapter 3). What is happening is that at increasing momentum transfer, the Cheshire Cat mechanism with the pion only starts to break down as momentum transfer increases. This breakdown is avoided, up to of course certain non-asymptotic momentum transfer, if the infinite tower of vector mesons is included in the meson sector as has been shown in holographic QCD. Quite analogously, as density increases way beyond the normal nuclear matter density the description of dense matter in terms of hadrons only must also start breaking down with quark degrees of freedom inevitably intervening at some point. The surprising finding in this volume is that this inevitable changeover of the degrees of freedom can be simulated for densities up to ~ 10 times the normal nuclear matter density by the changeover, when dense matter is treated as skyrmion matter, from skyrmions to half-skyrmions at a density $n_{1/2} \gtrsim 2n_0$ (Chapter 7). This is a topology change, not visible in standard nuclear EFT. There is an uncanny resemblance of this transition to the description of the hadrons-to-strongly interacting quark matter changeover, e.g., quarkyonic matter, invoked phenomenologically to account for the observed $\sim$ 2-solar mass stars. Now the striking feature of the half-skyrmion matter is that the bilinear quark condensate $\langle\bar{q}q\rangle$, considered as the order parameter of chiral symmetry, vanishes on the average when solitons are put on crystal lattice. The consequence is that there emerges parity-doubling with a chiral singlet piece $m_0 \neq 0$ remaining in the nucleon mass. It is nearly density-independent and is substantial, $m_0/m_N \sim (0.6 - 0.9)$, indicating a subtle origin of the nucleon mass. This topology change does not involve phase transition although the quark condensate averages to zero since the pion is still excited in the half-skyrmion "phase." It is our conjecture (Conjecture 1) that these two scenarios can be connected by the Cheshire Cat.

Incorporated into the EFT Lagrangian, the topology change has dramatic impacts on the properties of the "bare" parameters of *bs*HLS at $n \gtrsim n_{1/2}$. One of the most important is the "vector manifestation" (VM) of the ρ meson with the hidden gauge coupling g — and the ρ mass — going to zero as density increases, making hidden local symmetry "emerge" controlling the equation of state (EoS) of dense matter (Chapter 5). This VM is totally absent in EFTs without hidden local symmetry, so it is a distinctive signature of the role of a hidden local symmetry. This has a crucial effect on the nuclear tensor force and hence on the nuclear symmetry energy E_{sym} figuring in neutron-rich systems. The repulsive tensor force from the ρ exchange cancels a part of the attraction from the pion exchange, reducing the strength of the net tensor force which nearly vanishes at $n \sim 2n_0$. This feature, a robust prediction with *bs*HLS, is not present in low-order standard chiral perturbation theory. It offers an extremely simple explanation of the strongly suppressed Gamow–Teller matrix element in the C14 dating beta decay process (Chapter 8). What is significant in this result, as we will stress further below, is that the nearly vanishing Gamow–Teller matrix element at $n \sim n_0$ is brought about with the meson exchange currents to the Gamow–Teller operator completely ignored. Equally significant is that the presence of the VM fixed point — which is roughly estimated to be located at a high density, say, $n \gtrsim 10n_0$, where the explicit QCD degrees of freedom are expected to intervene — influences dramatically the behavior of the tensor force at the topology change density $n_{1/2}$, which is far from the VM fixed point. There the increasing ρ tensor gets drastically suppressed at $n_{1/2}$ so that the pion tensor strength, lost at increasing density, recovers its full strength above $n_{1/2}$. This produces a cusp in the symmetry energy at $n_{1/2}$ predicted in the skyrmion crystal model (Chapter 7). We take this as the strongest support for the crucial role topology plays in dense compact-star matter.

Another feature in our approach distinguishable from standard chiral effective theories is the role of the dilaton. Although a scalar of mass ~ 600 MeV as a local field has played important roles in nuclear physics, bringing in the dilaton as a pseudo-Nambu–Goldstone boson as we do in this volume is totally novel and naturally open to scrutiny. Given that whether an infrared (IR) fixed point associated with the dilaton is present or not in QCD for $N_f \sim 3$ is at present unestablished, the basic assumption we make is that scale symmetry, even if it is hidden or even absent in QCD in the vacuum, can emerge in highly correlated nuclear interactions. We are assuming that it makes sense to work near a putative infrared (IR) fixed point, although it may make no sense to be on the fixed point itself. This is similar in spirit to the VM fixed point at which hidden local symmetry, not present in QCD proper, is supposed to emerge. There is a good deal of indication that a scalar of mass ~ 600 MeV treated as a local field makes sense. In the way we implement the scalar degree of freedom as a dilaton, we have at our disposal a battery of low-energy theorems involving the scalar together with the pions as NG bosons. It has been incorporated into the hidden local symmetry setup in a scale-chiral symmetry scheme and a systematic scale-chiral symmetric expansion, extending the standard chiral EFT with generalized power counting with the departure from the IR fixed point as $O(p^2)$ in the power counting (Chapter 6). As it stands at present, the scale-invariant hidden local symmetric Lagrangian without (sHLS) and with (bsHLS) baryons is quite complicated with a large number of uncontrollable parameters. However if one works to the lowest order in scale symmetry (LOSS) with the symmetry breaking lumped into a dilaton potential, it can be drastically simplified. What results is somewhat like chiral Lagrangian where the symmetry breaking is put entirely in the quark mass term. In this volume, we limit our treatment to the LOSS order with going beyond the LOSS relegated to the future.

Now the dilaton figuring in sHLS or in bsHLS has an equally profound impact on the structure of nuclei and nuclear as well as dense matter as the hidden gauge fields do. But in a basically different way. Baryons generated as skyrmons in sHLS complement bsHLS where baryons figure as an explicit degree of freedom. The topological information gained in sHLS can be translated into the properties of the "bare" parameters of *bs*HLS. The first surprising observation is that both predict that the nucleon quasi-particle mass m_N^* goes proportional to the dilaton condensate $\langle\chi\rangle^*$ which approaches the chiral singlet mass m_0 at density $n \to n_{1/2}$. It is illuminating to see how the effective nucleon mass goes to a constant independent of density at $n \gtrsim n_{1/2}$. In the skyrmion description in *s*HLS with the solitons put on crystal lattice, as skyrmions fractionize into half-skyrmions with $\langle\bar{q}q\rangle_{\text{averaged}} \to 0$, the skyrmion mass goes to $\langle\chi\rangle \approx m_0$ (Chapter 7). This results from parity doubling in the nucleon structure. On the other hand, with the baryons explicitly incorporated, the mean-field approximation with *bs*HLS can be identified with the Fermi-liquid fixed point (Chapter 8). Thus one can write $m_L \sim \langle\chi\rangle^*\sqrt{g_A^L}$ where m_L is the Landau effective mass and g_A^L is the Landau g_A at the Fermi-liquid fixed point. One expects this in the "double large N limits": In the large N_c limit of QCD since $m_L \sim O(N_c)$, $\langle\chi\rangle \sim O(\sqrt{N_c})$ and $\sqrt{g_A} \sim O(\sqrt{N_c})$ *and* in the large $\tilde{N}$ limit where $\tilde{N} = k_F/\bar{\Lambda}$ with $\bar{\Lambda}$ being the cutoff relative to the Fermi surface. Now $g_A^L \to\sim 1$ more or less independently of density, the famous quenched g_A observed in nature. So the effective nucleon mass at $n \gtrsim n_{1/2}$ would go to $\langle\chi\rangle \sim m_0$.

It then follows some striking predictions with the *bs*HLS in both nuclear matter and dense compact star matter.

The first is the famous quenched g_A in nuclei. This has an important impact on double β decay processes. Closely connected to this is the hugely enhanced first forbidden beta decay process.

The second is the precocious conformal sound velocity in compact stars $v_s^2 = 1/3$.

The third is a possibly close link between the topology-change density $n_{1/2}$ and the tidal deformability measured in the gravitational waves coming from coalescing neutron stars GW170817.

Given the effective Lagrangian *bs*HLS with the "bare" parameters determined as described, we can resort to Wilsonian RG decimations to do the quantum theory with the Lagrangian. The strategy that takes into account high-order correlations is the well-known V_{lowk}RG that involves two decimations, i.e., "double decimation" (Chapters 4 and 8). What corresponds to Walecka-type mean-field approach is just a single decimation as the leading approximation to the double-decimation procedure. It is essentially a sort of energy density functional (EDF) analogous to the Kohn–Sham density functional in atomic and molecular physics, applied to strong interaction physics (Chapter 4). In this regard, what the IDD and DD_{induced} do is to implement what the EFT inherits from QCD by matching at the matching scale. The Hohenberg–Kohn functional obtained must therefore be closer to QCD in the sense of the Folk Theorem than the standard chiral EFT and the relativistic mean field theory currently popular in the literature. It could most certainly be formulated more elegantly and more powerfully than what's obtained in this volume. What we have achieved here should be considered as the first exploration.

Relying on the analysis of Matsui who arrives at Landau Fermi liquid structure with Walecka mean-field theory, we make the conjecture (Conjecture 2) that doing the mean-field calculation with the *bs*HLS encoded with the vacuum change in medium is equivalent to doing Landau Fermi-liquid fixed point theory. We assume that the dense system is at the Landau Fermi-liquid fixed point, hence we may ignore all $1/N$ suppressed terms where $N = k_F/\Lambda'$ with Λ' being the effective cutoff relative to the Fermi surface. Taking *bs*HLS in the leading-order scale symmetric (LOSS) limit, one can calculate the Gamow–Teller coupling constant in nuclear matter at its Landau Fermi-liquid fixed point. It comes out to be $g_A^L = g_A \times \left(1 - \frac{1}{3}\Phi\tilde{F}_1^\pi\right)^{-2}$ where $\Phi = f_\pi^*/f_\pi$ and $\tilde{F}_1^\pi$ is the in-medium pionic Fock contribution to the Landau parameter F_1 (Chapter 2). All the quantities involved here are density dependent. However it turns out that $g_A^L \approx 1$ nearly independent of density near n_0. This is a sort of universal result for Gamow–Teller transitions for all nuclear systems. Whether this result remains unaffected by higher-order correlations (i.e., $1/\tilde{N}$ corrections) could be checked by the double decimation V_{lowk}RG which has not yet been performed. As it stands, it offers the solution to the long-standing (four and half decades old) mystery of the quenched g_A in light and heavy nuclei and nuclear matter.

What's behind this surprisingly simple result can be understood from the beautiful result of the C14 dating, where the exchange currents to the Gamow–Teller operator are ignored completely. It suggests that (1) the exchange current corrections must be negligible *and* (2) the Gamow–Teller coupling g_A is invariant under the vacuum change, that is, unaffected by the change of the dilaton condensate caused by density. The first confirms the old "chiral filter" predicted in 1970's based on the double-soft-pion theorem (Chapter 2) and the second vindicates the LOSS approximation (Chapter 6) which predicts that the Gamow–Teller operator is invariant under scale transformation. Both are convincingly supported by recent "*ab initio*" quantum Monte Carlo calculations in light nuclei. What one learns from this is that the quenching of g_A from 1.27 in free space to 1 in medium nearly independent of density from light nuclei to nuclear matter, is the work of strong nuclear correlations dominated by the nuclear tensor force that is captured nearly entirely in the Landau Fermi-liquid structure at the fixed point. The factor of 4 enhancement of first-forbidden transitions, predicted by the theory and confirmed in experiments, gives another confirmation of the chiral filter mechanism which makes the unambiguous prediction based on the double-soft-pion theorem (Chapter 2).

A totally unexpected consequence of scale-chiral symmetry and topology in sHLS is found in compact stars. The V_{lowk}RG calculation with bsHLS globally gives the properties of stars in fair agreement with observations. The EoS so obtained agrees, nearly free of fine tuning, with nuclear matter properties at n_0 and accommodates well the recently found $\sim$ 2-solar mass stars. There is however a strikingly different prediction from all other calculations in the literature and that has to do with the sound velocity. It predicts that the sound velocity approaches and stays at the "conformal sound velocity" $v_s^2 = 1/3$ in the range of density $n \sim (3-7)n_0$.

To show how this phenomenon could come about, we recall the parity doubling due to which the dilaton condensate transmutes to the density-independent chiral scalar mass m_0. Next we look at the bsHLS in the mean field approximation interpreted as a Landau Fermi-liquid fixed point theory. In the LOSS approximation, the trace of the energy momentum tensor is given entirely in terms of the dilaton condensate $\langle\chi\rangle$. Therefore for $n \gtrsim n_{1/2}$, θ_μ^μ becomes independent of density, since the dilaton condensate goes to m_0. This is of course approximate since corrections to the LOSS approximation could depend on density. This then implies $\partial\theta_\mu^\mu/\partial n \approx 0$. It follows then that $v_s^2 \approx 1/3$ even though $\beta' \neq 0$ that signals scale anomaly (where β' is the anomalous dimension of the gluon energy-momentum tensor squared) and $\langle\chi\rangle \neq 0$. It has been verified that the same result is obtained in the double-decimation V_{lowk} approach — which includes some, if not all, of $1/\tilde{N}$ corrections. Then within the LOSS approximation, what we have here, taking place with $\langle\chi\rangle \neq 0$, is a robust result. In the limit $\langle\chi\rangle \to 0$ but with $\beta' \neq 0$, the scale symmetry is restored with the beta function going to zero. This is what is referred to as the dilaton limit fixed point. This can take place in medium, but most likely not in the vacuum, so it could be a phenomenon of "emergent symmetry." Whether this is a physically relevant observation or not is totally unclear at present. In QCD, one expects the sound velocity to reach the conformal from below at asymptotic density, but what's surprising here is that it sets in so precociously at relatively low density. If the conformal velocity were to set in at $\sim 3n_0$ and stay constant, then that would mean that standard nuclear EFT would breakdown. New physics must intervene to converge to density-independent conformal velocity. That new physics is provided here by the topology change which the Cheshire Cat encodes. This leads to the conjecture (Conjecture 3) that the skyrmion-half-skyrmion transition at $n_{1/2}$ could be equivalent to the transition from hadrons to strongly coupled quarks via Cheshire Cat mechanism. It would be interesting to formulate this in terms of gauge principle.

A potential breakthrough possibility for compact-star physics is offered by the recent observation of gravitational waves from the coalescing neutron stars GW170817. Among many fundamental issues associated with it, it could bring a valuable information on the EoS of dense compact-star matter, e.g., an information that could pin down in our description the topology change density $n_{1/2}$. Specifically the tidal deformability Λ seen in gravitational waves is known to be very sensitive to the nuclear symmetry energy E_{sym}. The prediction with the topology change taking place at $n = 2n_0$ in *bs*HLS is found to give a Λ slightly lower than what's presently quoted value, $\Lambda < 800$. If the tensor force were somewhat more attractive at $n \sim 2n_0$, then Λ could decrease. Here the cusp at $n_{1/2}$ can play precisely the role needed. If the topology change-density $n_{1/2}$ were increased from $2n_0$, then there would be rapid decrease of the tensor force at the density involved. A simple calculation with $n_{1/2}$ put at $2.6n_0$ shows indeed that Λ can be lowered considerably. This is consistent with the mechanism taking place already at $\sim n_0$ that accounts for the strongly suppressed C14 dating Gamow–Teller transition. The measurement of the precise value of Λ could then fix the density at which half-skrymions enter in *s*HLS, the most crucial information for the approach developed in this volume that no other observables could provide.

Only briefly touched on are two issues currently hotly debated in the literature, namely, the "requisite constraints" imposed on the EoS viable for high density and the strangeness problem in compact stars. The reason why little importance is given to those issues is that both do not play crucial roles in our approach. It is premature to claim that what we find is the correct or viable scenario. The future development, particularly in the experimental sector, will give the final verdict.

It is argued in the literature that the EoS for compact stars valid at large densities, $\sim (6-7)n_0$, must satisfy the "constraints" given at the normal nuclear matter density. The typical quantities involved are the symmetry energy and its derivatives "accurately" determined either empirically or theoretically at the normal nuclear matter density. Given an ultraviolet-complete fundamental theory, say, QCD, this should undoubtedly be feasible — and must be done. But it is doubtful that this axiom should hold in effective field theories. In the scenario developed in this volume, the topology change at $n_{1/2} \gtrsim 2n_0$ impacts changes in effective degrees of freedom within the same EFT framework, leading to the possibly "emergent" scale symmetry (and also flavor gauge symmetry) as seen in the "emergent" conformal sound velocity and similarly, in the hadron-quark continuity description. Even in EFT, approaches to finite nuclei, it is meaningless, for example, to impose the soft-pion theorems, the core of low-energy chiral EFT, as the constraints on the BPS nuclear model (Chapter 7) that gives an excellent description of certain aspects of heavy nuclei. The BPS structure applicable there could very well be "hidden" in the Lagrangian that has correct chiral symmetry that is manifested in different regimes. It may very well be that the BPS structure for the BPS nuclei is an emergent phenomenon in strong correlations involved.

The situation with the role of strangeness is of similar nature. Strangeness in dense matter, treated in EFT, is extremely subtle (Chapter 10), as is clear from the wide variety of phenomenological models with or without strange-quark degrees of freedom that can, with fiddled parameters, satisfactorily account for the observed massive stars. In the formulation of this volume, hidden symmetries — related via Cheshire Cat to QCD variables — are seen to intervene, with the net result to effectively banish both kaons and hyperons out of the range of density relevant to compact stars.

It is fair to say that a sufficiently reliable treatment of this subtle interplay of various degrees of freedom, which requires in the present framework a three-flavor V_{lowk}RG, is yet to obtain and that the verdict is not out yet.

Bibliography

Abbott, B.P. *et al.* [LIGO Scientific and Virgo Collaborations] (2017) GW170817: Observation of gravitational waves from a binary neutron star inspiral, *Phys. Rev. Lett.* **119**, *161101.*

Abrikosov, A.A., Gorkov, L.P. and Dzyaloshinsky, I.E. (1975) *Methods of Quantum Field Theory in Statistical Physics* (Dover, New York, 1975).

Adam, C., Sanchez-Guillen, J. and Wereszczynski, A. (2010) A Skyrme-type proposal for baryonic matter, *Phys. Lett. B* **691**, *105.*

Adam, C., Sanchez-Guillen, J. and Wereszczynski, A. (2010) A BPS Skyrme model and baryons at large N_c, *Phys. Rev. D* **82**, *085015.*

Adam, C. *et al.* (2013) Bogomolnyi-Prasad-Sommerfield Skyrme model and nuclear binding energies, *Phys. Rev. Lett.* **111** *no.23, 232501.*

Adam, C. *et al.* (2016) The Skyrme model in the BPS limit, *in The Multifaceted Skyrmion: 2nd Edition* (World Scientific, Singapore, 2016) ed. Rho, M and Zahed, I.

Adam, C., Sanchez-Guillen, J. and Wereszczynski, A. (2017) BPS submodels of the Skyrme model, *Phys. Lett. B* **769**, *362.*

Adami, C. and Brown, G.E. (1992) Order of the QCD transition and QCD sum rules, *Phys. Rev. D* **46**, *478.*

Adkins, G. S. Nappi, C. R. and Witten, E. (1983) Static properties of nucleons in the Skyrme model, *Nucl. Phys. B* **228**, *552.*

Adler, S. L., Collins, J. C. and Duncan, A. (1977) Energy-momentum-tensor trace anomaly in spin 1/2 quantum electrodynamics, *Phys. Rev. D* **15**, *1712.*

Adler, S.L. (1969) Axial vector vertex in spinor electrodynamics, *Phys. Rev.* **177**, *2426.*

Agbemava, S.E., Afanasjev, A.V., Ray, D. and Ring, P. (2014) Global performance of covariant energy density functionals: ground state observables of even-even nuclei and the estimate of theoretical uncertainties, *Phys. Rev. C* **89**, *054320.*

Akaishi, Y. and Yamazaki, T. (2016) High-density kaonic-proton matter (KPM) composed of $\Lambda^* \equiv K-p$ multiplets and its astrophysical connections, *Phys. Lett. B* **774**, *522.*

Akmal, A., Pandharipande, V.R. and Ravenhall, D.G. (1998) The equation of state of nucleon matter and neutron star structure, *Phys. Rev. C* **58**, *1804.*

Alexandrou, C. (2014) Nucleon structure from lattice QCD — recent achievements and perspectives, *EPJ Web Conf.* **73**, *01013.*

Alford, M.G., Rajagopal, K. and Wilczek, F. (1999) Color flavor locking and chiral symmetry breaking in high density QCD, *Nucl. Phys. B* **537**, *443.*

Antoniadis, J. *et al.* (2013), A massive pulsar in a compact relativistic binary, *Science* **340**, *6131.*

Aoki, S., *et al.* (2017) Review of lattice results concerning low-energy particle physics, *Eur. Phys. J. C* **77**, *no. 2, 112.*

Argaman, N. and Makov, G. (2000) Density functional theory — an introduction, *Amer. Phys.* **68**, *69.*

Atiyah, M.F., Hitchin, N.J., Drinfeld, V.G. and Manin, Yu.I. (1978) Construction of instantons, *Phys. Lett. A* **65**, *185.*

Atiyah, M.F. and Manton, N.S. (1989) Skyrmions from instantons, *Phys. Lett. B* **222**, *438.*

Atiyah, M. F. and Manton, N. S. (2016) Complex geometry of nuclei and atoms, *arXiv:1609.02816 [hep-th].*

Bahcall, J.N. (2000) Solar neutrino: What next?, *hep-ex/0002018.*

Baldo, M, Schuck, P. and Vinas, X. (2008) Kohn–Sham density functional approach to nuclear binding, *Phys. Lett. B* **663**, *390.*

Bando, M., Fujiwara, T and Yamawaki, K. (1988) Generalized hidden local symmetry and the A1 meson, *Prog. Theor. Phys.* **79**, *1140.*

Bando, M., Kugo, T., Uehara, S., Yamawaki, K. and Yanagida, T. (1985) Is ρ meson a dynamical gauge boson of hidden local symmetry?, *Phys. Rev. Lett.* **54**, *1215.*

Bando, M., Kugo, T. and Yamawaki, K. (1988) Nonlinear realization and hidden local symmetries, *Phys. Rept.* **164**, *217.*

Bardeen, W.A. and Hill, C.T. (1994) Chiral dynamics and heavy quark symmetry in a solvable toy field theoretic model, *Phys. Rev. D* **49**, *409.*

Bartolini, L., Bolognesi, S. and Proto, A. (2017) From the Sakai-Sugimoto model to the generalized Skyrme model, *arXiv:1711.03873 [hep-th],*

Bass, S.D. (2005) The spin structure of the proton, *Rev. Mod. Phys.* **77** *, 1257.*

Baym, G. and Chin, S. A. (1976) Landau theory of relativistic Fermi liquids, *Nucl. Phys. A* **262**, *527.*

Baym, G. and Chin, S.A. (1976) Can a neutron star be a giant MIT bag?, *Phys. Lett.* **62B**, *241.*

Baym, B., Pethick, C. and Sutherland, P. (1971) The ground state of matter at high densities: Equation of state and stellar models, *Astrophys. J.* **170**, *299.*

Battye, R.A., Manton, N.S. and Sutcliffe, P.M. (2017) Skrymions and nuclei, in [Rho and Zahed (2016)].

Beane, S.R. and van Kolck, U. (1994) The dilated chiral quark model, *Phys. Lett. B* **328**, *137.*

Bedaque, P. and Steiner, A.W. (2015) Sound velocity bound and neutron stars, *Phys. Rev. Lett.* **114**, *031103.*

Bedaque, P. and Steiner, A.W. (2015) Hypernuclei and the hyperon problem in neutron stars, *Phys. Rev. C* **92**, *no. 2, 025803.*

Bell, J.S. and Jackiw, R. (1969) A PCAC puzzle: $\pi^0 \to \gamma\gamma$ in the sigma model, *Nuovo Cim. A* **60**, *47.*

Bernard, V., Kaiser, N., Kambor, J. and Meissner, U.G. (1992) Chiral structure of the nucleon, *Nucl. Phys. B* **388**, *315.*

Bogner, S.K., Kuo, T.T.S. and Schwenk, A. (2003) Model independent low momentum nucleon interaction from phase shift equivalence, *Phys. Rept.* **386**, *1.*

Bogner, S.K., Kuo, T.T.S., Schwenk, A., Entem, D.R. and Machleidt, R. (2003) Towards a model independent low momentum nucleon nucleon interaction, *Phys. Lett. B* **576**, *265.*

Borasoy, B. (1999) Baryon axial currents, *Phys. Rev. D* **59**, *054021.*

Brodsky, S.J., de Teramond, G.F., and Deur, A. (2010) Nonperturbative QCD coupling and its β-function from light-front holography, *Phys. Rev. D* **81**, *096010.*

Brown, G.E. and Bethe, H. (1994) A scenario for a large number of low mass black holes in the galaxy, *Astrophys. J.* **423**, *659.*

Brown, G.E., Lee, C.H. and Rho, M. (2008) Recent developments on kaon condensation and its astrophysical implications, *Phys. Rept.* **462**, *1.*

Brown, G. E. and Machleidt, R. (1994) Strength of the ρ meson coupling to nucleons, *Phys. Rev. C* **50**, *1731.*

Brown, G.E. and Rho, M. (1988) The chiral bag, *Comments Nucl. Part. Phys.* **18**, *no. 1, 1.*

Brown, G.E. and Rho, M. (1991) Scaling effective Lagrangians in a dense medium, *Phys. Rev. Lett.* **66**, *2720.*

Brown, G.E. and Rho, M. (1996) From chiral mean field to Walecka mean field and kaon condensation, *Nucl. Phys. A* **596**, *503.*

Brown, G.E. and Rho, M. (2014) Double decimation and sliding vacua in the nuclear many body system, *Phys. Rept.* **396**, *1.*

Brown, G.E., Rho, M. and Weise, W. (1986) Phenomenological delineation of the quark–gluon structure from nucleon electromagnetic form-factors, *Nucl. Phys. A* **454**, *669.*

Brown, G.E., Thorsson, V., Kubodera, K. and Rho, M. (1992) A Novel mechanism for kaon condensation in neutron star matter, *Phys. Lett. B* **291**, *355.*

Buchheim, T., Kampfer, B. and Hilger, H. (2015) Algebraic vacuum limits of QCD condensates from in-medium projections of Lorentz tensors, *J. Phys. G* **43**, *055105.*

Buck, B. and Perez, S.M. (1983) New look at magnetic moments and beta decays of mirror nuclei, *Phys. Rev. Lett.* **50**, *1975.*

Callan Jr, C.G. and Klebanov, I.R. (1985) Bound state approach to strangeness in the Skyrme model, *Nucl. Phys. B* **262**, *365.*

Carbone, A., Polls, A., Providencia, C., Rios, A. and Vidana, I. (2014) Tensor force effects and high-momentum components in the nuclear symmetry energy, *Eur. Phys. J. A* **50**, *13.*

Carruthers, P. (1970) Broken scale invariance, *Phys. Rev. D* **2**, *2265.*

Casalbuoni, R. and Gatto, R. (1999) Effective theory for color flavor locking in high density QCD, *Phys. Lett. B* **464**, *111.*

Casher, A. (1979) Chiral symmetry breaking in quark confining theories, *Phys. Lett. B* **83**, *395.*

Catà, O., Crewther, R.J. and Tunstall, L.C. (2018) Crawling technicolor, *arXiv:1803.08513 [hep-ph].*

Caurier, E., Poves, A. and Zuker, A.P. (1995) Missing and quenched Gamow-Teller strength, *Phys. Rev. Lett.* **74**, *1517.*

Chemtob, M. and Rho, M. (1971) Meson exchange currents in nuclear weak and electromagnetic interactions, *Nucl. Phys. A* **163**, *1 Erratum: [Nucl. Phys. A* **212**, *628 (1973)].*

Chen, L.W. (2015) Symmetry energy systematics and its high density behavior, *EPJ Web Conf.* **88**, *00017.*

Chen, T., Frohlich, J. and Seifert M. (1995) Renormalization group methods: Landau-Fermi liquid and BCS superconductor, *cond-mat/9508063.*

Cohen, T.D., Furnstahl, R.J., Griegel, D.K. and Jin, X.M. (1999) QCD sum rules and applications to nuclear physics, *Prog. Part. Nucl. Phys.* **35**, *221.*

Collins, J. C., Duncan, A. and Joglekar, S. D. (1977) Trace and dilatation anomalies in gauge theories, *Phys. Rev. D* **16**, *438.*

Crewther, R.J. and L. C. Tunstall, L.C. (2015) $\Delta I = 1/2$ rule for kaon decays derived from QCD infrared fixed point, *Phys. Rev. D* **91**, *034016.*

Damgaard, P.H., Nielsen, H.B. and Sollacher, R. (1992) Smooth bosonization: The Cheshire cat revisited, *Nucl. Phys. B* **385**, *227.*

Damgaard, P.H., Nielsen, H.B. and Sollacher, R. (1994) Gauge symmetric approach to effective Lagrangians: The η' meson from QCD, *Nucl. Phys. B* **414**, *541.*

Danielewicz, P., Lacey, R. and Lynch, W.G. (2002) Determination of the equation of state of dense matter, *Science* **298**, *1592.*

Demorest, P. *et al.* (2010) Shapiro delay measurement of a two solar mass neutron star, *Nature* **467**, *1081.*

Detar, C. E., and Kunihiro, T. (1989) Linear σ model with parity doubling, *Phys. Rev. D* **39**, *2805..*

Diakonov, D. and Petrov, V.Y. (2000) Nucleons as chiral solitons, *In *Shifman, M. (ed.): At the frontier of particle physics, vol. 1* 359-41.*

Di Vecchia, P., Marotta, R. and Mojaza, M. (2017) Double-soft behavior of the dilaton of spontaneously broken conformal invariance, *JHEP 1709, 001.*

Doi, T. *et al.* (2017) Baryon interactions from lattice QCD with physical masses – Overview and $S = 0, -4$ sectors, *PoS LATTICE* **2016**, *110 [arXiv:1702.01600 [hep-lat]*

Dong, H., Kuo, T.T.S., Lee, H.K., Machleidt, R. and Rho, M. (2013) Half-skyrmions and the equation of state for compact-star matter, *Phys. Rev. C* **87**, *054332.*

Dong, H., Kuo, T.T.S., and Machleidt R. (2009) Neutron star and beta-stable ring-diagram equation of state with Brown–Rho scaling, *Phys. Rev. C* **80**, *065803.*

Dong, H., Kuo, T.T.S. and Machleidt, R. (2011) Low-momentum interactions with Brown–Rho–Ericson scalings and the density dependence of the nuclear symmetry energy, *Phys. Rev. C* **83**, *054002.*

Dong, H., Siu, L.W., Kuo, T.T.S. and Machleidt, R. (2010) Unitarity potentials and neutron matter at the unitary limit, *Phys. Rev. C* **81**, *034003.*

Drews, M. and Weise, W. (2017) Functional renormalization group studies of nuclear and neutron matter, *Prog. Part. Nucl. Phys.* **93**, *69.*

Drut, J. E. , Furnstahl, R.J. and Platter, L. (2010) Toward ab initio density functional theory for nuclei, *Prog. Part. Nucl. Phys.* **64**, *120.*

Dutra, M. *et al.* (2013) Relativistic mean-field models and nuclear matter constraints, *AIP Conf. Proc.* **1529**, *238 [arXiv:1303.2562 [nucl-th]].*

Ecker, G., Gasser, J., Leutwyler, H., Pich, A. and de Rafael, E. (1989) Chiral Lagrangians for massive spin 1 fields, *Phys. Lett. B* **223**, *425.*

Ecker, G., Gasser, J., Pich, A. and de Rafael, E. (1989) The role of resonances in chiral perturbation theory, *Nucl. Phys. B* **321**, *311.*

Ellis, J. (1970) Aspects of conformal symmetry and chirality, *Nucl. Phys. B* **22**, *478.*

Elliot-Ripley, M., Sutcliffe, P. and Zamaklar, M. (2016) Phases of kinky holographic nuclear matter, *JHEP* **1610**, *088.*

Engel, J. and Menèndez, J. J. (2017) Status and future of nuclear matrix elements for neutrinoless double-beta decay: A review, *Rept. Prog. Phys.* **80**, *no. 4, 046301.*

Fitzpatrick, A.L., Kachru, S. Kaplan, J. and Raghu, S. (2013) Non-Fermi liquid fixed point in a Wilsonian theory of quantum critical metals, *Phys. Rev. B* **88**, *125116.*

Fortin, M. *et al.* (2016) Neutron star radii and crusts: uncertainties and unified equations of state, *Phys. Rev. C* **94**, *no. 3, 035804.*

Freund, P. G. O and Nambu, Y. (1968) Scalar field coupled to the trace of the energy-momentum tensor, *Phys. Rev.* **174**, *1741.*

Friman, B. and Rho, M. (1996) From chiral Lagrangians to Landau Fermi liquid theory of nuclear matter, *Nucl. Phys. A* **606**, *303.*

Fubini, S. (1976) A new approach to conformal invariant field theories, *Nuovo Cim. A* **34**, *521.*

Fukano, H.S., Matsuzaki, S., Terashi, K. and Yamawaki, K. (2015) Conformal barrier and hidden local symmetry constraints: Walking technirhos in LHC diboson channels, *Nucl. Phys. B* **904**, *400.*

Fukushima, K. and Kojo, T. (2016) The quarkyonic star, *Astrophys. J.* **817**, *no. 2, 180.*

Furnstahl, R.J., Jin, X.M. and Leinweber, D.B. (1996) New QCD sum rules for nucleons in nuclear matter, *Phys. Lett. B* **387**, *253.*

Gasser, J. and Leutwyler, H. (1984) Chiral perturbation theory to one loop, *Annals Phys.* **158**, *142.*

Gasser, J. and Leutwyler, H. (1984) Chiral perturbation theory: Expansions in the mass of the strange quark, *Nucl. Phys. B* **250**, *465.*

Gelmini, G. and Ritzi, B. (1995) Chiral effective lagrangian description of bulk nuclear matter, *Phys. Lett. B* **357**, *431.*

Georgi, H. (1984) *Weak Interactions and Modern Particle Theory (Benjamin/Cummings, Menlo Park, 1984).*

Georgi, H. (1989) New realization of chiral symmetry, *Phys. Rev. Lett.* **63**, *1917.*

Georgi, H. (1990) Vector realization of chiral symmetry, *Nucl. Phys. B* **331**, *311.*

Glozman, L. Y. (2016) $SU(2N_F)$ symmetry of QCD at high temperature and its implications, *Acta Phys. Polon. Supp.* **10**, *3.*

Goldberger, W.D., Grinstein, B. and Skiba, W. (2008) Distinguishing the Higgs boson from the dilaton at the Large Hadron Collider, *Phys. Rev. Lett.* **100**, *111802.*

Goldhaber, A. S. and Manton, N. S. (1987) Maximal symmetry of the Skyrme crystal, *Phys. Lett. B* **198**, *231.*

Goldstone, J. and Jaffe, R.L. (1983) The baryon number in chiral bag models, *Phys. Rev. Lett.* **51**, *1518.*

Golterman, G. and Shamir Y. (2016) Low-energy effective action for pions and a dilatonic meson, *Phys. Rev. D* **94**, *no. 5, 054502.*

Golterman, G. and Shamir Y. (2018) The large-mass regime of the dilaton-pion low-energy effective theory, *arXiv:1805.00198 [hep-ph].*

Gorsky, A., Kopnin, P. N. and Krikun, A. (2014) Baryon as dyonic instanton-II. Baryon mass versus chiral condensate, *Phys. Rev. D* **89**, *no. 2, 026012.*

Gudnason, S.G. and Nitta, M. (2015) Fractional skyrmions and their molecules, *Phys. Rev. D* **91**, *no. 8, 085040.*

Harada, M., Lee, H. K., Ma, Y. L. and Rho, M. (2015) Inhomogeneous quark condensate in compressed skyrmion matter, *Phys. Rev. D* **91**, *no. 9, 096011.*

Harada, M., Kim, Y., Rho,M. and Sasaki, C. (2004) The pion velocity at chiral restoration and the vector manifestation, *Nucl. Phys. A* **730**, *379.*

Harada, M., Kugo, T. and Yamawaki, K. (1993) Proving the low-energy theorem of hidden local symmetry, *Phys. Rev. Lett.* **71**, *1299.*

Harada, M. and Ma, Y. L. (2013) Chiral partner structure of heavy baryons from the bound state approach with hidden local symmetry, *Phys. Rev. D* **87**, *056007.*

Harada, M., Matsuzaki, S. and Yamawaki, K (2010) Holographic QCD integrated back to hidden local symmetry, *Phys. Rev. D* **82**, *076010.*

Harada, H. and Sasaki, C. (2006) Dropping ρ and a_1 meson masses at chiral phase transition in the generalized hidden local symmetry, *Phys. Rev. D* **73**, *036001.*

Harada, M. and Yamawaki, K. (2001) Vector manifestation of the chiral symmetry, *Phys. Rev. Lett.* **86**, *757.*

Harada, M. and Yamawaki, K. (2003) Hidden local symmetry at loop: A New perspective of composite gauge boson and chiral phase transition, *Phys. Rept.* **381**, *1.*

Hashimoto, K., Sakai, T. and Sugimoto, S. (2008) Holographic baryons: static properties and form factors from gauge/string duality, *Prog. Theor. Phys.* **120**, *1093.*

Hata, H., Sakai, T., Sugimoto, S. and Yamato, S. (2007) Baryons from instantons in holographic QCD, *Prog. Theor. Phys.* **117**, *1157.*

Hell, T. and Weise, W. (2014) Dense baryonic matter: constraints from recent neutron star observations, *Phys. Rev. C* **90**, *045801.*

Hidaka, Y., Morimatsu, O. and Ohtani, M. (2006) Renormalization group equations in a model of generalized hidden local symmetry and restoration of chiral symmetry, *Phys. Rev. D* **73**, *036004.*

Hinderer, T. (2008) Tidal Love numbers of neutron stars, *Astrophys. J.* **677**, *1216.*

Hinderer, T., Lackey, B.D., Lang, R.N. and Read, S.J. (2010) Tidal deformability of neutron stars with realistic equations of state and their gravitational wave signatures in binary inspiral, *Phys. Rev. D* **81**, *123016.*

Holt, J.W., Brown, G.E. , Holt, J.D. and Kuo, T.T.S. (2007) Nuclear matter with Brown–Rho-scaled Fermi liquid interactions, *Nucl. Phys. A* **785**, *322.*

Holt, J.W., Brown, G.E., Kuo, T.T.S., Holt, J.D. and Machleidt, R. (2007) Shell model description of the C-14 dating beta decay with Brown–Rho-scaled NN interactions, *Phys. Rev. Lett.* **100**, *062501.*

Holt, J.W., Kaiser, N. and Weise, W. (2009) Chiral three-nucleon interaction and the C-14 dating beta decay, *Phys. Rev. C* **79**, *054331.*

Holt, J.W., Kaiser, N. and Weise, W, (2013) Nuclear chiral dynamics and thermodynamics, *Prog. Part. Nucl. Phys.* **73**, *35.*

Holt, J.W., Rho, M. and Weise, W. (2016) Chiral symmetry and effective field theories for hadronic, nuclear and stellar matter, *Phys. Rept.* **621**, *2.*

Holzwarth, G. and Schwesinger, B. (1986) Baryons in the Skyrme model, *Rept. Prog. Phys.* **49**, *825.*

Hong, D.K. (2000) An effective field theory of QCD at high density, *Phys. Lett. B* **473**, *118.*

Hong, D. K., Rho, M., Yee, H. U. and Yi, P. (2007) Chiral dynamics of baryons from string theory, *Phys. Rev. D* **76**, *061901.*

Hong, D. K., Rho, M., Yee, H. U. and Yi, P. (2007) Dynamics of baryons from string theory and vector dominance, *JHEP* **0709**, *063.*

Hong, D.K., Rho, M., Yee, H.Y. and Yi, P. (2008) Nucleon form-factors and hidden symmetry in holographic QCD, *Phys. Rev. D* **77**, *014030.*

Hong, D.K., Rho, M. and Zahed, I. (1999) Qualitons at high density, *Phys. Lett. B* **468**, *261.*

Horsley, R., Perlt, H., Rakow, P.E.L., Schierholz, G. and Schiller, A. (2014) The SU(3) beta function from numerical stochastic perturbation theory, *Phys. Lett. B* **728**, *1.*

Hotokezaka, K., Kyutoku, K., Sekiguchi, Y.I. and Shibata, M. (2016) Measurability of the tidal deformability by gravitational waves from coalescing binary neutron stars, *Phys. Rev. D* **93** *(2016) no. 6, 064082.*

Houghton, C. J., Manton, N. S. and Sutcliffe, P. M. (1998) Rational maps, monopoles and skyrmions, *Nucl. Phys. B* **510**, *507.*

Hyodo, T. and Weise, W. (2008) Effective anti-K N interaction based on chiral SU(3) dynamics, *Phys. Rev. C* **77**, *035204.*

Ichimura, M., Sakai, H. and Wakasa, T. (2006) Spin-isospin responses via (p,n) and (n,p) reactions, *Prog. Part. Nucl. Phys.* **56**, *446.*

Isham, C.J., Salam, A. and Strathdee, J.A. (1970) Broken chiral and conformal symmetry in an effective-lagrangian formalism, *Phys. Rev. D* **2**, *685.*

Jackson, A.D. and M. Rho, M. (1983) Baryons as chiral solitons, *Phys. Rev. Lett.* **51**, *751.*

Janssen, L. and He, Y.C. (2017) Critical behavior of the QED_3-Gross-Neveu model: duality and deconfined criticality, *Phys. Rev. B* **96**, *205113.*

Jenkins, E.E. and Manohar, A.V. (1990) Baryon chiral perturbation theory using a heavy fermion Lagrangian, *Phys. Lett. B* **255**, *558.*

Jido, D., Oka, M., and Hosaka, A. (2001) Chiral symmetry of baryons, *Prog. Theor. Phys.* **106**, *873.*

Jin, X.M. and Leinweber, D.B. (1995) Valid QCD sum rules for vector mesons in nuclear matter, *Phys. Rev. C* **52**, *3344.*

Johnson, K. (1975) The M.I.T. bag model, *Acta Phys. Polon.* **B6**, *865-919.*

Johnson, R.C. (1970) Simple derivation of the KSRF relation, *Nucl. Phys. B* **23**, *247.*

Kaiser, N., Fritsch, S. and Weise, W. (2002) Chiral dynamics and nuclear matter, *Nucl. Phys. A* **697**, *255.*

Kaplan, D.B. (1991) Qualitons, *Nucl. Phys. B* **351**, *137.*

Kaplan, D.B. and Nelson, A.E. (1988) Kaon condensation in dense matter, *Nucl. Phys. A* **479**, *273c.*

Kaplunovsky, V., Melnikov, D. and Sonnenschein, J. (2012) Baryonic popcorn, *JHEP* **1211**, *047.*

Kapusta, J. I. (1989) *Finite temperature field theory* (Cambridge University Press, Cambridge, 1989).

Kawarabayashi, K. and Suzuki, M. (1966) Partially conserved axial vector current and the decays of vector mesons, *Phys. Rev. Lett.* **16**, *255.*

Kerman, A.K. and Miller, L.D. (1974) Field theory methods for finite nuclear systems and the possibility of density isomerism, unpublished.

Kienle, P. and Yamazaki, T. (2004) Pions in nuclei, a probe of chiral symmetry restoration, *Prog. Part. Nucl. Phys.* **52**, *85.*

Kim, K., Lee, H.K. and Lee, J. (2016) Compact star matter: EoS with new scaling law, *arXiv:1607.03235 [nucl-th].*

Kim, K., Lee, H.K. and Rho, M. (2012) Triple layered compact star with strange quark matter, *Int. J. Mod. Phys. Conf. Ser.* **10**, *123.*

Klebanov, I.R. (1985) Nuclear matter in the Skyrme model, *Nucl. Phys. B* **262**, *133.*

Klebanov, I.R. (1989) Strangeness in the Skyrme model, *Cargese Lecture on Hadron and Hadronic Matter, Cargese, France.*

Kohn, W. and Sham, L.J. (1965) Self-consistent equations including exchange and correlation effects, *Phys. Rev.* **140**, *A1133.*

Kojo, T., Powell, P.D., Song, Y. and Baym, G. (2016) Phenomenological QCD equation of state for massive neutron stars, *Nucl. Phys. A* **956**, *821.*

Kolomeitsev, E.E., Lattimer, J.M., Ohnishi, A. and Tews, I. (2016) Unitary gas constraints on nuclear symmetry energy, *arXiv:1611.07133 [nucl-th].*

Komargodski, Z. (2011) Vector mesons and an interpretation of Seiberg duality, *JHEP* **1102**, *019.*

Krastev, P.G. and Li, B.A. (2018) Imprints of the nuclear symmetry energy on the tidal deformability of neutron stars, *arXiv:1801.04620 [nucl-th].*

Krause, A. (1990) Baryon matrix elements of the vector current in chiral perturbation theory, *Helv. Phys. Acta* **63**, *3.*

Kubodera, K., Delorme, J. and Rho, M. (1978) Axial currents in nuclei, *Phys. Rev. Lett.* **40**, *755.*

Kubodera, K. and Rho, M. (1991) Axial charge transitions in heavy nuclei and in-medium effective chiral Lagrangians, *Phys. Rev. Lett.* **67**, *3479.*

Kubodera, K. and Rho, M. (2011) Effective field theory and high-precision calculations of nuclear electroweak processes, in *From Nuclei to Stars: Festschrift in honor of Gerald E. Brown (World Scientific, Singapore) Ed. by S. Lee; arXiv:1011.4919 [nucl-th].*

Kugler, M. and Shtrikman, S. (1988) A new skyrmion crystal, *Phys. Lett. B* **208**, *491.*

Kugler, M. and Shtrikman, S. (1989) Skyrmion crystals and their symmetries, *Phys. Rev. D* **40**, *3421.*

Kunihiro, T. (1991) Quark number susceptibility and fluctuations in the vector channel at high temperatures, *Phys. Lett. B* **271**, *395.*

Kuo, T.T.S., Holt, J.W. and Osnes, E. (2016) Introduction to low-momentum effective interactions with Brown–Rho scaling and three-nucleon forces, *Phys. Scripta* **91**, *no. 3, 033009.*

Lattimer, J.M. and Lim, Y. (2013) Constraining the symmetry parameters of the nuclear interaction, *Astrophys. J.* **771**, *51.*

Lattimer, J.M. and Schutz, B.F. (2005) Constraining the equation of state with moment of inertia measurements, *Astrophys. J.* **629**, *979.*

Leader, E. (2016) The end of WHAT nucleon-spin crisis? *arXiv:1604.00305 [nucl-th].*

Lee, C.H., Brown, G.E., Min, D.P. and Rho, M. (1995) An effective chiral Lagrangian approach to kaon-nuclear interactions: Kaonic atom and kaon condensation, *Nucl. Phys. A* **585**, *401.*

Lee, H.J., Park, B.Y., Min, D.P., Rho, M. and Vento, V. (2003) A unified approach to high density: Pion fluctuations in skyrmion matter, *Nucl. Phys. A* **723** *(2003) 427.*

Lee, H.J., Park, B.Y., Rho, M. and Vento, V. (2004) The pion velocity in dense skyrmion matter, *Nucl. Phys. A* **741**, *161.*

Lee, H.K., Paeng, W.-G. and Rho, M. (2015) Scalar pseudo-Nambu-Goldstone boson in nuclei and dense nuclear matter, *Phys. Rev. D* **92**, *no. 12, 125033.*

Lee, H.K., Park, B.Y. and Rho, M. (2011) Half-skyrmions, tensor forces and symmetry energy in cold dense matter, *Phys. Rev. C* **83**, *025206 [Erratum-ibid. C* **84**, *059902 (2011)].*

Lee, H.K. and Rho, M. (2013) Hyperons and condensed kaons in compact stars, *arXiv:1301.0067 [nucl-th].*

Lee, H.K. and Rho, M (2014) Topology change and tensor forces for the EoS of dense baryonic matter, *Eur. Phys. J. A* **50**, *14.*

Lee, H.K. and Rho, M. (2014) Nuclear symmetry energy with strangeness in heavy-ion collisions, *Phys. Rev. C* **90**, *045201.*

Lee, H.K., Rho, M. and Sin, S.J. (1995) Renormalization group flow analysis of meson condensations in dense matter, *Phys. Lett. B* **348**, *290.*

Lee, T.D. and Wick, G.C. (1974) Vacuum stability and vacuum excitation in a spin 0 field theory, *Phys. Rev. D* **9**, *2291*: Abnormal nuclear states, in *Mesons in Nuclei* (North Holland, 1979) ed. Rho, M. and Wilkinson, D.H.

Lenaghan, J.T., Rischke, D.H. and Schaffner-Bielich, J. (2000) Chiral symmetry restoration at nonzero temperature in the $SU(3)_L \times SU(3)_R$ linear sigma model, *Phys. Rev. D* **62**, *085008*.

Li, B.A. and Chen, L.W. (2005) Nucleon-nucleon cross sections in neutron-rich matter and isospin transport in heavy-ion reactions at intermediate energies, *Phys. Rev. C* **72**, *064611*.

Li, X.Y, Lu, X.F., Wang, B., Sun, W.M. and Zong, H.S. (2009) Properties of cold dense nuclear matter based on a nonperturbative approach inspired by chiral perturbation theory, *Phys. Rev. C* **80**, *034909*.

Li, Y.L., Ma, Y.L. and Rho, M. (2017) Chiral-scale effective theory including a dilatonic meson, *Phys. Rev. D* **95**, *114011*.

Li, Y.L., Ma, Y.L. and Rho, M. (2018) The non-quenching of g_A in nuclei and emergent scale symmetry in dense baryonic matter, *arXiv:1804.00310 [nucl-th]*.

Li, Y.L., Wen, P.S., Ma, Y.L. and Rho, M. (2018) Scale-chiral effective field theory for nuclear interactions in the Veneziano limit, *arXiv:1802.08140 [nucl-th]*.

Lynn, B. W. (1993) Chiral $SU(2)_L \times SU(2)_R$ liquids: A theory of heavy nuclei and neutron stars, *Nucl. Phys. B* **402**, *281*.

Ma, Y.L., Harada, M., Lee, H.K., Oh, Y., Park, B.Y. and Rho, M. (2013) Dense baryonic matter in the hidden local symmetry approach: Half-skyrmions and nucleon mass, *Phys. Rev. D* **88**, *no. 1, 014016 [Erratum-ibid. D* **88**, *no. 7 (2013) 079904]*.

Ma, Y.L., Harada, M., Lee, H.K., Oh, Y., Park, B.Y. and Rho, M. (2014) Dense baryonic matter in conformally-compensated hidden local symmetry: Vector manifestation and chiral symmetry restoration, *Phys. Rev. D* **90**, *no. 3, 034015*.

Ma, Y.L., Lee, H.K., Paeng, W.G. and Rho, M. (2018) A pseudo-conformal equation of state in compact-star matter from topology change and hidden symmetries of QCD, *arXiv:1804.00305 [nucl-th]*.

Ma, Y. L., Oh, Y., Yang, G. S., Harada, M., Lee, H. K., Park, B. Y. and Rho, M. (2012) Hidden local symmetry and infinite tower of vector mesons for baryons, *Phys. Rev. D* **86**, *074025*.

Ma, Y.L. and Rho, M. (2017) Recent progress on dense nuclear matter in skyrmion approaches, *Sci. China Phys. Mech. Astron.* **60**, *no. 3, 032001*.

Ma, Y.L. and Rho, M. (2018) Scale-chiral symmetry, ω meson and dense baryonic matter, *Phys. Rev. D* **97**, *094017*.

Ma, Y. L., Yang, G. S., Oh, Y. and Harada, M. (2013) Skyrmions with vector mesons in the hidden local symmetry approach, *Phys. Rev. D* **87**, *no. 3, 034023*.

Ma, Y. L., Wang, Q. and Wu, Y. L. (2004) Hidden local symmetry and chiral effective theory for vector and axial-vector mesons, *Eur. Phys. J. C* **39**, *201*.

Machleidt, R. (1989) The meson theory of nuclear forces and nuclear structure, *Adv. Nucl. Phys.* **19**, *189*.

Machleidt, R. (1993) in *Computational Nuclear Physics 2–Nuclear Reactions* (Springer NY, 1993) ed. K. Langanke, J. A. Maruhn and S. E. Koonin.

Machleidt, R. and Sammarruca, F. (2016) Chiral EFT based nuclear forces: Achievements and challenges, *Phys. Scripta* **91**, *no. 8, 083007.*

Manohar, A. and Georgi, H. (1984) Chiral quarks and the nonrelativistic quark model, *Nucl. Phys. B* **234**, *189.*

Manton, N.S. (2017) Lightly bound skyrmions, tetrahedra and magic numbers, *arXiv:1707.04073 [hep-th].*

Mantovani Sarti, V. and Vento, V. (2014) The half-skyrmion phase in a chiral-quark model, *Phys. Lett. B* **728**, *323.*

Masuda, K. Hatsuda, T. and Takatsuka, T. (2016) Hadron-quark crossover and massive hybrid stars with strangeness, *Astrophys. J.* **764**, *12.*

Matsui, M (1981) Fermi liquid properties of nuclear matter in a relativistic mean - field theory, *Nucl. Phys. A* **370**, *365.*

Matsuzaki, S. and Yamawaki, K. (2014) Dilaton chiral perturbation theory: Determining the mass and decay constant of the technidilaton on the lattice, *Phys. Rev. Lett.* **113**, *082002.*

McLerran, L. (2010) Strongly interacting matter at high energy density, *Int. J. Mod. Phys. A* **25**, *5847.*

McLerran, L. and Pisarski, R.D. (2007) Phases of cold, dense quarks at large N_c, *Nucl. Phys. A* **796**, *83.*

Meissner, U.G. (2016) The long and winding road from chiral effective Lagrangians to nuclear structure, *Phys. Scripta* **91**, *no. 3, 033005.*

Meissner, U. G., Rakhimov, A. and Yakhshiev, U. T. (2000) The nucleon-nucleon interaction and properties of the nucleon in a $\pi\rho\omega$ soliton model including a dilaton field with anomalous dimension, *Phys. Lett. B* **473**, *200.*

Meng, J. (editor) 2016 *Relativistic Density Functional for Nuclear Structure* (World Scientific, Singapore, 2016).

Mermin, N.D. (1977) Surface singularities and superflow in ^{3}He-A, in *Quantum fluids and solids* ed. by Trickey, S.B., Adams, E.D. and Dufty, J.W. (Plenum, New York, 1977).

Mueller, E.J. (2018) Pseudogaps in strongly interacting Fermi gases, *arXiv: 1701.04838[cond-mat.quant-gas].*

Nadkarni, S., Nielsen, H.B. and Zahed, I. (1984) Bosonization relations for bag boundary conditions, *Nucl. Phys.* **B253**, *308.*

Nazarewicz, W., Reinhard, P.G., Satula, W. and Vretenar, D. (2011) Symmetry energy in nuclear density functional theory, *Eur. Phys. J. A* **50**, *20.*

Nielsen, H.B., Rho, M., Wirzba, A. and Zahed, I. (1991) Color anomaly in a hybrid bag model, *Phys. Lett.* **B269**, *389.*

Nielsen, H.B. and Wirzba, A. (1988) The Cheshire Cat Principle applied to hybrid bag models, *Proc. Les Houches Workshop (March-April 1987), Springer Proceedings in Physics, Vol. 26* (Springer, Berlin, 1988).

Nielsen, H.B. and Zahed, I. (2016) The Cheshire Cat principle from holography, in *The multifaceted skyrmion* (World Scientific, Singapore, 2016) ed. Rho, M and Zahed, I.

Nolte, R., Baumann, A., Rose, K.W. and Schumacher, M. (1986) Effect of exchange currents in E1 sum rule and orbital g -factor in ^{209}Bi, *Phys. Lett. B* **173**, *388.*

Nowak, M.A., Rho, M. and Zahed, I. (1993) Chiral effective action with heavy quark symmetry, *Phys. Rev. D* **48**, *4370.*

Nowak, M.A., Rho, M. and Zahed, I. (1996) *Chiral Nuclear Dynamics* (World Scientific, Singapore, 1996).

Olson, J.P., Warren, M., Meixner, M., Mathews, G.J., Lan, N.Q. and Dalhed, H.E. (2016) Generalized density functional equation of state for astrophysical simulations with 3-body forces and quark gluon plasma, *arXiv:1612.08992 [nucl-th].*

Ordonez, C., Ray, L. and van Kolck, U. (1994) Nucleon-nucleon potential from an effective chiral Lagrangian, *Phys. Rev. Lett.* **72**, *1982.*

Ordonez, C. and van Kolck, U. (1992) Chiral lagrangians and nuclear forces, *Phys. Lett. B* **291**, *459.*

Paeng, W.-G., Kuo, T.T.S., Lee, H.K. and Rho, M. (2015) Scale-invariant hidden local symmetry, topology change and dense baryonic matter, *Phys. Rev. C* **93**, *055203.*

Paeng, W.-G., Kuo, T.T.S., Lee, H.K., Ma, Y.L. and Rho, M. (2017) Scale-invariant hidden local symmetry, topology change and dense baryonic matter II, *Phys. Rev. D* **96**, *no. 1, 014031.*

Paeng, W.-G., Lee, H.K., Rho, M. and Sasaki, C. (2011) Dilaton-limit fixed point in hidden local symmetric parity doublet model, *Phys. Rev. D* **85**, *054022.*

Paeng, W.-G., Lee, H.K., Rho, M. and Sasaki, C. (2013) Interplay between ω-nucleon interaction and nucleon mass in dense baryonic matter, *Phys. Rev. D* **88**, *105019.*

Paeng, W.G., Lee, H.K., Rho, M. and Sasaki, C. (2011) Dilaton-limit fixed point in hidden local symmetric parity doublet model, *Phys. Rev. D* **85**, *054022.*

Paeng, W.G., Lee, H.K., Rho, M. and Sasaki, C. (2013) Interplay between ω-nucleon interaction and nucleon mass in dense baryonic matter, *Phys. Rev. D* **88**, *105019.*

Paeng, W.G. and Rho, M. (2015) Kaon condensation in baryonic Fermi liquid at high density, *Phys. Rev. C* **91**, *015801.*

Pagels, H. (1966) Energy-momentum structure form factors of particles, *Phys. Rev.* **144**, *1250.*

Pandharipande, V.R., Pethick, C.J. and Thorsson, V. (1995) Kaon energies in dense matter, *Phys. Rev. Lett.* **75**, *4567.*

Park, T.S. *et al.* (2003) Parameter free effective field theory calculation for the solar proton fusion and hep processes, *Phys. Rev. C* **67**, *055206.*

Park, B. Y., Kim, J. I. and Rho, M. (2009) Kaons in dense half-skyrmion matter, *Phys. Rev. C* **81**, *035203.*

Park, B.-Y., Min, D.P., Rho, M. and Vento, V. (2008) Atiyah–Manton approach to skyrmion matter, *Nucl. Phys. A* **707**, *381.*

Park, B. Y., Rho, M. and Vento, V. (2004) Vector mesons and dense skyrmion matter, *Nucl. Phys. A* **736**, *129.*

Park, B.Y., Rho, M. and Vento, V. (2008) The role of the dilaton in dense skyrmion matter, *Nucl. Phys. A* **807**, *28.*

Pastore, S., Baroni, A. Carlson, J., Gandolfi, S., Pieper, S.C., Schiavilla, R. and Wiringa, R.B. (2018) Quantum Monte Carlo calculations of weak transitions in $A = 6$–10 nuclei, *Phys. Rev. C* **97**, *022501.*

Patrignani, C. *et al.* (2016) (Particle Data Group), *Chin. Phys. C, 40, 100001.*

Pelaez, J.R. (2015) From controversy to precision on the sigma meson: a review on the status of the non-ordinary $f_0(500)$ resonance, *Phys. Rept.* **658**, *1.*

Peskin, M.E. and Schroeder, D.V. (1990) *Introduction to quantum field theory* (Westview Press, UK, 1995).

Piarulli, M. *et al.* (2018) Light-nuclei spectra from chiral dynamics, *Phys. Rev. Lett.* **120**, *052503.*

Polchinski, J. (1992) Effective field theory and the Fermi surface *In *Boulder 1992, Proceedings, Recent directions in particle theory* 235-274, and Calif. Univ. Santa Barbara - NSF-ITP-92-132 (92,rec.Nov.) 39 p. (220633) Texas Univ. Austin - UTTG-92-20 (92,rec.Nov.) 39 p.*

Radha, P.B., Dean, D.J., Koonin, S.E., Langanke, K. and Vogel, P. (1997) Gamow-Teller strength distributions in f - p shell nuclei, *Phys. Rev. C* **56**, *3079.*

Radice, D., Bernuzzi, S., Del Pozzo, W., Roberts, L.F. and Ott, C.D. (2017) Probing extreme-density matter with gravitational wave observations of binary neutron star merger remnants, *Astrophys. J.* **842**, *no. 2, L10.*

Radice, D., Perego, A. and Zappa, F. (2018) GW170817: Joint constraint on the neutron star equation of state from multimessenger observations, *Astrophys. J.* **852**, *no.2, L29.*

Read, J.S. *et al.* (2013) Matter effects on binary neutron star waveforms, *Phys. Rev. D* **88**, *044042.*

Regge, T. and Wheeler, J.A. (1957) Stability of a Schwarzschild singularity, *Phys. Rev.* **108**, *1063.*

Rho, M. (1974) Quenching of axial-vector coupling constant in beta-decay and pion-nucleus optical potential, *Nucl. Phys. A* **231**, *493.*

Rho, M. (1991) Exchange currents from chiral Lagrangians, *Phys, Rev. Lett.* **66**, *1275.*

Rho, M. (2008) *Chiral Nuclear Dynamics: From Quarks to Nuclei to Compact Stars* (World Scientific, Singapore, 2008)

Rho, M. (2014) A new topological "twist" to BR scaling, *Nucl. Phys. A* **928**, *144.*

Rho, M. (2016) In search of a pristine signal for (scale-)chiral symmetry in nuclei, *New Physics:Sae Mulli* **66**.

Rho, M. (2016) Probing the source of proton mass by "unbreaking" scale-chiral symmetry, *arXiv:1610.07012 [nucl-th].*

Rho, M., Goldhaber, A.S. and Brown, G.E. (1983) Topological soliton bag model for baryons, *Phys. Rev. Lett.* **51**, *747.*

Rho, M., Sin, S.J. and Zahed, I. (2010) Dense QCD: A holographic dyonic salt, *Phys. Lett. B* **689**, *23.*

Rho, M. and Zahed, I. (1986) The Wess–Zumino term in the chiral bag model, *Phys. Lett.* **B182**, *274.*

Rho, M. and Zahed. I. (ed) (2016) *The multifaceted skyrmion: 2nd edition* (World Scientific, Singapore).

Riazuddin and Fayyazuddin (1966) Algebra of current components and decay widths of rho and K* mesons, *Phys. Rev.* **147**, *1071.*

Ritz, R. et al. (2013) Formation of a topological non-Fermi liquid in MnSi, *Nature* **497**, *231.*

Runge, E. and Gross, E.K.U. (1984) Density-functional theory for time-dependent systems, *Phys. Rev. Lett.* **52**, *997.*

Ryttov, T.A. and Shrock, R. (2016) Scheme-independent series expansions at an infrared zero of the beta function in asymptotically free gauge theories, *Phys. Rev. D* **94**, *125005.*

Sakai, T. and Sugimoto, S. (2005) Low energy hadron physics in holographic QCD, *Prog. Theor. Phys.* **113**, *843.*

Sasaki, C., Lee, H.K., Paeng, W.G. and Rho, M. (2011) Conformal anomaly and the vector coupling in dense matter, *Phys. Rev. D* **84**, *034011.*

Sasano, M. *et al.* (2011) Gamow–Teller transition strengths from Ni-56, *Phys. Rev. Lett.* **107**, *202501.*

Schäfer, T. and Wilczek, F. (1999) Continuity of quark and hadron matter, *Phys. Rev. Lett.* **82**, *3956.*

Scherer, S. (2003) Introduction to chiral perturbation theory, *Adv. Nucl. Phys.* **27**, *277.*

Schroder, H.C. *et al.* (2001) The pion nucleon scattering lengths from pionic hydrogen and deuterium, *Eur. Phys. J. C* **21**, *473.*

Schwinger, J. (1967) Chiral dynamics, *Phys. Lett.* **24B**, *473.*

Schwinger, J. (1970) Particles and Sources, *Conf. Proc. V* **1**, *209.*

Serot, B.D. and Walecka, J.D. (1997) Recent progress in quantum hadrodynamics, *Int. J. Mod. Phys. E* **6**, *515.*

Serot, B.D. and Walecka, J.D. (1986) The relativistic nuclear many body problem, *Adv. Nucl. Phys.* **16**, *1.*

Serra, J. (2013) A Higgs-like dilaton: Viability and implications, *EPJ Web Conf.* **60**, *17005.*

Shankar, R. (1994) Renormalization group approach to interacting fermions, *Rev. Mod. Phys.* **66**, *129.*

Shapiro, S.L. and Teukolsky, S.A. (1983) *Black holes, white dwarfs and neutron stars* (Wiley, New York 1983).

Shaposhnikov, M. and Zenhausern, D. (2009) Quantum scale invariance, cosmological constant and hierarchy problem, *Phys. Lett. B* **671**, *162.*

Siu, L.-W., Holt, J.W., Kuo, T.T.S. and Brown, G.E. (2009) Low-momentum NN interactions and all-order summation of ring diagrams of symmetric nuclear matter, *Phys. Rev. C* **79**, *054004.*

Siu, L.W., Kuo, T.T.S. and Machleidt, R. (2008) Low-momentum ring diagrams of neutron matter at and near the unitary limit, *Phys. Rev. C* **77**, *034001.*

Skyrme, T.H.R. (1959) The effective nuclear potential, *Nucl. Phys.* **9**, *615.*

Skyrme, T.H.R. (1962) A unified field theory of mesons and baryons, *Nucl. Phys.* **31**, *556.*

Son, D.T. and Stephanov, M.A. (2002) Pion propagation near the QCD chiral phase transition, *Phys. Rev. Lett.* **88**, *202302.*

Son, D.T. and Stephanov, M.A. (2004) QCD and dimensional deconstruction, *Phys. Rev. D* **69**, *065020.*

Song, C., Brown, G.E., Min, D.P. and Rho, M. (1997) Fluctuations in "BR scaled'"chiral Lagrangians, *Phys. Rev. C* **56**, *2244.*

Song, C., Min, D.-P. and Rho, M. (1998) Thermodynamic properties of effective chiral Lagrangians with Brown–Rho scaling, *Phys. Lett. B* **424**, *226.*

Song, H.Q., Yang, S.D. and Kuo, T.T.S. (1987) Infinite order summation of particle-particle ring diagrams in a model-space approach for nuclear matter, *Nucl. Phys. A* **462**, *491 (1987).*

Song, Y. and Baym, G. (2017) Generalized Nambu–Goldstone pion in dense matter: a schematic NJL model, *arXiv:1703.08236 [nucl-th].*

Speicher, C., Dreizler, R.M. and Engel, E. (1992) Density functional approach to quantum hadrondynamics: Theoretical foundations and construction of extended Thomas–Fermi models, *Ann. of Phys.* **213**, *312.*

Strominger, A. (2017) Lectures on the infrared structure of gravity and gauge theory, *arXiv:1703.05448.*

Suenaga, D., He, B.R., Ma, Y.L. and Harada, M. (2015) Mass degeneracy of heavy-light mesons with chiral partner structure in the half-Skyrmion phase, *Phys. Rev. D* **91**, *036001.*

Suhonen, J.T. (2017) Value of the axial-vector coupling strength in β and $\beta\beta$ decays: A Review, *Front. in Phys.* **5**, *55.*

Sutcliffe, P. (2010) Skyrmions, instantons and holography, *JHEP* **1008**, *019.*

Sutcliffe, P. (2011) Skyrmions in a truncated BPS theory, *JHEP* **1104**, *045.*

Tews, I., Carlson, J., Gandolfi, S. and Reddy, S. (2018) Constraining the speed of sound inside neutron stars with chiral effective field theory interactions and observations, *arXiv:1801.01923 [nucl-th].*

't Hooft, G. (1974) A planar diagram theory for strong interactions, *Nucl. Phys. B* **72**, *461.*

't Hooft, G. (1976) Symmetry breaking through Bell–Jackiw anomalies, *Phys. Rev. Lett.* **37**, *8.*

't Hooft, G. (1976) Computation of the quantum effects due to a four-dimensional pseudoparticle, *Phys. Rev. D* **14**, *3432 Erratum: [Phys. Rev. D* **18**, *2199 (1978)].*

't Hooft, G. (1980) Naturalness, chiral symmetry, and spontaneous chiral symmetry breaking, *Recent Developments in Gauge Theories* (Plenum Press, 1980) ed. 't Hooft, G.

Thorne, K.S. (1998) Tidal stabilization of rigidly rotating, fully relativistic neutron stars, *Phys. Rev. D* **58**, *124031.*

Thorne, K.S. and Campolattaro, A. (1967) *Astrophys. J.* **149**, *591.*

Todd, T.B.G. and Piekarewicz, J. (2003) Relativistic mean field study of neutron rich nuclei, *Phys. Rev. C* **67**, *044317.*

Tolos *et al.* (2012), Implications for compact stars of a soft nuclear equation of state from heavy-ion data, *PoS NICXII , 036.*

Tolos, L., Centelles, M. and Ramos, A. (2017) Equation of state for nucleonic and hyperonic neutron stars with mass and radius constraints, *Astrophys. J.* **834**, *no. 1, 3.*

Tsang, M.B. *et al.* (2009) Constraints on the density dependence of the symmetry energy, *Phys. Rev. Lett.* **102**, *122701.*

Vafa, C. and Witten, E. (1984) Restrictions on symmetry breaking in vector-like gauge theories, *Nucl. Phys. B* **234**, *173.*

Valiev, M, and Fernando, G.W, (1997) Generalized Kohn–Sham density functional theory via effective action formalism, *arXiv: cond-mat/9702247.*

Veneziano, G. (1976) Some aspects of a unified approach to gauge, dual and Gribov theories, *Nucl. Phys. B* **117**, *519*.

Vento, V. (2017) Skyrmions at high density, *Int. J. Mod. Phys. E* **26**, *1740029*.

Walecka, J. D. (1974) A theory of highly condensed matter, *Annals Phys.* **83**, *491*.

Wang, Q., Kuang, Y. P., Xiao, M. and Wang, X. L. (2000) Derivation of the effective chiral Lagrangian for pseudoscalar mesons from QCD, *Phys. Rev. D* **61**, *054011*.

Warburton, E.K. (1991) Mesonic enhancement of the weak axial vector current evaluated from beta decay in the lead region, *Phys. Rev. Lett.* **66**, *1823*.

Watanabe, H. and Murayama, H. (2014) Effective Lagrangian for nonrelativistic systems, *Phys. Rev. X* **4**, *031057*.

Watanabe, H. and Vishwanath, A. (2014) Criterion for stability of Goldstone modes and Fermi liquid behavior in a metal with broken symmetry, *Proc. Nat. Acad. Sci.* **111**, *16314*.

Weigel, H., Schwesinger, B. and Holzwarth, G. (1986) Exotic baryon number $B = 2$ states in the SU(2) Skyrme model, *Phys. Lett.* **168B**, *321*.

Weinberg, S. (1967) Precise relations between the spectra of vector and axial vector mesons, *Phys. Rev. Lett.* **18**, *507*.

Weinberg, S. (1979) Phenomenological Lagrangians, *Physica A* **96**, *327*.

Weinberg, S, (1990) Mended symmetries, *Phys. Rev. Lett.* **65**, *1177*; Unbreaking symmetries, *Salamfestschrift* (World Scientific, Singapore 1994) ed. by Ali, Ellis, Randjbar-Daem.

Weinberg, S. (1990) Nuclear forces from chiral Lagrangians, *Phys. Lett. B* **251**, *288*.

Weinberg, S. (1991) Effective chiral Lagrangians for nucleon-pion interactions and nuclear forces, *Nucl. Phys. B* **363**, *3*.

Weinberg, S. (1997) What is quantum field theory, and what did we think it is? *In *Boston 1996, Conceptual foundations of quantum field theory* 241-251 [hep-th/9702027]*.

Weinberg, S. (2001) *The quantum theory of fields, Vol. II, Modern applications* (Cambridge University Press.)

Weinberg, S. (2016) Effective field theory, past and future, *Int. J. Mod. Phys. A* **31**, *no.06, 1630007*.

Wess, J. and Zumino, B. (1971) Consequences of anomalous Ward identities, *Phys. Lett.* **37B**, *95*.

Wilczek, F. (2006) The origin of mass, *Mod. Phys. Lett. A* **21**, *701*.

Wilkinson, D.H. (1978) Symmetries and nuclei, in *Nuclear Physics with Heavy Ions and Mesons* (North-Holland, Amsterdam, 1978) ed. Balian, R., Rho, M. and Ripka, G.

Witten, E. (1979) Baryons in the $1/N$ expansion, *Nucl. Phys. B* **160**, *57*.

Witten, E. (1983) Global aspects of current algebra, *Nucl. Phys. B* **223**, *422*.

Yamawaki, K. (2016) Origin of mass — Horizons expanding from the Nambu's theory, *PTEP* **2016**, *no.6, 06A107*.

Yamawaki, K. (2017) Hidden local symmetry and beyond, *Int. J. Mod. Phys. E* **26**, *1740032*.

Yamawaki, K. (2018) Dynamical gauge boson of hidden local symmetry within the Standard Model, *arXiv:1803.07271 [hep-ph].*

Zahed, I. and Brown, G.E. (1986) The Skyrme model, *Phys. Rept.* **142**, *1.*

Zaks, A. (1985) Derivation of the Skyrme-Witten Lagrangian from QCD. *Nucl. Phys. B* **260**, *241.*

Zarembo, K. (2002) Possible pseudogap phase in QCD, *JETP Lett.* **75**, *59 (2002) [Pisma Zh. Eksp. Teor. Fiz.* **75**, *67 (2002)].*

Zhang, P., Kimm, K., Zou, L. and Cho, Y.M. (2017) Re-interpretation of Skyrme Theory: New topological structures, *arXiv:1704.05975 [hep-th].*

Index